Electrochemical Transformation of Renewable Compounds

Editors

Zhiqun Lin
Georgia Institute of Technology
Atlanta, GA, United States

Xueqin Liu
Engineering Research Center of Nano-Geomaterials of Ministry of Education
China University of Geosciences
Wuhan, PR China

Zhen Li
Engineering Research Center of Nano-Geomaterials of Ministry of Education
China University of Geosciences
Wuhan, PR China

Yanqiu Huang
Engineering Research Center of Nano-Geomaterials of Ministry of Education
China University of Geosciences
Wuhan, PR China

CRC Press
Taylor & Francis Group
Boca Raton London New York

CRC Press is an imprint of the
Taylor & Francis Group, an **Informa** business

A SCIENCE PUBLISHERS BOOK

Cover credit: Image used on the cover designed by the second editor, Xueqin Liu.

First edition published 2023
by CRC Press
6000 Broken Sound Parkway NW, Suite 300, Boca Raton, FL 33487-2742

and by CRC Press
4 Park Square, Milton Park, Abingdon, Oxon, OX14 4RN

Library of Congress Cataloging-in-Publication Data (applied for)

ISBN: 978-0-367-34432-0 (hbk)
ISBN: 978-1-032-38259-3 (pbk)
ISBN: 978-0-429-32678-3 (ebk)

DOI: 10.1201/9780429326783

Typeset in Times New Roman
by Radiant Productions

Preface

Electrochemical transformation plays a key role in the green energy revolution processes that have been developed to meet the worldwide energy consumption requirements and avoid potential environmental damage. *Electrochemical Transformation of Renewable Compounds* presents the basic fundamentals of different electrochemical transformations for clean energy and places a significant emphasis on the key developments of various electrochemical processes using the state-of-the-art materials. Written by electrochemical energy scientists who have worked on the application of electrocatalysis in environmental and energy area, this book provides comprehensive coverage of main electrochemical transformation processes, including oxygen evolution, hydrogen generation, oxygen reduction, carbon dioxide reduction, nitrogen reduction, methanol oxidation, urea oxidation and ammonia oxidation. To develop the effective and low cost electrocatalysts for electrochemical transformation processes, various kinds of materials with different structures and properties from nano to giant have been explored that includes earth abundant electrocatalysts, transition metal phosphide, carbon based electrocatalysts, metal based heterogeneous electrocatalysts, transition metal dichalcogenides, self-supported nanoarrays, single-atom electrocatalysts and metal-organic frameworks derived electrocatalysts. The opportunities, challenges and prospects of different electrochemical transformation are also discussed by the scientists who are keen on electrochemical energy.

Specifically, Chapter 1 provides a comprehensive review of each chapter in order to give readers an idea of the content of this book. The bifunctional electrocatalysts with the ability to simultaneously reduce overpotential and increase reaction rate of hydrogen evolution reaction and oxygen evolution reaction are covered in Chapter 2. Chapter 3 focuses on the metal-organic frameworks-based electrolytes for oxygen evolution reaction, oxygen reduction reaction and CO_2 reduction reaction, while methanol and urea oxidation electrocatalysts used in the fuel cells are given in Chapter 4. Chapter 5 summarizes the latest developments for various typical carbon-based materials for oxygen reduction reaction. In Chapter 6, the authors discuss the recent advancement of metal-based heterogeneous electrocatalysts applied for CO_2 reduction. Chapter 7 gives an overview of earth-abundant electrocatalysts for oxygen evolution reaction. In Chapter 8, the recent advancement of self-support nanoarrays for a wide range of energy-conversion processes are the topic in focus. Chapter 9 focuses on the single-metal-atom electrocatalysts covering the unique features and mechanistic origins of electrocatalytic activity.

This book acts as a resource for researchers and graduate students who are keen to know about the development of electrochemistry, material sciences, energy and environment because it gives a comprehensive review of renewable compounds produced by electrochemical process.

Xueqin Liu
Wuhan

Contents

||||||||||||||||||||||||||||||||||||||||||||||||||||||||||||||||||||||||||||||||||||||||||||||||||||||||||||||||||||||||||||||||||

1

Electrochemical Transformation of Renewable Compounds
A Perspective

Xueqin Liu,[1] *Zhen Li,*[1] *Yanqiu Huang*[1] and *Zhiqun Lin*[2,*]

Introduction

To meet the increasing worldwide energy requirements and avoid associated environmental pollution, the search for clean and renewable energy sources has attracted more and more public attention. Electrochemical transformation plays a key role in the development of sustainable energy resources, which can be used to generate power, store energy and synthesise chemicals.

Electrochemical Transformation of Renewable Compounds presents the basic fundamentals of different electrochemical transformations for clean energy and places a significant emphasis on the key developments of various electrochemical processes using the state-of-the-art materials. Written by electrochemical energy scientists who have worked on the application of electrocatalysis in environmental and energy area, this book provides comprehensive coverage of main electrochemical transformation processes, including oxygen (O_2) evolution, hydrogen (H_2) generation, oxygen reduction, carbon dioxide (CO_2) reduction, nitrogen (N_2) reduction, methanol oxidation, urea oxidation and ammonia oxidation. To develop the effective and low cost electrocatalysts for electrochemical transformation processes, various kinds of materials with different structures and properties from nano to giant have been explored that includes earth abundant electrocatalysts, transition metal phosphide, carbon based electrocatalysts, metal based heterogeneous electrocatalysts, transition metal dichalcogenides, self-supported nanoarrays, single-atom electrocatalysts and

---
[1] Engineering Research Center of Nano-Geomaterials of Ministry of Education, Faculty of Materials Science and Chemistry, China University of Geosciences, Wuhan 430074, PR China.
[2] Georgia Institute of Technology, Atlanta, GA, United States.
* Corresponding author: zhiqun.lin@mse.gatech.edu

metal-organic frameworks derived electrocatalysts. The opportunities, challenges and prospects of different electrochemical transformation are also discussed. Due to the comprehensive review of renewable compounds produced by electrochemical process, we hope this book will act as a resource for researchers and graduate students who are keen to know about the development of electrochemistry, material sciences, energy and environment. In order to give readers a concise summary of the content, an overview of each chapter included in this book is provided in the following sections.

In Chapter 2 by Yang Yang, the text focuses on bifunctional electrocatalysts, which have the ability to simultaneously reduce overpotential and increase reaction rate of the two half-reactions (hydrogen evolution reaction and oxygen evolution reaction), for overall water splitting. A brief introduction of the fundamental physical or chemical processes involved in overall water splitting is provided first. Then, noble metal and noble metal-free electrocatalytic materials for overall water splitting in acidic, neutral, and basic media three different electrolytes are specially discussed. Finally, the possible trends of future developments of bifunctional electrocatalysts are also proposed.

In Chapter 3 by Wei et al., the authors summarize recent developments in electrochemical energy conversion based on metal-organic frameworks (MOFs) derived electrocatalysts, which have the advantages of controllable morphology and structure, unique physical and chemical properties, and excellent electrochemical performance. MOFs-derived carbon-based hybrid nanomaterials, including metal bimetallic and alloy, metal chalcogenides, metal phosphides and nitrides, as well as carbides and single-atom materials, as electrocatalysts for hydrogen evolution reaction will be given first. Then the chapter focuses on the MOFs-derived carbon-based hybrid nanomaterials for oxygen evolution reaction, oxygen reduction reaction and CO_2 reduction reaction. Fabrication methods of MOF-derived porous carbon electrocatalysts are highlighted as well.

In Chapter 4 by Cai et al., methanol and urea oxidation electrocatalysts used in the fuel cells are the topic in focus. Some basic concepts and principles about fuel cell and alternative fuels are briefly introduced first. Then electrocatalysts for methanol oxidation reaction (MOR) in direct methanol fuel cell, including Pt-based and Pt-free MOR catalysts are discussed in detail. In addition, the mechanism and nickel-based catalysts for urea oxidation reaction (UOR) are also briefly introduced. At last, some prospects are put forward for the future development of MOR and UOR catalysts.

In Chapter 5 by Chen et al., the authors present the promising application of carbon-based materials in oxygen reduction reaction (ORR). A brief introduction of the main mechanism of the ORR is discussed first. Summarization of the latest developments for various typical carbon-based materials including platinum group metal-based carbon materials, transition metal-based carbon materials, and metal-free carbon materials, applied in ORR is then introduced in detail. Finally, challenges and perspectives for effective carbon-based ORR electrocatalysts are proposed.

In Chapter 6 by Ye et al., the latest developments of metal-based heterogeneous electrocatalysts, including molecular systems, single-atom and nanostructured catalysts, applied for CO_2 reduction are comprehensively reviewed. The fundamentals and evaluation parameters of metal-based heterogeneous catalysts for CO_2 reduction are indicated first. The factors for the performance of molecular catalyst are also discussed, especially for the intrinsic effect from the catalyst structure and the extrinsic effect from the immobilization strategy. Finally, the chapter discusses the main strategies employed for the design of single-atom and nanostructured catalysts.

In Chapter 7 by Zhao et al., earth-abundant electrocatalysts for oxygen evolution reaction (OER) are the topic in focus. Reaction mechanisms and evaluation criteria for the OER are summarized first. Then earth-abundant electrocatalysts for OER, including transition metal compounds (metal and alloys, oxides, hydroxides and oxyhydroxides, sulfides, selenides, nitrides, phosphides, and other compounds), heteroatom doped carbon nanomaterials, and their hybrids are discussed in detail. At last, some prospects are put forward for the future development for rationally designing OER electrocatalysts.

In Chapter 8 by Wang et al., the authors summarize the recent advancement of self-support nanoarrays for a wide range of energy-conversion processes. The preparation methods, including chemical bath deposition, hydro/solvothermal methods, electrochemical deposition, template transcription, MOF-derived synthesis and chemical vapor deposition, for the different categories of nanoarrays electrocatalysts are introduced first. Then the application of nanoarrays electrocatalysts in the field of electrocatalysis, such as oxygen reduction reaction, oxygen evolution reaction, hydrogen evolution reaction, CO_2 reduction reaction, N_2 reduction reaction and urea oxidation reaction are discussed in detail. Finally, the authors conclude the challenges and prospects on the promising electrocatalytic performance and valuable innovation directions of architectural arrays in the energy conversion field.

In Chapter 9 by Cui et al., the overview of single-metal-atom electrocatalysts covering the unique features and mechanistic origins of electrocatalytic activity are indicated first. The advanced synthetic strategies and characterization techniques for the development of single-metal-atom electrocatalysts are then performed. Subsequently, the most recent advances of single-metal-atom electrocatalysts and their potential applications for reactions involved in clean energy conversion technologies, including oxygen reduction reaction (ORR), oxygen evolution reaction (OER), hydrogen evolution reaction (HER), and carbon dioxide reduction reaction (CO_2RR) are briefly outlined. Particular attentions are paid to the structure-property correlations and intrinsic origins of electrocatalytic activity. Finally, the future perspectives with regard to the challenges and opportunities awaiting this particular research field are offered and discussed.

2

Bifunctional Electrocatalysts for Overall Water Splitting

Yang Yang

॥॥॥

2.1 Fundamentals of overall electrochemical water splitting

2.1.1 The mechanism of overall water splitting

Overall water splitting (OWS) involves two half-cell reactions, namely 2-electron transfer during the hydrogen evolution reaction (HER) at the cathode and a 4-electron couple during the oxygen evolution reaction (OER) at the anode. Both the HER and OER comprise multistep reaction pathways and exhibit sluggish reaction kinetics in different media (acidic, neutral, and alkaline electrolytes). The proposed mechanisms for the HER and OER in universal pH ranges is follows, shown in Table 2.1.[1,2] There are two main mechanisms proposed for the HER of the cathode in all electrolytes: the Volmer-Heyroskỳ and Volmer-Tafel reaction path ways. In acidic media, protons from the electrolyte are reduced to hydrogen, while in neutral and alkaline electrolyte, water is reduced to hydrogen and hydroxyl ions. The OER of the anode involves three reaction intermediates: *O, *OOH, and *OH. In acidic and neutral media, water is oxidised to oxygen and hydrogen ions. Hydroxyl ions are oxidized to water and oxygen in alkaline media. Identifying the mechanism of the two half-reaction assists in investigating the OWS reaction in more detail.

Key Laboratory of Chemical Additives for China National Light Industry, College of Chemistry and Chemical Engineering, Shaanxi University of Science and Technology, Xi'an 710021, People's Republic of China.
Email: yyang399@sust.edu.cn

Table 2.1. The mechanism of the HER, OER, and OWS.

Reactions	Acidic media (pH < 7)	Neutral media (pH = 7)	Alkaline media (pH > 7)
HER	Volmer step: $H^+ + * + e^- \rightarrow H^*$ Heyrosky step: $H^+ + H^+ + e^- \rightarrow H_2$ Tafel step: $H^* + H^* \rightarrow H_2$ Overall reaction: $2H^+ + 2e^- \rightarrow H_2$	Volmer step: $H_2O + * + e^- \rightarrow H^* + OH^-$ Heyrosky step: $H^* + H_2O + e^- \rightarrow H_2 + OH^- + *$ Tafel step: $H^* + H^* \rightarrow H_2$ Overall reaction: $2H_2O + 2e^- \rightarrow 2OH^- + H_2$	Volmer step: $H_2O + * + e^- \rightarrow H^* + OH^-$ Heyrosky step: $H^* + H_2O + e^- \rightarrow H_2 + OH^- + *$ Tafel step: $H^* + H^* \rightarrow H_2$ Overall reaction: $2H_2O + 2e^- \rightarrow 2OH^- + H_2$
OER	$H_2O + * \rightarrow {}^*OH + H^+ + e^-$ ${}^*OH \rightarrow {}^*O + H^+ + e^-$ $H_2O + {}^*O \rightarrow {}^*OOH + H^+ + e^-$ ${}^*OOH \rightarrow * + O_2 + H^+ + e^-$ Overall reaction: $2H_2O \rightarrow 4H^+ + O_2 + 4e^-$	$H_2O + * \rightarrow {}^*OH + H^+ + e^-$ ${}^*OH \rightarrow {}^*O + H^+ + e^-$ $H_2O + {}^*O \rightarrow {}^*OOH + H^+ + e^-$ ${}^*OOH \rightarrow * + O_2 + H^+ + e^-$ Overall reaction: $2H_2O \rightarrow 4H^+ + O_2 + 4e^-$	$OH^- + * \rightarrow {}^*OH + e^-$ $OH^- + {}^*OH \rightarrow {}^*O + H_2O + e^-$ $OH^- + {}^*O \rightarrow {}^*OOH + e^-$ $OH^- + {}^*OOH \rightarrow * + O_2 + H_2O + e^-$ Overall reaction: $4OH^- \rightarrow O_2 + 2H_2O + 4e^-$
OWS	$2H_2O \rightarrow 2H_2 + O_2$		

2.1.2 Parameters for overall water splitting

Bifunctional electrocatalysts for OWS play a key role in accelerating the OWS reaction kinetics. To evaluate whether a catalyst has a beneficial effect on OWS, several important parameters should be considered in the initial screening process. These parameters include overpotential, Tafel slope, Faradaic efficiency, and stability.

Overpotential: The overpotential (η) is a key parameter to assess the catalytic activity of bifunctional electrocatalysts in OWS. The theoretical value of the equilibrium potential (E_{theory}) for OWS is 1.23 V *vs.* reversible hydrogen evolution (RHE) whereas an additional potential, which is higher than the E_{theory}, is applied to drive OWS as the sluggish kinetics of the HER and OER causes polarisation of the electrode. The difference in the OWS cell potential (two-electrode configuration) from the equilibrium potential is termed as the overpotential: $\eta = E_i - E_{theory}$, where E_i is the actual potential at a defined current density. Clearly, the overpotential reflects the extent of electrode polarisation in the OWS. Therefore, reducing the degree of polarisation assists in decreasing the overpotential.

Determining E_i is the key to calculating η. The polarization curve was used to obtain the value of E_i at a defined current density. Polarisation curves are also called current-potential curves, and are obtained by cyclic voltammetry (CV) and linear sweep voltammetry (LSV) measurements. The general method involves establishing the corresponding potential (E_i) at a defined current density based on the LSV or CV curves (Fig. 2.1a). The oxidation peak, which has a higher current density, appears in the CV curves if the catalyst contains Ni or Co (Fig. 2.1b). This influences the calculation of E_i, which affects the overpotential. In this case, the CV curves are usually split into two sections: the "oxidation curves" and "reduction curves". The reduction curves are used to calculate E_i as it does not contain an oxidation peak. E_i is constant when the potential on the E-T curves stabilises. Generally, a bifunctional catalyst that requires an overpotential in the range of 200–300 mV over the full pH

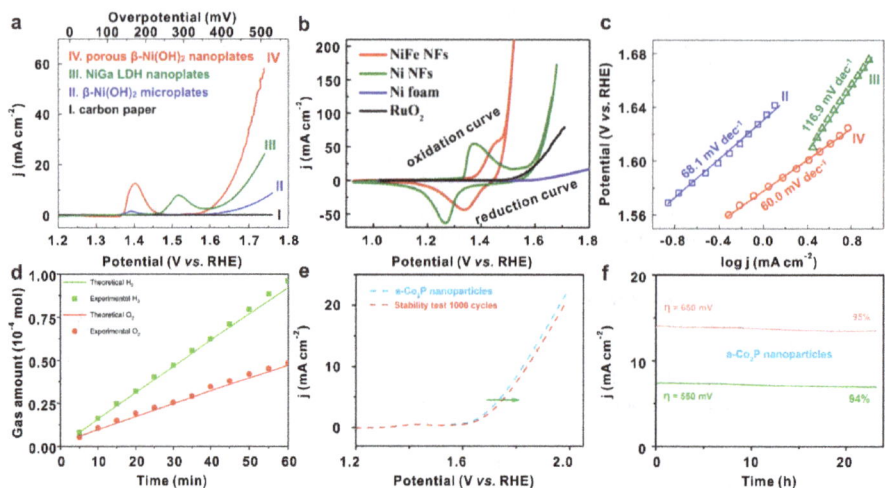

Figure 2.1. (a) Linear sweep voltammetry (LSV) curves of various catalysts. Reproduced with permission from Ref. 3. Copyright 2015, American Chemical Society. (b) Cyclic voltammetry (CV) curves plot of Ni-based catalysts. Reproduced with permission from Ref. 4. Copyright 2019, American Chemical Society. (c) Tafel slopes derived from LSV curves in panel (a). Reproduced with permission from Ref. 3. Copyright 2015, American Chemical Society. (d) Faradic efficiency measurements using gas chromatography. Reproduced with permission from Ref. 5. Copyright 2019, Nature Publishing Group. (e) LSV curves of a-Co$_2$P nanoparticles before and after 1000 CV cycles. (f) Chronoamperometry measurement of a-Co$_2$P nanoparticles at various potentials. (e and f) Reproduced with permission from Ref. 6. Copyright 2017, American Chemical Society.

range is considered an excellent catalyst for OWS; however, only a few catalysts can reduce the overpotential to below 200 mV.

Tafel slope: The Tafel slope is an important parameter to understand the electrode kinetics of OWS. The Tafel equation is defined as $\eta = a + b \log j$, where η, a, b, and j are the overpotential, exchange current density, Tafel slope, and current density, respectively. The Tafel plot (η–$\log j$) is obtained from the CV and LSV curves (Fig. 2.1c). An excellent bifunctional electrocatalyst for OWS should exhibit a small Tafel slope and large current density.

Faradaic efficiency (FE): FE is a quantitative parameter used in OWS. This parameter represents the electron transfer efficiency in a single-cell system with a two-electrode configuration to improve the OWS. It is the ratio of the experimental and theoretical amounts of gas produced. The quantity of theoretically produced gas is calculated through integration from the chronoamperometric or chronopotentiometric analysis. The experimental amount of gas produced is determined by gas chromatography or the water-gas displacement method (Fig. 2.1d).

Stability: Electrochemical stability is another crucial parameter that is used to assess the ability of an electrocatalyst for long-term operation under working conditions. Generally, two methods are used to test the electrochemical durability of an electrocatalyst for OWS. The first is by performing CV analysis for thousands of cycles at a constant scan rate (Fig. 2.1e), while the second is by performing chronoamperometry or chronopotentiometry measurements (Fig. 2.1f). The electrocatalyst is exposed to a constant potential or current density over several hours to days. The ideal catalyst always exhibits negligible changes in overpotential or current density after long-term operation.

2.2 Bifunctional electrocatalysts for overall water splitting

Bifunctional electrocatalysts for OWS are catalysts that have the ability to simultaneously reduce overpotential and increase reaction rate of the two half-reactions (HER and OER) in the same electrolyte solution. Figure 2.2 shows a schematic diagram of a single cell with dual-function catalysts for water splitting. Bifunctional electrocatalysts are classified into three types based on their chemical composition: noble metal, noble metal-free, and metal-free electrocatalysts. In this section, we present the recent advances in noble metal and noble metal-free bifunctional electrocatalysts for OWS in different electrolytes (acidic, neutral, and basic media).

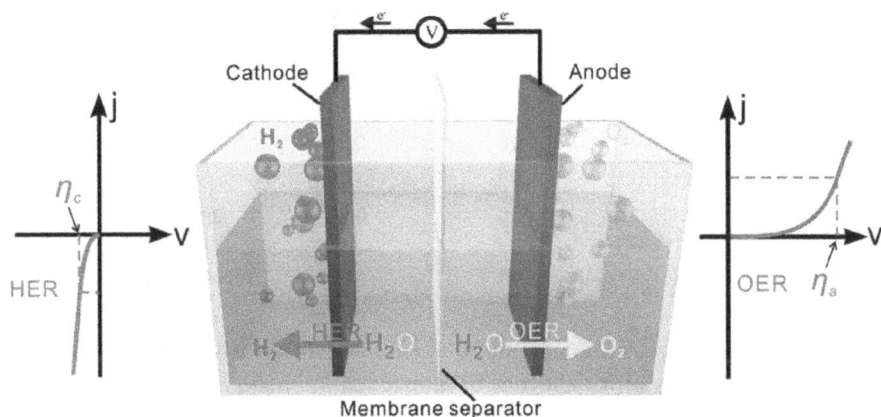

Figure 2.2. Scheme of overall water splitting in a single cell with bifunctional catalysts on cathode and anode. Reproduced with permission from Ref. 7. Copyright 2020, American Chemical Society.

2.2.1 Bifunctional electrocatalysts for overall water splitting in acidic media

2.2.1.1 Noble metal-based bifunctional electrocatalysts

OWS requires a large overpotential to lower the reaction energy barrier and accelerate the reaction rate. Noble metal-based catalysts that contain precious metal (e.g., Pt, Rh, Ru, and Ir) can achieve this aim. It has recently been reported that noble metal-based bifunctional electrocatalysts exhibit excellent OWS performances in acidic electrolytes (Table 2.2). Luo et al. synthesized extremely small Ir wavy nanowires (NWs), with diameters of 1.7 nm, using a coprecipitation method. The Ir NWs were applied as the anode and cathode in two-electrode cell system. The Ir NWs can bestow an overpotential of 390 mV at a catalytic current density of 10 mA cm^{-2} in 0.1 M HClO$_4$ electrolyte, which is much lower than the overpotential achieved with an Ir nanoparticle (NP) electrode ($\sim$ 500 mV). This is due to the unique morphology and

Table 2.2. The OWS Performance of the reported bifunctional electrocatalysts in acidic media.

Catalysts	Electrolytes	Current density (j/mA cm^{-2})	Overpotential (η/mV)	Stability	References
Ir nanowires	0.1 M HClO$_4$	10	390	40000 s @10 mA cm^{-2}	8
Ir-NSG	0.1 M HClO$_4$	10	190	24 h @10 mA cm^{-2}	9
IrW-ND	0.5 M H$_2$SO$_4$	10/50	250/310	8 h @10 mA cm^{-2}	10
IrAg NTs	0.5 M H$_2$SO$_4$	10	320	1000 cycles	11
RhCu NTs	0.5 M H$_2$SO$_4$	10/20	410/510	30 h @1.64 V	12
Ru$_3$Ni$_3$NAs	0.5 M H$_2$SO$_4$	10	280	10 h @ 5mA cm^{-2}	13
IrNi$_{0.57}$Fe$_{0.82}$/C	0.5 M HClO$_4$	10	410	20000 s @10 mA cm^{-2}	14
Co-RuIr	0.1 M HClO$_4$	10	290	25 h @10 mA cm^{-2}	15
Co-doped RuO$_2$ NWs	0.5 M H$_2$SO$_4$	10	307	12 h @ 1.537 V	16
RuIrO$_x$	0.5 M H$_2$SO$_4$	10	220	24 h @1.45V	17
N-WC	0.5 M H$_2$SO$_4$	30	470	-	21
Co-MoS$_2$	0.5 M H$_2$SO$_4$	10	670	5000 s @10 mA cm^{-2}	24
PMFCP	0.5 M H$_2$SO$_4$	10	520	10000 cycles	25

structure and large electrochemical surface area of the Ir wavy NWs.[8] The Liu group reported the synthesis of Ir nanoclusters embedded in N and S co-doped graphene (Ir-NSG) for OWS under all pH conditions. Ir-NSG exhibited an outstanding OWS performance in a two-electrode configuration, which only required an overpotential of 190 mV to achieve a current density of 10 mA cm^{-2} in 0.1 M HClO$_4$ solution (Fig. 2.3a). The overpotential of the Ir-NSG ‖ Ir-NSG electrode was lower than that of the benchmark Ir/C ‖ Pt/C electrode on a carbon fibre paper electrode. Its good catalytic performance was ascribed to the tuneable and favourable electronic state of the Ir sites coordinated to both N and S in the Ir-NSG bifunctional electrocatalyst; this was based on density functional theory (DFT) simulations and *in situ* X-ray absorption spectroscopy (Fig. 2.3b and c).[9]

Chemical composition modulation plays a key role in producing excellent bifunctional electrocatalysts for OWS. To reduce the amount of precious metals used and improve the intrinsic activity of noble metal-based bifunctional electrocatalysts toward OWS, noble metals are alloyed with various earth-abundant metals to further optimise their overall performance. Guo et al. synthesised an IrW alloy with a nanodendritic structure (IrW-ND) using a colloidal-chemical method. The IrW/C ‖ IrW/C electrode demonstrated promising OWS performance, which achieved current densities of 10 and 50 mA cm^{-2} at overpotentials of 250 and 310 mV in 0.5 M H$_2$SO$_4$ solution. This is superior to the benchmark Pt/C ‖ Ir/C

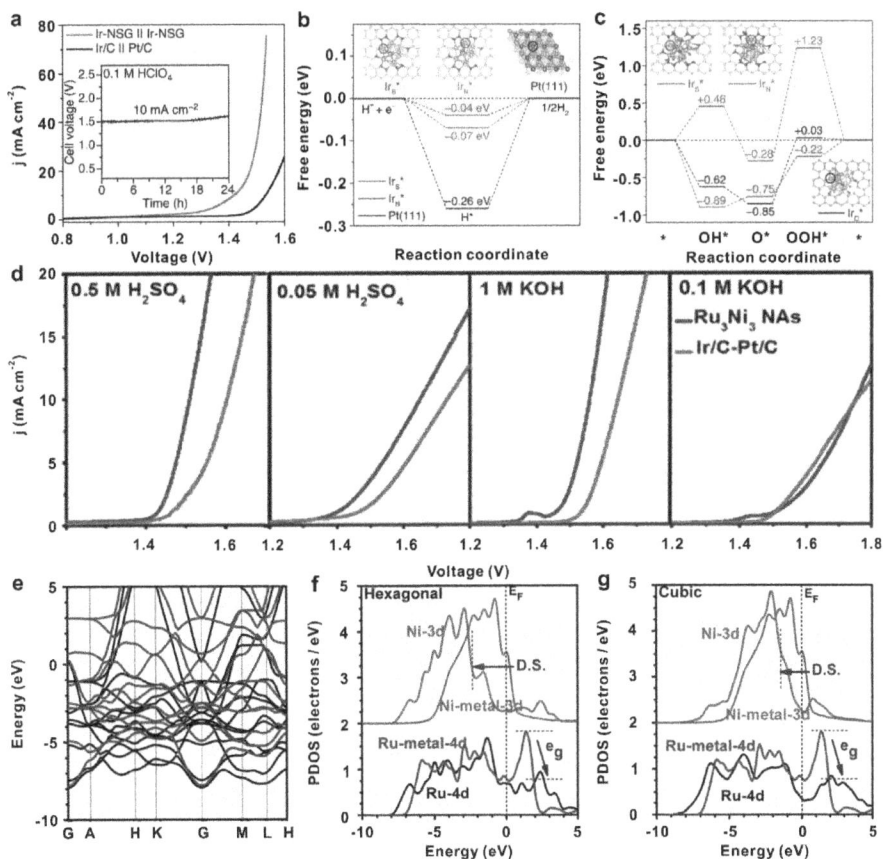

Figure 2.3. (a) LSV curves of Ir-NSG ‖ Ir-NSG electrode in 0.1 M HClO$_4$, inset shows the corresponding chronopotentiometric curves at 10 mA cm^{-2}. (b-c) DFT calculation of the predicted free-energy diagrams for HER at U = 0 V and OER at U = 1.23 on various metal sites. (a–c) Reproduced with permission from Ref. 9. Copyright 2020, Nature Publishing Group. (d) LSV curves for the OWS of the Ru$_3$Ni$_3$NAs ‖ Ru$_3$Ni$_3$NAs electrode in different electrolytes. (e) The bandstructure and (f-g) PDOS of Ru-Ni alloy. (d–g) Reproduced with permission from Ref. 13. Copyright 2019, Elsevier.

electrode (320 and 390 mV at 10 and 50 mA cm^{-2}, respectively). DFT calculations revealed that the IrW-ND bifunctional electrocatalysts could optimize the H and OH binding energies for the HER and lower the binding energy of the OER intermediates, therefore enhancing the intrinsic activity of IrW-ND towards OWS in acidic electrolytes.[10] Huang et al. successfully synthesised bimetallic IrAg nanotubes (IrAg NTs) with hollow nanostructures *via* a solvothermal and selective etching method. The IrAg NTs were used as an excellent bifunctional electrocatalyst for OWS displaying an extremely low overpotential of 320 mV at 10 mA cm^{-2} in 0.5 M H$_2$SO$_4$ solution.[11] Cao et al. prepared defect-rich RhCu nanotubes (RhCu NTs) containing Rh$_3$Cu$_1$ alloy, using a mixed-solvent method, which exhibited a high intrinsic activity and considerable stability for OWS at all pH values. The RhCu NTs/CP ‖ RhCu NTs/ CP electrode achieved a catalytic current density of 10 and 20 mA cm^{-2} in 0.5 M H$_2$SO$_4$ electrolyte at low voltages of 1.64 and 1.74 V, respectively. DFT calculations suggested that the Rh_3Cu_1 alloy on the surface of the RhCu NTs was the active site centre for improvement in the OWS performance, as it can weaken the adsorption energy of atomic H and O and expedite dissociation of the water molecules. Moreover, the RhCu NTs exhibited a porous and hollow structure that exposed more active sites to OWS.[12] Yang et al. reported 3D core-shell nanosheet assemblies (NAs), consisting of an ultrathin Ru shell and RuNi alloy core, as bifunctional electrocatalysts for OWS using a universal pH range. The $Ru_3Ni_3NAs \parallel Ru_3Ni_3NAs$ electrode required an overpotential of 280 mV at 10 mA cm^{-2} in 0.5 M H_2SO_4 (Fig. 2.3d), which is much lower than that of the benchmark Ir/C $\parallel$ Pt/C electrode (370 mV). DFT results show that due to the mutually restricted D-band interaction, the binding of Ru/Ni and H/O is weakened, and the formation of O-O and H-H is promoted, leading to excellent performance (Fig. 2.3e–g).[13] These alloying strategies are not limited to two types of metals; the introduction of a third metal may further promote the OWS performance of the alloyed catalyst. Fu et al. synthesised trimetallic NiFeIr (NPs) using a colloidal strategy; ultrasmall NPs with an average diameter of 2.2 nm were produced. The $IrNi_{0.57}Fe_{0.82}/C \parallel IrNi_{0.57}Fe_{0.82}/C$ electrode required an overpotential of 410 mV at 10 mA cm^{-2} in 0.5 M $HClO_4$ solution. The high performance was attributed to the ultra-small size and monodisperse distribution of the NPs and the strong synergistic effect among Ir, Ni, and Fe.[14] Qiao et al. used a series of transition metals (M = Co, Ni, and Fe) as dopants in a RuIr alloy to prepare a trimetallic alloy through a co-reduction polyol method. The Co-RuIr nanocrystals displayed considerable OWS performance in acidic environments. The Co-RuIr $\parallel$ Co-RuIr electrode required a low cell voltage of 1.52 V at 10 mA cm^{-2} in nitrogen-saturated 0.1 M $HClO_4$ solution, which is superior to that of the commercial Pt/C $\parallel$ RuO_2 electrode. This result is owing to the ability of transition metals to tune the electronic structure and binding strength of reaction intermediates during the OWS process.[15]

Furthermore, doping noble metal oxides with transition metals is an effective avenue to promote its intrinsic activity towards OWS. Wang et al. doped ultrathin RuO_2 NWs with transition metals (Fe, Co, and Ni) to form a network structure by combining a wet-chemical method with an annealing strategy; the structure presented an abundance of defects and grain boundaries. The Co-doped RuO_2 NWs $\parallel$ Ni-doped RuO_2 NWs electrode required an overpotential of 307 mV at 10 mA cm^{-2} in 0.5 M H_2SO_4, which is superior to the commercial Ir/C $\parallel$ Pt/C electrode (412 mV). The DFT simulation results suggested that the outstanding OWS performance could be ascribed to the optimised adsorption energy of the reaction intermediates after doping.[16] Li et al. reported a highly active and durable $RuIrO_x$ (x $\geq$ 0) nano-netcage catalyst formed during electrochemical testing by *in situ* etching which removed amphoteric ZnO from the RuIrZnOx hollow nanobox (Fig. 2.4a). The $RuIrO_x \parallel RuIrO_x$ electrode required a cell voltage of 1.45 V at 10 mA cm^{-2} in 0.5 M H_2SO_4 solution (Fig. 2.4b–d). The high activity was ascribed to great exposure of the active sites of the porous nano-netcage, which significantly increased the electrochemical surface area during OWS.[17]

2.2.1.2 Noble metal-free bifunctional electrocatalysts

In principle, transition metal-based compounds can be utilised as bifunctional electrocatalysts for the HER and OER in acidic media. However, the low corrosion

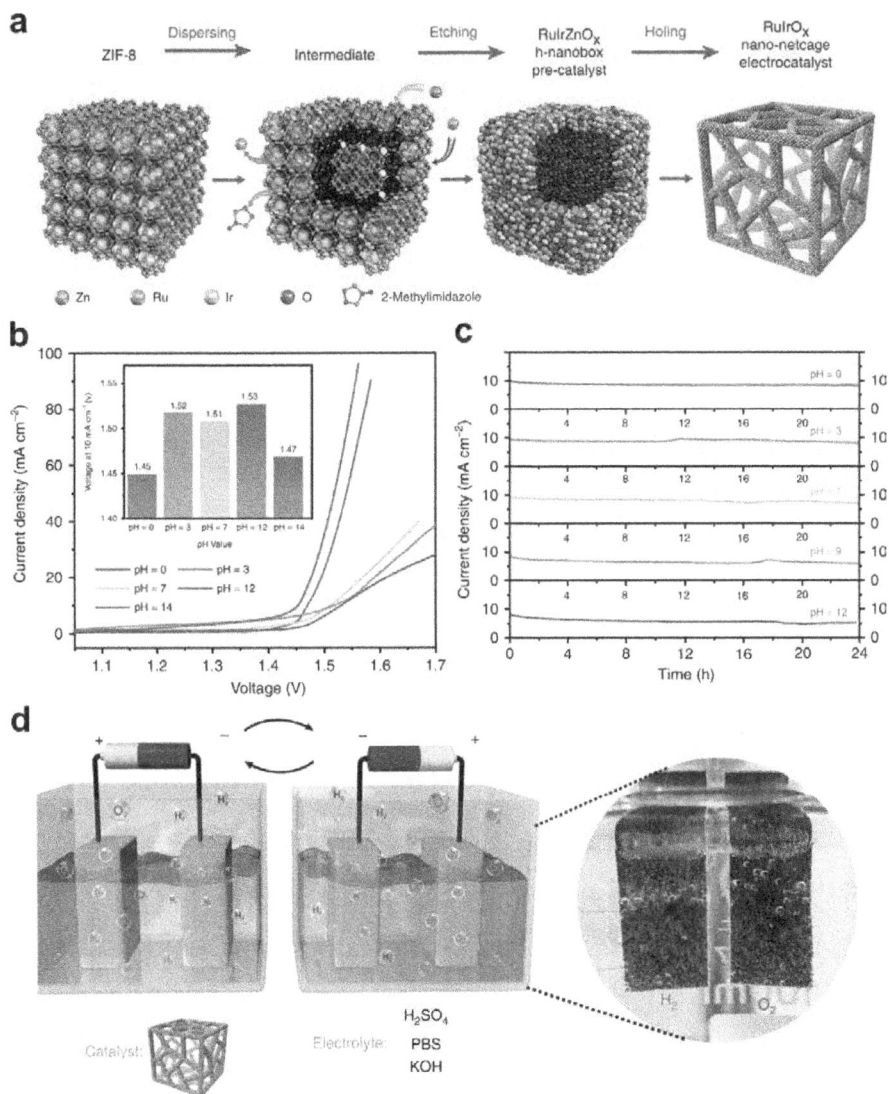

Figure 2.4. Synthetic scheme and OWS performance of the RuIrO$_x$ nano-netcages. (a) Schematic illustration of the synthetic procedure. (b) LSV curves and the voltages at 10 mA·cm^{-2} (inset) of the RuIrO$_x$ nanonetcages. (c) The chronoamperometric curves of the RuIrO$_x$ nano-netcages for 24 h at different pH values. (d) Schematic illustration of RuIrO$_x$ nano-netcages for the OWS with an AA battery. (a–d) Reproduced with permission from Ref. 28. Copyright 2017, Wiley-VCH.

resistance of transition metal species under acidic conditions is a significant challenge, based on analysis of the Pourbaix diagram. The delicate design of electrocatalysts through the use of metal chalcogenides, phosphides, and carbides is a potential solution for the application of these compounds in OWS in acidic electrolytes.[18–20] Transition metal-based chalcogenides, phosphides, and carbides are promising candidates for catalyzing OWS, mainly because the S/Se/P/C atoms in these catalysts can moderately bond with the reaction intermediates, thereby providing a reaction interface with proton and water acceptor sites.[21–23] Sun et al. prepared a superaerophobic nitrogen-doped tungsten carbide (N-WC) nanoarray electrode using hydrothermal and chemical vapour deposition methods (Fig. 2.5a). The N-WC ∥ N-WC electrode exhibited a significant OWS performance in 0.5 M H_2SO_4, where the current density reached values up to 30 mA cm^{-2} before 1.7 V *vs.* RHE (Fig. 2.5b). The high intrinsic activity of these electrodes was attributed to the following reasons: first, nitrogen doping optimised the surface energy level and thus boosted the reaction kinetics. Second, the nanoarray structure provided a large effective surface area and facilitated the evolution of H_2 and O_2.[21] Xiong et al. synthesised Co-doped MoS_2 (Co-MoS_2) nanosheets using a hydrothermal method. The Co-MoS_2 ∥ Co-MoS_2 electrode exhibited a cell voltage of 1.90 V to reach 10 mA cm^{-2} in 0.5 M H_2SO_4 (Fig. 2.5c), which is larger than that of a Pt/C ∥ RuO_2 electrode. Although the catalytic activity of the Co-MoS_2 electrode is lower than that of the commercial electrode, Co doping can optimize the conductivity and active sites of MoS_2, providing a new method for designing highly active bifunctional OWS electrocatalysts.[24] Luo et al. reported porous manganese-doped FeP/$Co_3(PO_4)_2$ (PMFCP) nanosheets on carbon cloth, which were utilized as both the anode and cathode to produce a PMFCP ∥ PMFCP electrode. This electrode required a low cell voltage of 1.75 V and achieved a current density of 10 mA cm^{-2} in 0.5 M H_2SO_4 solution.[25]

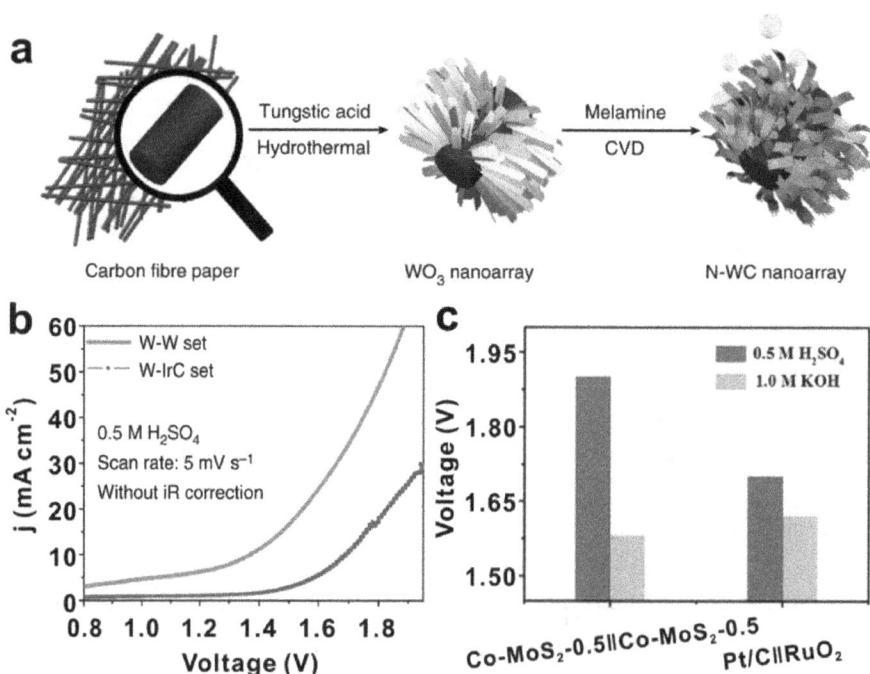

Figure 2.5. (a) Schematic illustration of WO_3 nanoarray on CFP. (b) LSV curves of N-WC ∥ N-WC electrode (noted as W-W set) for OWS. (a and b) Reproduced with permission from Ref. 21. Copyright 2018, Nature Publishing Group. (c) The cell voltage of Co-MoS_2 ∥ Co-MoS_2 electrode in 1 M KOH and 0.5 M H_2SO_4 solutions. Reproduced with permission from Ref. 24. Copyright 2018, Royal Society of Chemistry.

2.2.2 Bifunctional electrocatalysts for overall water splitting in neutral media

OWS in neutral electrolytes is closely related to several environmentally friendly renewable energy technologies and devices, such as microbial electrolysis of wastewater and direct electrolysis of seawater. Bifunctional electrocatalysts with highly intrinsic activity that can promote water splitting in neutral media have attracted widespread attention; an example of this is noble metal-based bifunctional electrocatalysts. Liu et al. reported that their Ir-NSG ‖ Ir-NSG electrode displayed excellent OWS performance in 1.0 M phosphate buffer saline (PBS) solution, affording a current density of 10 mA cm^{-2} at an overpotential of 300 mV for the OWS.[26] Li et al. showed that their RuIrO$_x$ ‖ RuIrO$_x$ electrode only required a low overpotential of 270 mV at 10 mA cm^{-2} in 1.0 M PBS solution for OWS.[27] However, the commercial application of these catalysts is limited by their low reserves and high price. Therefore, there is a need to design and synthesise non-noble metal based bifunctional electrocatalysts for OWS in neutral media. Currently, non-noble metal based bifunctional electrocatalysts that have been reported consist mainly of transition metal-based phosphides, chalcogenides, oxides, and nitrides. Table 2.3 summarizes the catalytic performances of the reported bifunctional electrocatalysts for OWS under neutral condition.

Table 2.3. The OWS performance of the reported bifunctional electrocatalysts in neutral media.

Catalysts	Electrolytes	Current density (j/mA cm^{-2})	Overpotential (η/mV)	Stability	References
Ir-NSG	1.0 M PBS	10	300	24 h @10 mA cm^{-2}	26
RuIrO$_x$	1.0 M PBS	10	270	24 h @10 mA cm^{-2}	27
CoP NA/CC	1.0 M PBS	2	370	35 h @10 mA cm^{-2}	28
np-Co$_9$S$_4$P$_4$	1.0 M PBS	10	440	24 h @10 mA cm^{-2}	29
PMFCP	1.0 M PBS	10	590	30000 s @5 mA cm^{-2}	25
Ni(S$_{0.5}$Se$_{0.5}$)$_2$	1.0 M PBS	10	640	12 h @10 mA cm^{-2}	30
MoS$_2$/Co$_9$S$_8$/ Ni$_3$S$_2$/Ni	1.0 M PBS	10	570	20 h @1.80 V	31
CoO/CoSe$_2$	0.5 M PBS	10	950	10 h @10 mA cm^{-2}	32
S-NiFe$_2$O$_4$/NF	1.0 M PBS	10	720	24 h @1.95 V	34
CoO/Co$_4$N/NF	1.0 M PBS	10/42.3	560/770	50 h @10 mA cm^{-2}	36

2.2.2.1 Transition metal-based phosphides

Transition metal-based phosphides are widely utilised as effective bifunctional electrocatalysts for OWS in neutral solution. For example, Sun et al. synthesized CoP nanosheet arrays on carbon cloth (CoP NA/CC) using α-Co(OH)$_2$ NA/CC as a precursor catalyst. The CoP NA/CC ‖ CoP NA/CC electrode reached a current density of 2 mA cm^{-2} at an overpotential of 370 mV in 1.0 M PBS electrolyte. These results demonstrate that CoP NA/CC is a promising candidate among earth-abundant electrocatalysts for OWS.[28] Tan et al. reported the use of nanoporous Co$_9$S$_4$P$_4$

(np-Co$_9$S$_4$P$_4$) pentlandite, which was prepared by phosphatising nano-sized Co$_9$S$_8$. The np-Co$_9$S$_4$P$_4 \parallel$ np-Co$_9$S$_4$P$_4$ electrode only required a voltage of 1.67 V *vs.* RHE to achieve a current density of 10 mA cm^{-2} in 1.0 M PBS (Fig. 2.6a), which is superior to the noble metal-based Pt/C $\parallel$ IrO$_2$ electrodes (490 mV). Furthermore, this couple displayed excellent stability, with the applied potential only increasing $\sim$ 0.02 V after 24 h of chronopotentiometric operation (Fig. 2.6b).[29] Luo et al. designed and

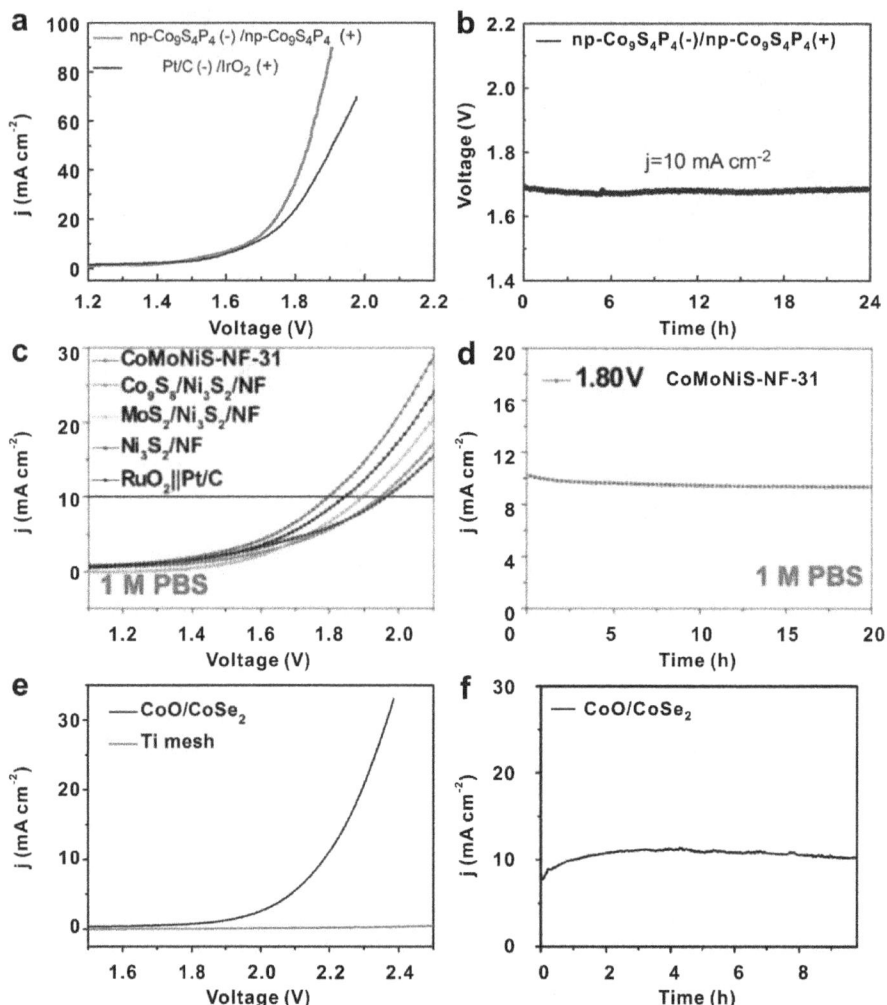

Figure 2.6 (a) OWS performance of the np-Co$_9$S$_4$P$_4 \parallel$ np-Co$_9$S$_4$P$_4$ electrode. (b) Chronopotentiomeric curves of the np-Co$_9$S$_4$P$_4 \parallel$ np-Co$_9$S$_4$P$_4$ electrode at 10 mA cm^{-2}. (a and b) Reproduced with permission from Ref. 29. Copyright 2019, American Chemical Society. (c) LSV curves of CoMoNiS-NF-xy serial electrodes for OWS. (d) Chronoamperometric curve of CoMoNiS-NF-31 at constant 1.80 V *vs.* RHE. (c and d) Reproduced with permission from Ref. 31. Copyright 2019, American Chemical Society. (e) LSV curve of CoO/CoSe$_2 \parallel$ CoO/CoSe$_2$ electrode for OWS. (f) Chronoamperometric curve of CoO/CoSe$_2 \parallel$ CoO/CoSe$_2$ electrode at constant 2.18 V *vs.* RHE. (e and f) Reproduced with permission from Ref. 32. Copyright 2016, Wiley-VCH.

synthesised porous Mn-doped $FeP/Co_3(PO_4)_2$ (PMFCP) nanosheets on carbon cloth, which was utilized as anodes and cathodes for OWS in neutral electrolyte. The PMFCP ‖ PMFCP electrode required a cell voltage of 1.82 V *vs.* RHE to reach 10 mA cm^{-2} in 1.0 M PBS. This electrode also maintained a high OWS stability after 30,000 s of operation.[25]

2.2.2.2 Transition metal-based chalcogenides

Transition metal-based sulphides and selenides are extensively utilized as bifunctional electrocatalysts for OWS. Pan et al. synthesised a series of hierarchical nickel sulphoselenide ($NiS_{2(1-x)}Se_{2x}$) hollow/porous spheres using a simple hydrothermal and selenisation method. The atomic ratio of Se and S in the $NiS_{2(1-x)}Se_{2x}$ catalysts can be controlled by regulating the selenisation levels of the NiS_2 hollow spheres. This is favourable for tuning the electronic structure of nickel sulphoselenide and improving its intrinsic catalytic activity. $Ni(S_{0.5}Se_{0.5})_2$ displayed the best intrinsic activity towards the HER and OER compared to other $NiS_{2(1-x)}Se_{2x}$ compositions. The $Ni(S_{0.5}Se_{0.5})_2$ ‖ $Ni(S_{0.5}Se_{0.5})_2$ electrode required an applied voltage of 1.87 V at a current density of 10 mA cm^{-2} in 1.0 M PBS, outperforming the commercial Pt/C ‖ IrO_2 electrode. This couple also exhibited high electrochemical stability; the current density barely changed over 12 h at 1.87 V vs. RHE.[30] Another strategy for fabricating effective bifunctional electrocatalysts is to directly integrate the HER and OER catalyst into one system. Kanatzidis et al. reported a hierarchical nano assembly of $MoS_2/Co_9S_8/Ni_3S_2/Ni$ as a highly active bifunctional electrocatalyst for OWS. They synthesised various CoMoNiS-NF-xy containing different Co/Mo ratios by using a facile hydrothermal strategy. Compared to catalysts containing a different Co/Mo ratio, CoMoNiS-NF-31 exhibited the best HER and OER performance under neutral conditions. The CoMoNiS-NF-31 ‖ CoMoNiS-NF-31 electrode required an applied voltage of 1.80 V *vs.* RHE at 10 mA cm^{-2} in 1.0 M PBS (Fig. 2.6c-d). The outstanding electrocatalytic activity of this electrode was ascribed to the synergistic effects of Co_9S_8, which are the active phases during the OWS process.[31] Li et al. loaded CoO domains in $CoSe_2$ nanobelts on the surface of a Ti mesh constructed $CoO/CoSe_2$ heterostructure as an OWS bifunctional electrocatalyst in neutral media. The $CoO/CoSe_2$ ‖ $CoO/CoSe_2$ electrode only required an applied voltage of 2.18 V at 10 mA cm^{-2} in 0.5 M PBS (Fig. 2.6e). Additionally, this electrode demonstrated superior durability (10 h at 10 mA cm^{-2}) and ~ 100% FE (Fig. 2.6f). This excellent OWS performance can be attributed to two features: (1) the integrated three-dimensional porous electrode possesses a large active surface area for OWS, promoting electron and mass transfer, and (2) the synergistic effect between metallic CoO and $CoSe_2$ affords superactive sites.[32] Zhang et al. reported a core-shell $MoS_2@CoO$ electrocatalyst for OWS in neutral solution. This $MoS_2@CoO$ electrode required an overpotential of 325 and 176 mV in 1.0 M PBS for the OER and HER, respectively. Although the OWS performance of the two-electrode configuration ($MoS_2@CoO$ ‖ $MoS_2@CoO$) was not further investigated in neutral media, this bifunctional electrocatalyst boosted both the HER and OER intrinsic activities in parallel, making it promising for seawater electrolysis.[33]

2.2.2.3 Transition metal-based oxides

Transition metal-based oxides are widely used as bifunctional electrocatalysts for OWS in basic environments. Recently, the modification and fabrication of transition-metal-based oxides for OWS in neutral media has attracted significant attention. Qiao et al. synthesized sulphur-doped $NiFe_2O_4$ on nickel foam (S-$NiFe_2O_4$/NF) using a facile thiourea-assisted electrodeposition and calcination strategy. The S-$NiFe_2O_4$/NF ‖ S-$NiFe_2O_4$/NF electrode achieved 10 mA cm^{-2} at the applied voltage of 1.95 V in 1.0 M PBS, which outperformed the PtC/NF ‖ PtC/NF and NF ‖ NF electrodes. This favourable OWS performance was attributed to a sufficient number of active sites and good electrical and mass transfer of the integrated 3D electrode.[34] Zhang et al. synthesised Co_3O_4 and $Co(OH)_2$ ultrathin nanoplate arrays (UNA) using a simple electrodeposition method without using templates or surfactants. Compared with the IrO_2 ‖ Pt/C electrode, the Co_3O_4 UNA ‖ $Co(OH)_2$ UNA electrode exhibited superior OWS performance in 1.0 M PBS based on polarisation curve analysis.[35]

2.2.2.4 Transition metal-based nitrides

Transition metal-based nitrides are extensively applied in OWS in acidic and basic solutions based on their good electrical conductivity and anti-corrosion properties. However, the key to applying transition metal-based nitrides as bifunctional electrocatalysts in neutral media is to improve their intrinsic activity towards the OER. Jiang et al. fabricated heterostructured CoO/Co_4N on a Ni foam substrate (CoO/Co_4N/NF), in which CoO and Co_4N were active for the OER and HER, respectively. The CoO/Co_4N/NF ‖ CoO/Co_4N/NF electrode achieved current densities of 10 and 42.3 mA cm^{-2} at overpotentials of 560 and 770 mV in 1.0 M PBS solution, respectively. Chronoamperometric measurements demonstrated that this electrode has outstanding stability after 50 h of operation. The good OWS performance was attributed to two factors. First, the high-index crystal plane (111) between Co_4N nanowhiskers and CoO facilitated electron transportation. Second, the integrated 3D electrode, with a porous structure, possess a large surface area, and acts as current collector.[36]

2.2.3 Bifunctional electrocatalysts for overall water splitting in basic media

Compared with neutral and acidic water splitting, the development of bifunctional electrocatalysts for OWS in basic electrolyte has attracted more attention. This is because most OER catalysts can prevent material dissolution in an alkaline electrolyte and thus expand the range of the OWS electrocatalysts. The recent advances in OWS electrocatalysts applied under basic conditions are summarised and listed in Table 2.4.

2.2.3.1 Noble metal-based bifunctional electrocatalysts

The chemical composition of noble metal-based electrocatalysts mainly include Pt, Ru, and Ir, which have unique electronic structures and high intrinsic activities for the HER and OER.[8,37] Although RuO_2, IrO_2, and Pt/C are the benchmark electrocatalysts for the OER and HER, their intrinsic activities for the two half-reactions can be further improved to produce a promising bifunctional electrocatalyst for OWS.

Table 2.4. The OWS Performance of the reported bifunctional electrocatalysts in basic media.

Catalysts	Electrolytes	Current density (j/mA cm^{-2})	Overpotential (η/mV)	Stability	References
RuO$_2$/N-C	1.0 M KOH	10	304	12 h @1.55 V	38
Co-Pt/C NAs	1.0 M KOH	10	310	120 h @1.7 V	39
IrRh NAs	1.0 M KOH	10	340	20 h @10 mA cm^{-2}	40
RuS$_2$	1.0 M KOH	10	297	20 h @10 mA cm^{-2}	41
NiRu@MWCNTs	1.0 M KOH	10	280	48 h @10 mA cm^{-2}	42
CoO$_x$–RuO$_2$/NF	1.0 M KOH	10	260	48 h @1.5 A cm^{-2}	43
Ir-NiCo LDH	1.0 M KOH	10	220	200 h @10 mA cm^{-2}	44
NiVRu(Ir) LDH	1.0 M KOH	10	190	300 h @10 mA cm^{-2}	5
SNCF-NRs	1.0 M KOH	10	450	30 h @ 10 mA cm^{-2}	47
S-LCF	1.0 M KOH	10	440	7 h @ 1.78 V	48
Co(OH)$_2$/SFM	1.0 M KOH	10	370	68 h @ 10 mA cm^{-2}	49
CoP-PBSCF	1.0 M KOH	10	440	12 h @ 10 mA cm^{-2}	50
STFN/CNT-700	1.0 M KOH	10	570	30 h @ 10 mA cm^{-2}	51
MoS$_2$-Ni$_3$S$_2$ HNRs/NF	1.0 M KOH	10	270	48 h @ 1.53 V	52
CuCo$_2$S$_4$/NF	1.0 M KOH	100	420	50 h @ 1.50 V	53
P$_{9.03\%}$-(Ni,Fe)$_3$S$_2$/NF	1.0 M KOH	10	310	15 h @ 1.80 V	54
Ni$_{0.33}$Co$_{0.67}$S$_2$ NWs/CC	1.0 M KOH	10	340	30 h @ 1.65 V	55
CoMoS$_4$/Ni$_3$S$_2$	1.0 M KOH	10	338	10 h @ 10 mA cm^{-2}	56
CoSe$_2$/CF	1.0 M KOH	10	400	30 h @ 20 mA cm^{-2}	57
CoNiSe$_2$/NF	1.0 M KOH	10	361	20 h @ 10 mA cm^{-2}	58
Ni$_2$P	1.0 M KOH	10	240	100 h @ 1.47 V	59
MoP	1.0 M KOH	10	390	20 h @ 10/20 mA cm^{-2}	60
Cu$_3$P/g-C$_3$N$_4$	1.0 M KOH	10	310	35 h @10 mA cm^{-2}	61

Table 2.4 contd. ....

...Table 2.4 contd.

Catalysts	Electrolytes	Current density $(j/mA\ cm^{-2})$	Overpotential (η/mV)	Stability	References
CoP/SPNF	1.0 M KOH	30	317	24 h @40 mA cm^{-2}	62
Fe-CoP UNSs/NF	1.0 M KOH	10	230	50 h @30 mA cm^{-2}	63
Ni-Fe-P@NC/NF	1.0 M KOH	10	240	100 h @50 mA cm^{-2}	64
V-FeNi$_2$P/NF	1.0 M KOH	10	340	48 h @ 1.57 V	65
Mo-NiCoP	1.0 M KOH	10	380	48 h @ 10,20, and 50 mA cm^{-2}	66
Ni$_2$P	1.0 M KOH	10	350	24 h @10 mA cm^{-2}	67
$(Ni_{0.75}Fe_{0.25})_2P$	1.0 M KOH	10	311	30 h @10 mA cm^{-2}	68
$(Ni_{0.66}Fe_{0.33})_2P$	1.0 M KOH	10	380	-	69
NiCo$_2$O$_4$/Ni$_2$P	1.0 M KOH	10	360	50 h @ 10 and 50 mA cm^{-2}	70
Ni$_2$P/NF	1.0 M KOH	10	400	24 h @10 mA cm^{-2}	67
P-NiFe	1.0 M KOH	10	280	100 h @10 mA cm^{-2}	71
Ni$_2$P@NiFeAlO$_x$	1.0 M KOH	10	290	24 h @10 mA cm^{-2}	72
Ni$_2$P/Ni/NF	1.0 M KOH	10	260	40 h @10 and 20 mA cm^{-2}	73
NiCoP	1.0 M KOH	50	540	20 h @2 V	74
NiCoP@NiMn LDH	1.0 M KOH	10	289	50 h @100 mA cm^{-2}	75
NiFe-NiMo/Ni-P	1.0 M KOH	10	280	200 h @1.75 V	76
NiFeP@C	1.0 M KOH	47	370	25 h @10 mA cm^{-2}	77
Ni-Fe-P-B@CFP	1.0 M KOH	10	350	60 h @10, 30, 50, and 100 mA cm^{-2}	78
NiMn$_{1.5}$PO$_4$/NF	1.0 M KOH	10	250	50 h @10 mA cm^{-2}	79
NiMoP$_2$-Ni$_2$P/CC	1.0 M KOH	10	250	20 h @10 mA cm^{-2}	80
Ni$_x$P$_y$-325	1.0 M KOH	10	340	60 h @1.57 V	81
Ni$_2$Mn$_1$P	1.0 M KOH	10	370	20 h @1.6 V	82
Co$_{0.6}$Fe$_{0.4}$P-1.125	1.0 M KOH	10	340	120 h @1.65 V	83

Table 2.4 contd. ....

...Table 2.4 contd.

Catalysts	Electrolytes	Current density $(j/mA\ cm^{-2})$	Overpotential (η/mV)	Stability	References
Co$_2$P/Mo$_2$C/ Mo$_3$Co$_3$C@C	1.0 M KOH	10	510	-	84
Er doped CoP	1.0 M KOH	10	350	25 h @ 1.58, 1.63, and 1.71 V	85
10%Cr-CoP/30% Fe-CoP	1.0 M KOH	100	440	-	86
CoFeP/CF	1.0 M KOH	10	265	60 h @1.65 V	87
Co-Mo-P-O	1.0 M KOH	10	340	80 h @1.65 V	88
CoP@NPCSs	1.0 M KOH	10	413	20 h @ 1.643 V	89
CoP/MXene	1.0 M KOH	10	330	24 h @ 1.69 V	90
CoP NFs	1.0 M KOH	10	420	30 h @1.7 V	91
CoP$_3$/NiMoO$_4$-NF-2	1.0 M KOH	10	340	12 h @1.68 V	92
CoPS	1.0 M KOH	10	360	50 h @ 1.65 V	93
MoP/Ni$_2$P/NF	1.0 M KOH	10	320	15 h @10 mA cm^{-2}	94
CoVP@CC	1.0 M KOH	10	380	-	95
FeCoP$_x$S$_y$	1.0 M KOH	10	360	10 h @ 1.67 V	96
FeP/Ni$_2$P	1.0 M KOH	10	190	36 h @ 30 and 100 mA cm^{-2}	97
Fe-Ni$_2$P/MoSx/ NF	1.0 M KOH	10	380	36 h @ 10 mA cm^{-2}	98
Fe$_x$-NiCoP	1.0 M KOH	10	380	25 h @ 10 mA cm^{-2}	99
H-CoP@NC	1.0 M KOH	10	380	24 h @ 1.72 V	100
N-CoO@CoP	1.0 M KOH	100	560	50 h @ 50 mA cm^{-2}	101

Xu et al. synthesized highly dispersed ultra-small RuO$_2$ NPs anchored *in situ* to an N-doped carbon matrix (RuO$_2$/N–C) using a molecule-assisted pyrolysis and subsequent oxidation method (Fig. 2.7a). The RuO$_2$/N–C ‖ RuO$_2$/N–C electrode required an overpotential of only 304 mV to reach 10 mA cm^{-2} in 1.0 M KOH solution (Fig. 2.7b), which is lower than that of the Pt/C ‖ RuO$_2$ electrode (381 mV). Moreover, this electrode maintained a high performance at cell voltage of 1.55 V *vs.* RHE after 12 h of operation (Fig. 2.7c), which suggests that the electrode has long-term stability in basic media. The considerable activity of the electrode was ascribed to the synergistic effect between the active RuO$_2$ ultra-small NPs and the N-C substrate based on DFT calculation results.[38] Zhang et al. successfully synthesised a series of open hollow Co-Pt bimetallic nanoclusters incorporated into nitrogen- doped carbon nanoflake arrays grown on flexible carbon cloth (Co-Pt/C NAs) by a carbonisation treatment and galvanic replacement reaction using Co-MOFs as the initial precursor.

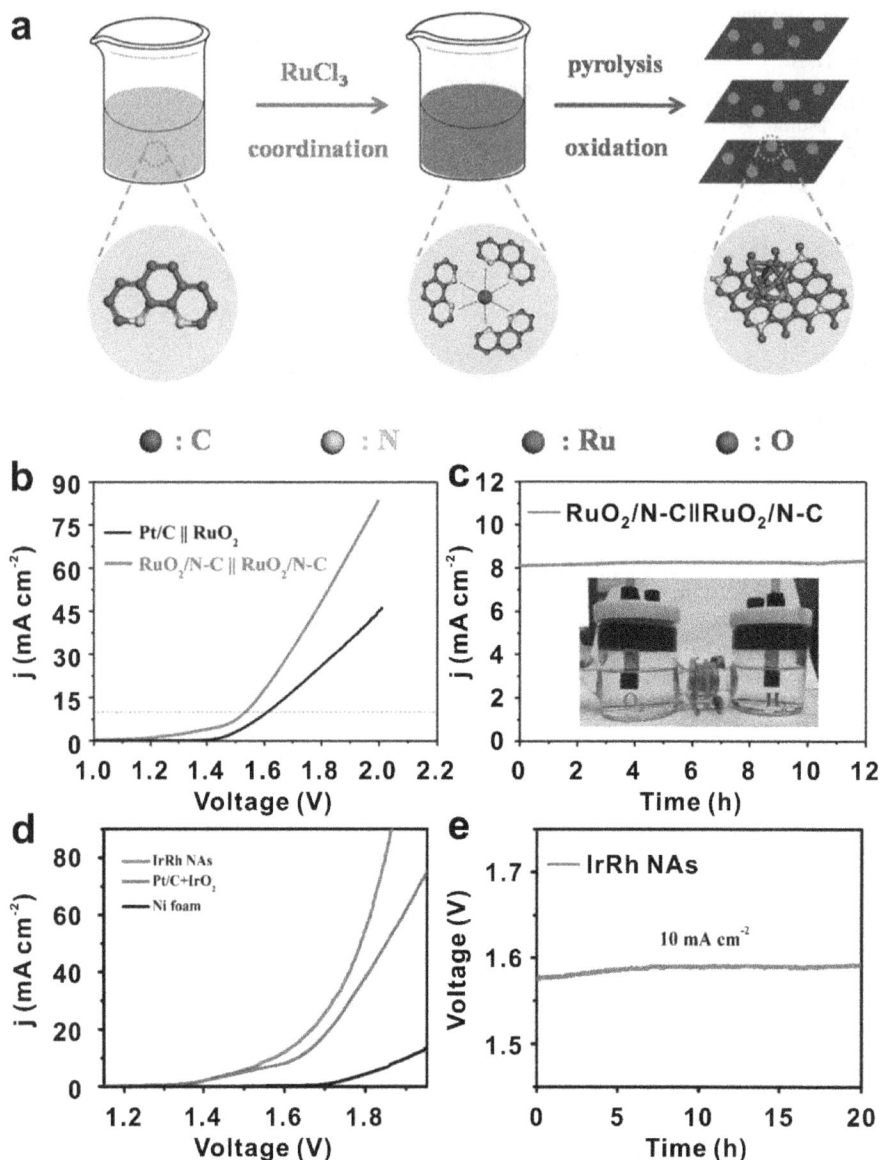

Figure 2.7. (a) Scheme for the preparation of the RuO$_2$/N–C Catalyst. (b) LSV curves for OWS of RuO$_2$/N–C ‖ RuO$_2$/N–C electrode. (c) Chronoamperometric curve of RuO$_2$/N–C ‖ RuO$_2$/N–C electrode at constant voltage of 1.55 V *vs.* RHE. (a–c) Reproduced with permission from Ref. 38. Copyright 2018, American Chemical Society. (d) LSV curves of IrR NAs ‖ IrR NAs electrode. (e) Chronopotentiomeric curve of the IrRh NAs ‖ IrRh NAs electrode at the constant 10 mA cm^{-2} for the OWS. (d and e) Reproduced with permission from Ref. 40. Copyright 2019, American Chemical Society.

The Co-Pt/C-10 sample showed the best performance for the OER and HER in 1.0 M KOH solution. The Co-Pt/C-10 ‖ Co-Pt/C-10 electrode reached a catalytic current density of 10 mA cm^{-2} with a low overpotential of 310 mV in 1.0 M KOH solution, which is lower than that of the commercial Ir/C ∥ Pt/C electrode. Furthermore, this two-electrode system exhibited good stability with only small changes in the current density after 120 h of operation at 1.7 V. DFT calculation results demonstrated that the outstanding performance of the catalyst resulted from the synergistic effect between Co and Pt which accelerated water dissociation and shifted the d-band centre.[39] Wang et al. prepared IrRh nanosheet assemblies (IrRh NAs) by a facile solution growth method using $IrCl_3$ and $RhCl_3$ precursors. The IrRh NAs ∥ IrRh NA electrode required an applied voltage of 1.57 V *vs.* RHE to reach a current density of 10 mA cm^{-2} in 1.0 M KOH electrolyte (Fig. 2.7d), which is superior to that of the Pt/C ∥ IrO_2 (1.63 V) and NF ∥ NF (1.90 V) electrodes. Furthermore, this electrode exhibits considerable electrochemical stability, with slight changes in the voltage after 20 h of operation at 10 mA cm^{-2} (Fig. 2.7e). The excellent performance of the IrRh NAs toward OWS was mostly attributed to its beneficial chemical composition and unique morphology.[40] Zhu et al. reported a pyrite-type RuS_2, with disorder and defects, using a low-temperature sulphurisation strategy. The RuS_2-500/CP ∥ RuS_2-500/CP electrode only require an applied voltage of 1.527 V to achieve 10 mA cm^{-2} in 1.0 M KOH solution, outperforming the Pt/C/CP ∥ RuO_2/CP and RuO_2/CP ∥ RuO_2/CP electrodes. This electrode also maintained its great stability with negligible performance decay after 20 h of testing at 10 mA cm^{-2}. This significant activity was ascribed to the unique disordered structure and abundant defects of the catalyst, which increased the number of active sites and tuned the electronic structure of the RuS_2 electrocatalysts based on experimental and DFT calculation results.[41]

Combining precious metal with earth-abundant catalysts to form an integrated bifunctional electrocatalyst is another effective strategy to reduce the cost of noble metal-based electrocatalysts and enhance the intrinsic activities of the electrocatalysts. Zhang et al. synthesised NiRu alloy NPs anchored on multi-walled carbon nanotubes (NiRu@MWCNTs) using a plasma reduction method (Fig. 2.8a). The NiRu@MWCNTs ∥ NiRu@MWCNTs electrode needed a cell voltage of 1.51 V at 10 mA cm^{-2} in 1.0 M KOH solution (Fig. 2.8b), which is superior to those Pt/C ∥ RuO_2, Ni@MWCNTs ∥ Ni@MWCNTs, and Ru@MWCNTs ∥ Ru@MWCNTs electrodes. This electrode exhibited excellent durability, and the current density remained at 97.7% after 48 h of chronopotentiometry operation at 10 mA cm^{-2} (Fig. 2.8c). DFT simulations demonstrated that the metal-modified defective and non-defective carbon sites act as the HER and OER active sites, respectively.[42] Yu et al. successfully synthesised an amorphous CoO_x anchored crystalline RuO_2 nanosheet catalyst on Ni foam (CoO_x–RuO_2/NF) using a hydrothermal and annealing strategy. The CoO_x–RuO_2/NF ∥ CoO_x–RuO_2/NF electrode only needed 1.49 V to reach 10 mA cm^{-2} for OWS in 1.0 M KOH solution, outperforming the Pt/C/NF ∥ IrO_2/C/NF electrode. This electrode showed satisfactory stability, and the potential showed no noticeable changes after 48 h of galvanostatic operation at 1500 mA cm^{-2}. The great activity of CoO_x–RuO_2/NF was owing to a sufficient number of active sites provided by the combination of the amorphous CoOx and crystalline RuO_2.[43] Fan et al. doped atomic Ir into a NiCo layered double hydroxide (LDH) to form an iridium-doped NiCo LDH (Ir-NiCo LDH) using solvothermal and galvanic displacement methods. The Ir-NiCo LDH ∥ Ir-NiCo LDH electrode required an extremely small overpotential of 220 mV at a current density of 10 mA cm^{-2} in 1.0 M KOH solution, which is much lower

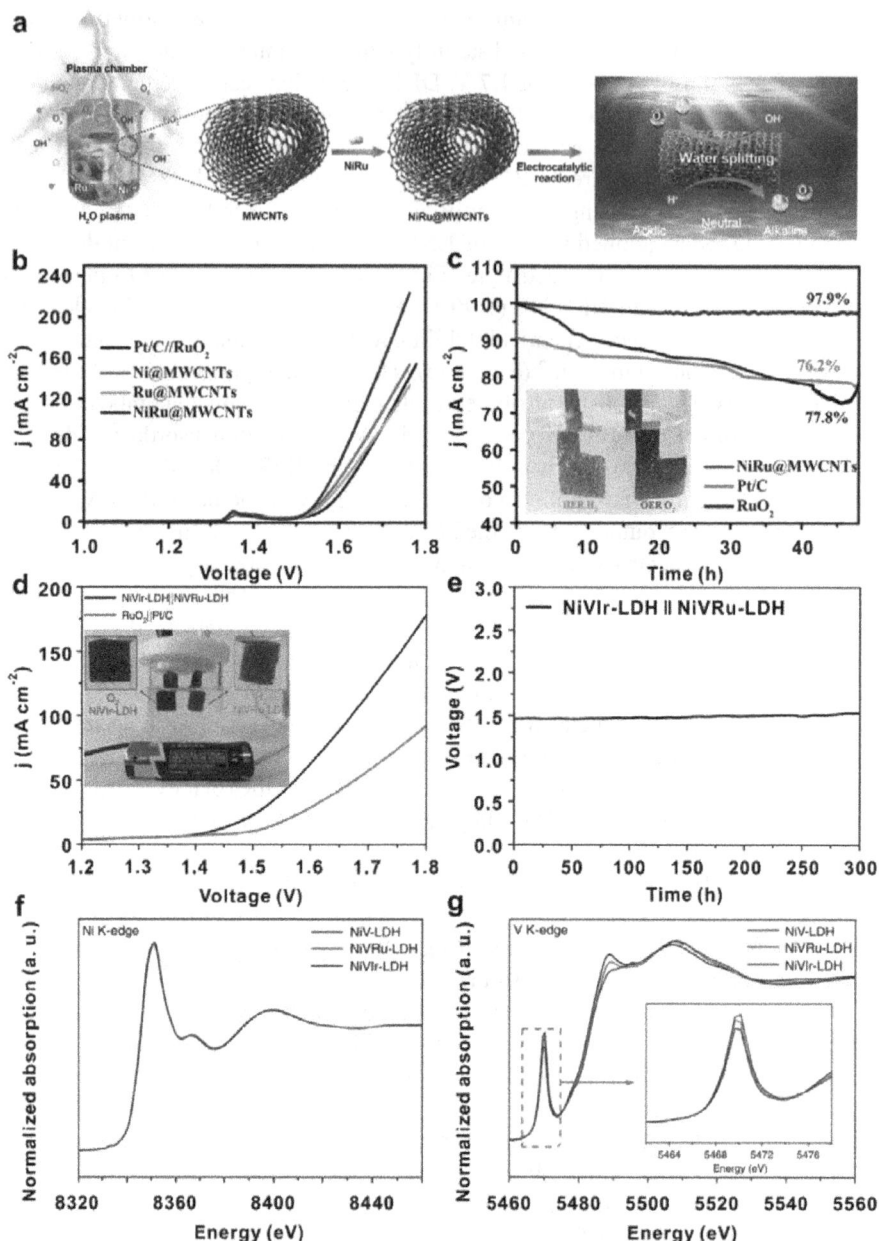

Figure 2.8. (a) Scheme of the synthesis of NiRu@MWCNTs electrocatalyst. (b) LSV curves of various electrodes. (c) Choronoamperometric curves of the Ni@MWCNTs ‖ Ni@MWCNTs electrode at 10 mA cm⁻². (a–c) Reproduced with permission from Ref. 42. Copyright 2020, American Chemical Society. (d) LSV curves of NiVIr-LDH ‖ NiVRu-LDH electrode for OWS. (e) Chronopotentiomeric curve of the NiVIr-LDH ‖ NiVRu-LDH electrode toward OWS at constant 10 mA cm⁻². XANES spectra of trimetallic NiVRu-LDH for (f) Ni K-edge and (g) V K-edge. (d–g) Reproduced with permission from Ref. 5. Copyright 2019, Nature Publishing Group.

than that required by the NiCo LDH || NiCo LDH, Ir NPs/NiCo LDH || Ir NPs/NiCo LDH, and Pt/C|| Ir/C electrodes. Additionally, the long-term durability results indicated that the potential did not clearly change after 200 h of operation at 10 mA cm^{-2}. The excellent stability was ascribed to (1) Ir incorporated into the crystal lattice of the NiCo LDH which avoids corrosion from the harsh alkaline environment and (2) the Ir-NiCo LDH defects that can tune the electronic structures and afford more efficient anchoring sites, stabilising the heteroatom sites.[44] Yang et al. introduced noble metals (Ru and Ir) into a NiV LDH to modulate the electronic structure which accelerated the reaction kinetics and improved the intrinsic activity of the catalyst towards OWS. The NiVIr-LDH || NiVRu-LDH electrode only required a cell voltage of 1.42 V *vs.* RHE at 10 mA cm^{-2} in 1.0 M KOH electrolyte (Fig. 2.8d), which is superior to that of the benchmark RuO$_2$ || Pt/C electrode (1.50 V). In addition, this electrode exhibited excellent long-term catalytic stability, with negligible changes in the potential after 300 h of continuous current operation (Fig. 2.8e). This outstanding performance was attributed to three factors: (1) the synergistic effects among Ni, V, and Ru (Ir) optimised the electronic structure of the NiVIr LDH and NiVRu LDH, based on X-ray absorption near-edge structure spectroscopy analysis (Fig. 2.8f). (2) The Ir or Ru doping tuned the local coordination environments around the Ni and V sites, and generated distorted octahedral V vacancies, based on Raman spectra and extended X-ray absorption fine structure results (Fig. 2.8g). (3) DFT simulation results demonstrated that Ru or Ir incorporation could optimise the adsorption energy of the reaction intermediates in the rate-determining step of the HER and accelerate the entire kinetic process for the OER, thus boosting the OWS performance.[5]

2.2.3.2 Transition-metal-based oxides

Perovskite-type oxides have been studied comprehensively as representative noble metal-free-based oxide OWS electrocatalysts under alkaline conditions owing to their earth abundance, tuneable chemical composition and structure, and excellent intrinsic activities. As displayed in Fig. 2.9a, primitive perovskite oxides have the general formula ABO$_3$, where A represents an alkali/rare earth metal cation and B is located in the centre of the octahedron structure and coordinated to O at the corners of the octahedron. The OWS activity of perovskite oxides can be easily optimised by modifying its chemical composition.

Notably B-sites are usually deemed as the active centres, because the electronic structure of the B-site is closely related to the reaction intermediates. In addition to primitive perovskites, other novel perovskite-type oxides exhibit good OWS performance.[46] Liu et al. synthesised SrNb$_{0.1}$Co$_{0.7}$Fe$_{0.2}$O$_{3-\delta}$ perovskite nanorods (SNCF-NRs) using facile electrospinning strategies. The SNCF-NR || SNCF-NR electrode required an overpotential of 450 mV at 10 mA cm^{-2} and exhibited excellent stability in 1.0 M KOH solution. Chronopotentiometry measurements demonstrated that the voltage did not noticeably change after 30 h of operation at 10 mA cm^{-2} in 1.0 M KOH. The excellent performance was ascribed to the increased electrochemical surface area and faster charge transfer rate.[47]

To further improve the intrinsic OWS activity of perovskites in alkaline electrolytes, hybrid catalysts are constructed based on the synergistic effect between

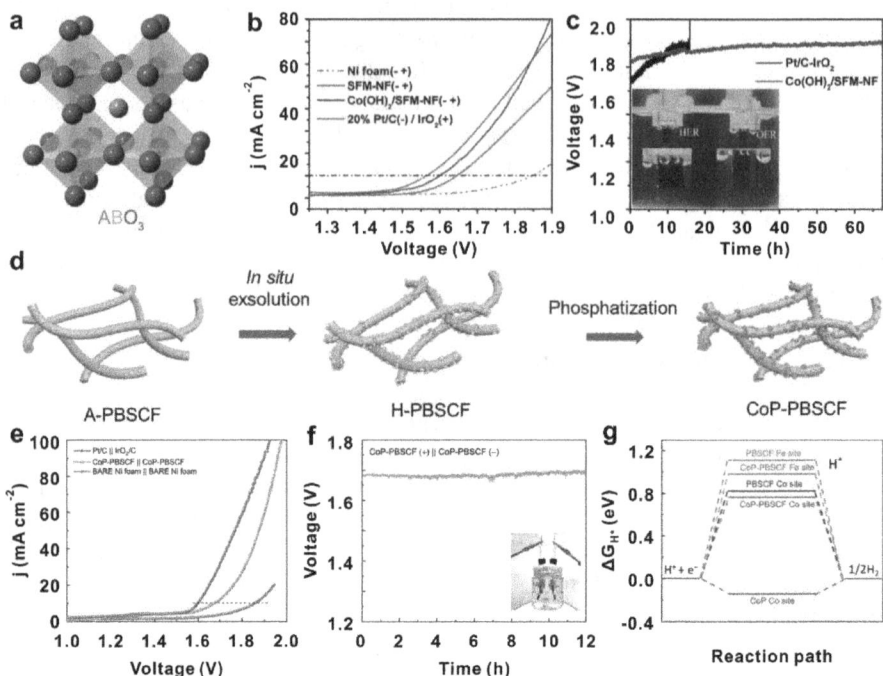

Figure 2.9. (a) Crystal structure of a primitive perovskite oxide. Reproduced with permission from Ref. 45. Copyright 2017, Royal Society of Chemistry. (b) LSV curves of $Co(OH)_2$/SFM-NF ∥ $Co(OH)_2$/SFM-NF electrodes for OWS. (c) Chronopotentiometric curves of $Co(OH)_2$/SFM-NF ∥ $Co(OH)_2$/SFM-NF electrode toward OWS at fixed 10 mA cm⁻². (b and c) Reproduced with permission from Ref. 49. Copyright 2020, Royal Society of Chemistry. (d) Scheme of the synthesis for CoP-PBSCF. (e) LSV curves of the OWS on the CoP-PBSCF ∥ CoP-PBSCF electrodes. (f) Chronopotentiometric curve of CoP-PBSCF ∥ CoP-PBSCF toward OWS at 10 mA cm⁻². (g) HER free energy diagrams of Co-sites on the surface of CoP-PBSCF. (d–g) Reproduced with permission from Ref. 50. Copyright 2019, Royal Society of Chemistry.

the heterogeneous interfaces. For example, Li et al. constructed a CoS_2 pinned $LaFe_{0.8}Co_{0.2}O_3$ (S-LCF) perovskite oxide by *in situ* etching and sulphurisation methods, which tuned the electronic structure of the B-site (Co) and improved the intrinsic activity of the catalyst towards OWS. The S-LCF ∥ S-LCF electrode required an overpotential of 440 mV at 10 mA cm⁻² in 1.0 M KOH, outperforming a Ni foam ∥ Ni foam electrode (650 mV). This electrode also displayed good stability, as the current density showed no apparent change after 7 h of tests at 1.78 V in nitrogen-saturated basic solution. The excellent performance and stability of the S-LCF ∥ S-LCF electrode are mainly owing to the strong coupling effect between the CoS_2 NP and THE LCF perovskite.[48] He et al. deposited amorphous $Co(OH)_2$ on the surface of $Sr_2Fe_{1.5}Mo_{0.5}O_{6-\delta}$ (SFM) ($Co(OH)_2$/SFM) using atomic layer deposition method. The $Co(OH)_2$/SFM ∥ $Co(OH)_2$/SFM electrode only required a low overpotential of 370 mV at a current density of 10 mA cm⁻² and displayed an excellent durability after 68 h of operation at 10 mA cm⁻² in 1.0 M KOH solution (Fig. 2.9b-c). The considerable performance is mainly owing to the synergistic effect between the amorphous $Co(OH)_2$ nanoflakes and SFM nanofibres.[49] Zhang et al.

prepared CoP socketed on perovskite $PrBa_{0.5}Sr_{0.5}Co_{1.5}Fe_{0.5}O_{5+\delta}$ (CoP-PBSCF) with O deficiencies using exsolution and post-growth strategies (Fig. 2.9d). The CoP-PBSCF ‖ CoP-PBSCF electrode required an overpotential of 440 mV at 10 mA cm^{-2} in 1.0 M KOH solution (Fig. 2.9e), which is slightly higher than that of the noble metal-based Pt/C ‖ IrO$_2$/C electrolyte (370 mV). Furthermore, this electrode was stable in the alkaline electrolyte, with no remarkable change in the cell voltage after 12 h of testing at a constant current density of 10 mA cm^{-2} (Fig. 2.9f). This enhanced performance was ascribed to the optimal electronic structure of the catalyst (Fig. 2.9g).[50] Shao et al. deposited carbon nanotubes (CNTs) on the surface of a $SrTi_{0.1}Fe_{0.85}Ni_{0.05}O_{3-\delta}$ perovskite at 700°C (STFN/CNT-700) using a facile chemical vapour deposition (CVD) strategy. The STFN/CNT-700 ‖ STFN/CNT-700 electrode required an overpotential of 570 mV at 10 mA cm^{-2} and exhibited excellent stability after 30 h of galvanostatic measurements. Its excellent performance, mainly owing to the synergistic effect between the CNTs and STFN, improved its electronic transfer rate and electrochemical active surface area (ECSA).[51]

2.2.3.3 Transition metal-based chalcogenides

Numerous transitional metal-based chalcogenides, including sulphides and selenides, such as Ni, Cu, Co, Mo, Fe, and alloys, have been explored for their excellent OWS performance under alkaline conditions. Yang et al. reported the successful preparation of MoS2-Ni3S2-bonded heterojunction nanorods on nickel foam (MoS2-Ni3S2 HNRs/NF) (Fig. 2.10a). This electrode required a low overpotential of 270 mV at 10 mA cm^{-2} and exhibited excellent durability for approximately 48 h in 1.0 M KOH solution (Fig. 2.10b and c). The excellent performance of the electrode was ascribed to the highly active heterointerfaces and good electron and mass transport.[52] Czioska et al. successfully synthesised $CuCo_2S_4$ nanowire arrays on Ni foam (CuCo$_2$S$_4$/NF) by hydrothermal and sulphidation strategies using Na$_2$S as asulphur precursor. The CuCo$_2$S$_4$/NF ‖ CuCo$_2$S$_4$/NF electrode required an extremely low cell voltage of 1.65 V at 100 mA cm^{-2} in 1.0 M KOH solution. This electrode also demonstrated remarkable stability, with a slight degradation after 50 h operation at 1.50 V.[53] Liu et al. doped Fe into (Ni,Fe)$_3$S$_2$ nanosheet arrays on Ni foam (P-(Ni,Fe)$_3$S$_2$/NF) by a simultaneous phosphorisation and sulphuration approach using $NaH_2PO_2 \cdot H_2O$ (phosphorus source) and 2-mercaptoethanol (sulphur source) as reaction precursors. The P$_{9.03\%}$-(Ni,Fe)$_3$S$_2$/NF sample displayed a better HER and OER performance compared to the other samples which contained different phosphorus contents. The P$_{9.03\%}$-(Ni,Fe)$_3$S$_2$/NF ‖ P$_{9.03\%}$-(Ni,Fe)$_3$S$_2$/NF electrode required an overpotential of 310 mV at 10 mA cm^{-2} and exhibited excellent durability after 15 h of electrolysis at 1.80 V in 1.0 M KOH. Its remarkable performance was attributed to P incorporation, which can optimise the absorption energy and improve the conductivity.[54] Li et al. prepared NiCo disulphide nanowires grown *in situ* on a carbon cloth (Ni$_{0.33}$Co$_{0.67}$S$_2$ NWs/CC) by hydrothermal and sulphuration approaches (Fig. 2.10d). The Ni$_{0.33}$Co$_{0.67}$S$_2$ NWs/CC ‖ Ni$_{0.33}$Co$_{0.67}$S$_2$ NWs/CC electrode required a low overpotential of only 340 mV achieved at 10 mA cm^{-2} and exhibited good stability after 30 h of electrolysis at 1.65 V in 1.0 M KOH solution (Fig. 2.10e and f). The

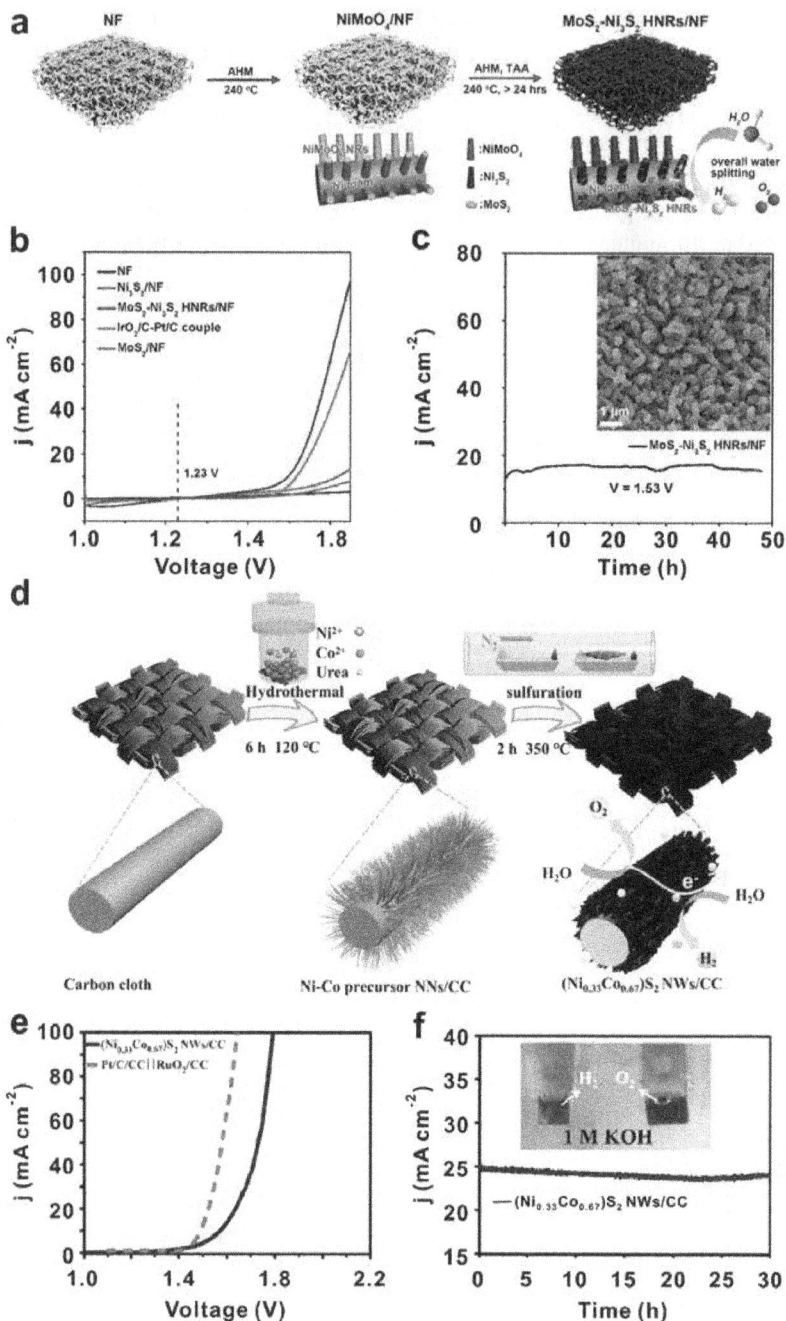

Figure 2.10. MoS_2–Ni_3S_2 HNRs/NF composites for (a) scheme illustration of the preparation, (b) LSV curves and (c) chronoamperometric tests at 1.53 V *vs.* RHE of MoS_2–Ni_3S_2 HNRs/NF for OWS. (a–c) Reproduced with permission from Ref. 52. Copyright 2017, American Chemical Society. $(Ni_{0.33}Co_{0.67})S_2$ NWs/CC electrodes for (d) illustration of the preparation procedure, (e) LSV curves, and (f) chronoamperometric curve operated at 1.65 V *vs.* RHE. (d–f) Reproduced with permission from Ref. 55. Copyright 2018, American Chemical Society.

synergistic effect between the NiCo bisulphides and the 3D substrate with high conductivity results in an excellent performance.[55]

In addition to the above reports, Li et al. fabricated a porous $CoMoS_4/Ni_3S_2$ heterostructure on a Ni foam through a self-templating route using $CoMoO_4$ as precursor. The $CoMoS_4/Ni_3S_2 \parallel CoMoS_4/Ni_3S_2$ electrode required an overpotential of only 338 mV to achieve 10 mA cm^{-2} and exhibited good durability after 10 h of operation at a constant current density of 10 mA cm^{-2} in 1.0 M KOH solution. The excellent performance of the electrode was attributed to the synergistic effect between the $CoMoS_4$ nanosheets and Ni_3S_2 NPs.[56] Sun et al. synthesised $CoSe_2$ NPs grown *in situ* on carbon paper ($CoSe_2/CF$) through co-deposition and selenisation strategies. The $CoSe_2/CF \parallel CoSe_2/CF$ electrode required an overpotential of 400 mV to achieve a current density of 10 mA cm^{-2} in 1.0 M KOH solution, outperforming the benchmark Pt/C/CF $\parallel$ RuO$_2$/CF electrode. Chronopotentiometric measurements showed that the potential did not noticeably degrade after 30 h of tests at a constant current density of 20 mA cm^{-2}. The outstanding performance and stability of the electrode was ascribed to the unique synthetic approach and 3D NF substrate which has a high surface area and electronic conductivity.[57] The Tan group grew ultrathin $CoNiSe_2$ nanorods on NF and fabricated an integrated 3D electrode ($CoNiSe_2/NF$). The $CoNiSe_2/NF \parallel CoNiSe_2/NF$ electrode required a low cell voltage of 1.591 V at 10 mA cm^{-2} and displayed good durability after 20 h of operation at 10 mA cm^{-2} in 1.0 M KOH electrolyte. Its remarkable performance compared to $CoSe_2$ and $NiSe_2$ mixtures was ascribed to the electronic structure change of the ternary electrode.[58]

2.2.3.4 Transition metal-based phosphides

Transition metal-based phosphides are promising candidates for OWS under in alkaline conditions because of their good electronic conductivity and high electronegativity of the phosphorus sites. Numerous transition metal-based phosphides have been studied using metal phosphating and doping methods. Transition metal-based phosphides are characterised into three categories based on the type and number of metals used: monometallic, bimetallic, and trimetallic phosphides.

Monometallic phosphides are extensively utilised as highly active bifunctional electrocatalysts for OWS under basic conditions. For example, Nouraiz et al. synthesised the ultrathin Ni_2P nanosheets through microwave-assisted and phosphorisation strategies using NaH_2PO_2 as the phosphorus precursor (Fig. 2.11a). The $Ni_2P \parallel Ni_2P$ electrode required an overpotential of only 240 mV at 10 mA cm^{-2} and the electrolyte is 1.0 M KOH solution (Fig. 2.11b), and that is superior to the benchmark $RuO_2 \parallel$ Pt/C electrode. The electrode also exhibited excellent electrochemical stability there, at a constant potential of 1.47 V *vs.* RHE, with no noticeable fluctuation in the current density after 100 h of operation (Fig. 2.11c). The outstanding performance of the electrode can be ascribed to its low thickness, rough surface, abundant defects, and porous morphology.[59] Lu et al. fabricated a self-supported hierarchical electrode by phosphorising MoO_2 nanoflakes on Ni foam to form porous MoP nanoflakes. The MoP/NF $\parallel$ MoP/NF electrode required an overpotential of 390 mV, which was achieved at 10 mA cm^{-2} in 1.0 M KOH solution; this compares well to RuO$_2$/NF $\parallel$ Pt/C/NF electrode. Stability measurements demonstrated that the potential did not change after 20 h of operation at 10 and 20 mA cm^{-2}.

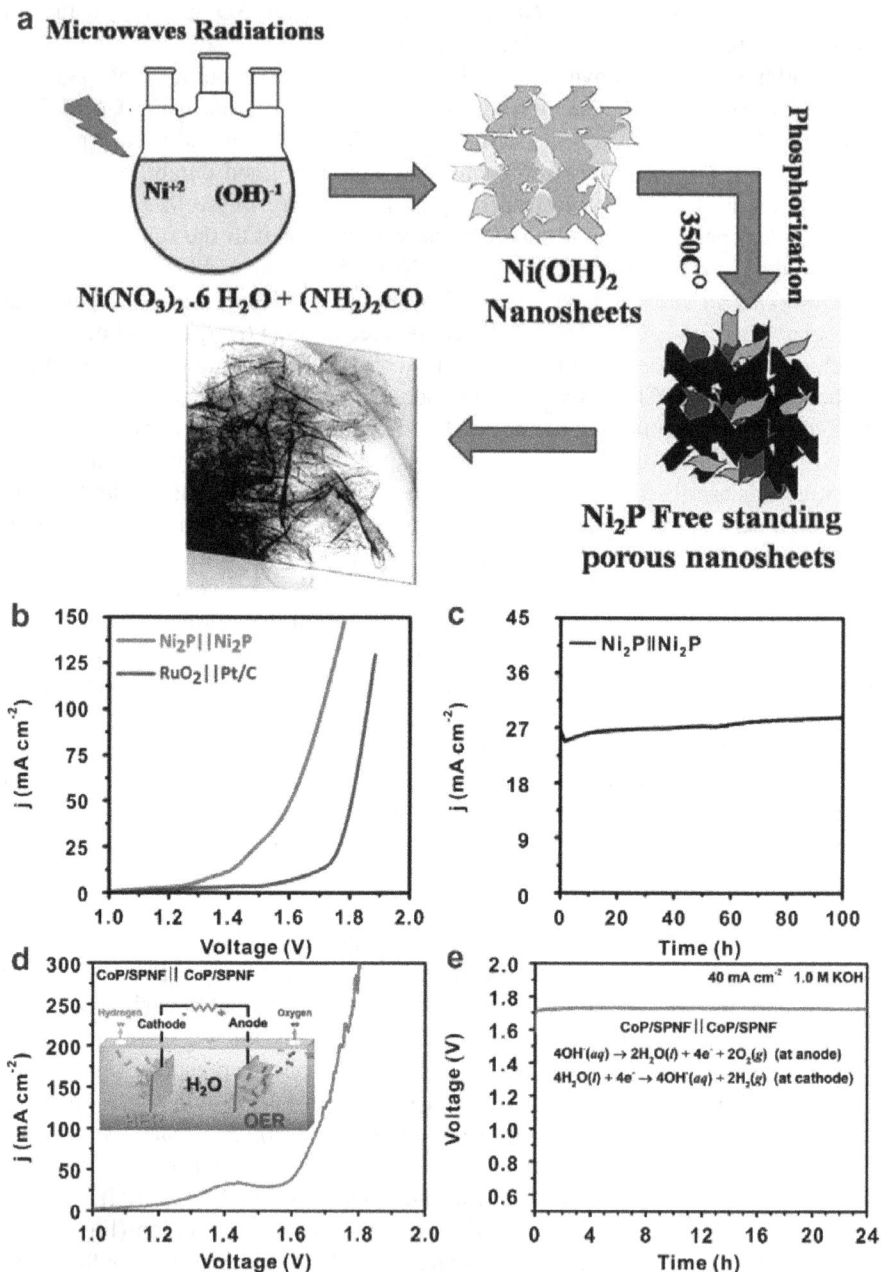

Figure 2.11. (a) Illustration of the synthesis procedure of Ni_2P nanosheets. (b) LSV curves of the $Ni_2P \parallel Ni_2P$ electrode for OWS. (c) Chronoamperometric curve of the $Ni_2P \parallel Ni_2P$ electrode at 1.47 V *vs.* RHE. (a–c) Reproduced with permission from Ref. 59. Copyright 2020, Royal Society of Chemistry. (d) LSV curve of the CoP/SPNF $\parallel$ CoP/SPNF electrode for OWS. (e) Chronopotentiometry curve of CoP/SPNF $\parallel$ CoP/SPNF at 40 mA cm^{-2}. (d and e) Reproduced with permission from Ref. 62. Copyright 2020, American Chemical Society.

Its excellent performance was attributed to the porous nanoflake array morphology of the catalyst and the synergistic effect between MoP and the surface oxide.[60] Ghosh et al. prepared Cu_3P on g-C_3N_4 via chemical vapour deposition using red phosphorous as the P source. The Cu_3P/g-C_3N_4 || Cu_3P/g-C_3N_4 electrode required an overpotential of 310 mV to afford current density of 10 mA cm^{-2} in 1.0 M KOH solution. This electrode also exhibited good durability, with no clear increase in the potential after 35 h of operation at 10 mA cm^{-2}.[61] Cao et al. reported CoP porous microscale nanoflake arrays grown on surface-phosphatised Ni foam (CoP/SPNF) by controllable structural transformation engineering. The CoP/SPNF||CoP/SPNF electrode only required an overpotential of 317 mV at 30 mA cm^{-2} in 1.0 M KOH solution (Fig. 2.11d). Galvanostatic measurements demonstrated that the potential did not change after 24 h of testing at 40 mA cm^{-2} (Fig. 2.11e). The good OWS performance was mostly attributed to the porous superstructure arrays, and this structure exposed more electrochemical active sites. Besides, it also increased the contact between catalysts and the basic media, thus accelerating electron and mass transfer.[62]

To improve the intrinsic activity of monometallic phosphide electrocatalysts, metal doping is used in the design and preparation of efficient bimetallic and trimetallic phosphides for OWS. For example, Li et al. successfully doped Fe into CoP to form 2.3 nm ultrathin Fe-Co nanosheets on Ni foam (Fe-CoP UNS/NF) by high-temperature phosphidation treatment using CoFe oxide and NaH_2PO_2 H_2O as precursors. The Fe-CoP UNSs/NF || Fe-CoP UNS/NF electrode only required an overpotential of 230 mV, which was reached at a current density of 10 mA cm^{-2} in 1.0 M KOH solution, outperforming a Pt/C/NF || RuO_2 NP/NF electrode (380 mV). Chronopotentiometry measurements showed that the applied potential was extremely stable over 50 h at 30 mA cm^{-2}. The excellent activity of Fe-CoP UNS/NF was attributed to the synergistic effect among Fe, Co, and P.[63] Selomulya et al. prepared Ni-Fe-P NPs arched on nitrogen-doped carbon on 3D Ni foam (Ni-Fe-P@NC/NF) by anion exchange and phosphidation strategies using $K[NiFe(CN)_6]$ (PBA) as the precursor. The Ni-Fe-P@NC/NF || Ni-Fe-P@NC/NF electrode delivered a low overpotential of 240 mV achieved at 10 mA cm^{-2}, with outstanding stability over 100 h in 1.0 M KOH solution. The excellent performance was ascribed to the abundant active sites in the catalyst and the synergistic effect among the Ni-Fe-P NPs, nitrogen-doped carbon, and 3D Ni foam substrate.[64] Suo et al. synthesised a hydrangea-like vanadium-doped $FeNi_2P$ electrocatalyst on Ni foam (V-$FeNi_2P$/NF) using hydrothermal and phosphorisation methods (Fig. 2.12a). The V-$FeNi_2P$/NF || V-$FeNi_2P$/NF electrode required a low applied voltage of 1.57 V at 10 mA cm^{-2} in 1.0 M KOH solution (Fig. 2.12b) that was outperforming the Pt/C/NF || RuO_2/NF electrode. Chronoamperometry measurements demonstrated the current density changed slightly after 48 h of testing test at 1.57 V (Fig. 2.12c). The high performance and stability of the electrode was ascribed to (1) the electron-rich state of Ni in the material which is conducive to absorption of the reaction intermediates, based on the d-band centre theory; (2) the obtained open-ended structure which improves mass transfer and provides sufficient active sites; and (3) V doping which improved the conductivity of V-$FeNi_2P$/NF, and subsequently

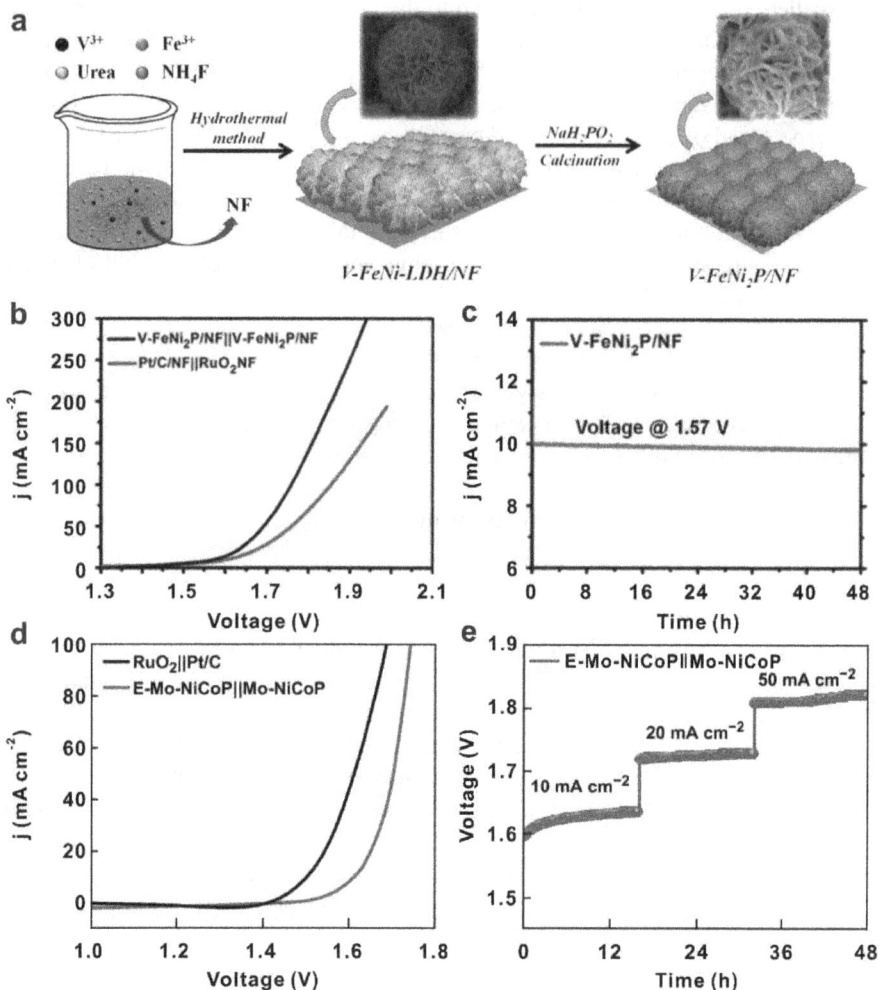

Figure 2.12. (a) Scheme of the preparation of V-FeNi$_2$P/NF. (b) LSV curves of the V-FeNi$_2$P/NF || V-FeNi$_2$P/NF electrode. (c) Chronoamperometry curve of the V-FeNi$_2$P/NF || V-FeNi$_2$P/NF electrode toward OWS at 10 mA cm^{-2}. (a–c) Reproduced with permission from Ref. 65. Copyright 2020, American Chemical Society. (d) LSV curves of the E-Mo-NiCoP || Mo-NiCoP electrode for OWS. (e) Chronopotentiometry curves at 10, 20, and 50 mA cm^{-2}. (d and e) Reproduced with permission from Ref. 66. Copyright 2019, Springer.

enhanced electron transfer.[65] Qi et al. prepared molybdenum-doped NiCoP nanosheet arrays (Mo-NiCoP) by hydrothermal and phosphation strategies using Na$_2$MoO$_4$ 6H$_2$O, CoCl$_2$ 6H$_2$O, NiCl$_2$ 6H$_2$O, and NaH$_2$PO$_2$ H$_2$O as precursors. They further converted the Mo-NiCoP nanosheet arrays to core-branched Mo-(Ni, Co) OOH arrays (denoted as E-Mo-NiCoP) by *in situ* electrochemical activation because of the instability of metal phosphides under alkaline OER conditions. Subsequently, they constructed a two-electrode system to study its OWS performance, using E-Mo-NiCoP and Mo-NiCoP as the anode and cathode, respectively. The E-Mo-NiCoP || Mo-NiCoP electrode required an applied cell voltage of 1.61 V achieved at 10 mA cm^{-2} in 1.0 M KOH solution (Fig. 2.12d), which is lower than that of a RuO$_2$ ‖ Pt/C electrode. Furthermore, this electrode showed long-term durability, with a slight change in the voltage after 48 h of operation at constant current densities of 10, 20, and 50 mA cm^{-2} (Fig. 2.12e). The excellent OWS activity was mainly attributed to the incorporation of Mo that effectively modulated the electronic structure of NiCoP, and subsequently enhanced the boosted intrinsic activity of each electrochemically active site.[66]

2.3 Summary and future perspectives

This chapter summarises the recent progress in bifunctional electrocatalysts, including noble metal-based oxides and alloys, transition metal-based oxides, chalcogenides, nitrides, and phosphides, for overall water splitting in acidic, neutral, and alkaline electrolyte solutions. All the representative work discussed herein is compiled in Table 2.2–2.4. Despite the exciting achievements, several challenges still exist which need to be addressed to improve the intrinsic activity and stability of bifunctional elecrocatalysts for OWS in different electrolyte solutions. (1) The reaction mechanism of OWS with bifunctional electrocatalysts needs to be further investigated. In particular, surface restructuring and determination of the active sites in the OWS process in all types of electrolytes requires elucidation. Therefore, innovative computational simulation methods and *in situ/operando* characterisation (X-ray absorption spectroscopy, Raman spectroscopy, ambient-pressure X-ray photoelectron spectroscopy, and electrochemical atomic force microscopy) are highly recommended to elucidate the active sites and mechanism of OWS by theoretical and experimental approaches. (2) The corrosion resistance of bifunctional electrocatalysts needs to be improved via electrocatalyst design. Most metal-based electrocatalysts dissolve to some extent in acidic or alkaline solutions. Vacancy engineering is favoured to prevent corrosion. Furthermore, the development of metal-free electrocatalysts is a new avenue to improve anticorrosion. (3) It still needs further improvement of the design strategy for bifunctional electrocatalysts. The nature of the two distinctive active sites for the HER and OER of OWS renders it challenging to design a single electrocatalyst that is active for both reactions. Most of the prevalent modification strategies still largely depend on pre-requisite knowledge of existing materials and trial-and-error; few materials have been discovered based on rational design and fundamental understanding of the catalytic mechanism and the targeted electrochemical reactions. Developing an understanding of the fundamental nature of the electrochemical performance evaluations benefits the discovery of new bifunctional electrocatalysts with high intrinsic activities. Finally, we hope that this chapter can boost the development of bifunctional catalysts for water splitting in all types of electrolytes.

Acknowledgements

This work was supported by grants from the National Science Foundation of Shaanxi Province of China (Grant No. 2020JQ-706) and Scientific Research Foundation for Ph.D. of Shaanxi University of Science and Technology (Grant No. 2019QNBJ-04).

References

[1]   Guo, Y. N., T. Park, J. W. Yi, J. Henzie, J. Kim, Z. L. Wang, B. Jiang, Y. Bando, Y. Sugahara, J. Tang and Y. Yamauchi. 2019. Nanoarchitectonics for transition-metal-sulfide-based electrocatalysts for water splitting. *Adv. Mater.*, 31: 1807134.

[2]   Peng, J., W. Dong, Z. Wang, Y. Meng, W. Liu, P. Song and Z. Liu. 2020. Recent advances in 2D transition metal compounds for electrocatalytic full water splitting in neutral media. *Mater. Today Adv.*, 8: 100081.

[3]   Liang, H., L. Li, F. Meng, L. Dang, J. Zhuo, A. Forticaux, Z. Wang and S. Jin. 2015. Porous two-dimensional nanosheets converted from layered double hydroxides and their applications in electrocatalytic water splitting. *Chem. Mater.*, 27: 5702–5711.

[4]   Tao, J. Y., Y. J. Zhang, S. P. Wang, G. Wang, F. Hu, X. J. Yan, L. F. Hao, Z. J. Zuo and X. W. Yang. 2019. Activating three-dimensional networks of Fe@Ni nanofibers via fast surface modification for efficient overall water splitting. *ACS Appl. Mater. Inter.*, 11: 18342–18348.

[5]   Wang, D. W., Q. Li, C. Han, Q. Q. Lu, Z. C. Xing and X. R. Yang. 2019. Atomic and electronic modulation of self-supported nickel-vanadium layered double hydroxide to accelerate water splitting kinetics. *Nat. Commun.*, 10: 3899.

[6]   Xu, K., H. Cheng, L. Liu, H. Lv, X. Wu, C. Wu and Y. Xie. 2017. Promoting active species generation by electrochemical activation in alkaline media for efficient electrocatalytic oxygen evolution in neutral media. *Nano Lett.*, 17: 578–583.

[7]   Zhu, J., L. S. Hu, P. X. Zhao, L. Y. S. Lee and K.-Y. Wong. 2020. Recent advances in electrocatalytic hydrogen evolution using nanoparticles. *Chem. Rev.*, 120: 851–918.

[8]   Fu, L. H., F. L. Yang, G. Z. Cheng and W. Luo. 2018. Ultrathin Ir nanowires as high-performance electrocatalysts for efficient water splitting in acidic media. *Nanoscale*, 10: 1892–1897.

[9]   Wang, Q. L., C. Q. Xu, W. Liu, S. F. Hung, H. B. Yang, J. J. Gao, W. Z. Cai, H. M. Chen, J. Li and B. Liu. 2020. Coordination engineering of iridium nanocluster bifunctional electrocatalyst for highly efficient and pH-universal overall water splitting. *Nat. Commun.*, 1: 4246.

[10]   Lv, F., J. R. Feng, K. Wang, Z. P. Dou, W. Y. Zhang, J. H. Zhou, C. Yang, M. C. Luo, Y. Yang, Y. J. Li, P. Gao and S. J. Guo. 2018. Iridium-tungsten alloy nanodendrites as pH-universal water-splitting electrocatalysts. *ACS Cent. Sci.*, 4: 1244–1252.

[11]   Zhu, M. W., Q. Shao, Y. Qian and X. Q. Huang. 2019. Superior overall water splitting electrocatalysis in acidic conditions enabled by bimetallic Ir-Ag nanotubes. *Nano Energy*, 56: 330–337.

[12]   Cao, D., H. X. Xu and D. J. Cheng. 2020. Construction of defect-rich RhCu nanotubes with highly active Rh_3Cu_1 alloy phase for overall water splitting in all pH values. *Adv. Energy Mater.*, 10: 1903038.

[13]   Yang, J., Q. Shao, B. L. Huang, M. Z. Sun and X. Q. Huang. 2019. pH-universal water splitting catalyst: Ru-Ni nanosheet assemblies. *iScience*, 11: 492–504.

[14]   Fu, L. H., G. Z. Cheng and W. Luo. 2017. Colloidal synthesis of monodisperse trimetallic IrNiFe nanoparticles as highly active bifunctional electrocatalysts for acidic overall water splitting. *J. Mater. Chem. A*, 5: 24836–24841.

[15]   Shan, J. Q., T. Ling, K. Davey, Y. Zheng and S.-Z. Qiao. 2019. Transition-metal-doped RuIr bifunctional nanocrystals for overall water splitting in acidic environments. *Adv. Mater.*, 31: 1900510.

[16]   Wang, J., Y. J. Ji, R. G. Yin, Y. Y. Li, Q. Shao and X. Q. Huang. 2019. Transition metal-doped ultrathin RuO_2 networked nanowires for efficient overall water splitting across a broad pH range. *J. Mater. Chem. A*, 7: 6411–6416.

[17]   Baggio, J. A., J. Freeman, T. R. Coyle, T. T. Nguyen, D. Hancock, K. E. Elpers, S. Nabity, H. J. F. Dengah II and D. Pillow. 2019. The importance of cognitive diversity for sustaining the commons. *Nat. Commun.*, 10: 875.

[18]   Ghadge, S. D., O. I. Velikokhatnyi, M. K. Datta, P. M. Shanthi, S. S. Tan, K. Damodaran and P. N. Kumta. 2019. Experimental and theoretical validation of high efficiency and robust electrocatalytic response of one-dimensional (1D) $(Mn,Ir)O_2$:10F nanorods for the oxygen evolution reaction in PEM-based water electrolysis. *ACS Catal.*, 9: 2134–2157.

[19] Joo, J., T. Kim, J. Lee, S. I. Choi and K. Lee. 2019. Morphology-controlled metal sulfides and phosphides for electrochemical water splitting. *Adv. Mater.*, 31: 1806682.

[20] Shi, Y. M. and B. Zhang. 2016. Correction: Recent advances in transition metal phosphide nanomaterials: synthesis and applications in hydrogen evolution reaction. *Chem. Soc. Rev.*, 45: 1781–1781.

[21] Han, N., K. R. Yang, Z. Y. Lu, Y. J. Li, W. W. Xu, T. F. Gao, Z. Cai, Y. Zhang, V. S. Batista, W. Liu and X. M. Sun. 2018. Nitrogen-doped tungsten carbide nanoarray as an efficient bifunctional electrocatalyst for water splitting in acid. *Nat. Commun.*, 9: 924.

[22] Jiao, Y., Y. Zheng, M. Jaroniec and S. Z. Qiao. 2015. Design of electrocatalysts for oxygen and hydrogen-involving energy conversion reactions. *Chem. Soc. Rev.*, 44: 2060–2086.

[23] Yu, J., Q. Q. Li, Y. Li, C. Y. Xu, L. Zhen, V. P. Dravid and J. S. Wu. 2016. Ternary metal phosphide with triple-layered structure as a low-cost and efficient electrocatalyst for bifunctional water splitting. *Adv. Funct. Mater.*, 26: 7644.

[24] Xiong, Q. Z., X. Zhang, H. J. Wang, G. Q. Liu, G. Z. Wang, H. M. Zhang and H. J. Zhao. 2018. One-step synthesis of cobalt-doped MoS_2 nanosheets as bifunctional electrocatalysts for overall water splitting under both acidic and alkaline conditions. *Chem. Commun.*, 54: 3859–3862.

[25] Liu, H. X., X. Y. Peng, X. J. Liu, G. C. Qi and J. Luo. 2019. Porous Mn-doped FeP/ $Co_3(PO_4)_2$ nanosheets as efficient electrocatalysts for overall water splitting in a wide pH range. *Chem. Sus. Chem.*, 12: 1334–1341.

[26] Wang, Q. L., C. Q. Xu, W. Liu, S. F. Hung, H. B. Yang, J. J. Gao, W. Z. Cai, H. M. Chen, J. Li and B. Liu. 2020. Coordination engineering of iridium nanocluster bifunctional electrocatalyst for highly efficient and pH-universal overall water splitting. *Nat. Commun.*, 11: 4246.

[27] Zhuang, Z. W., Y. Wang, C. Q. Xu, S. J. Liu, C. Chen, Q. Peng, Z. B. Zhuang, H. Xiao, Y. Pan, S. Q. Liu, R. Yu, W. C. Cheong, X. Cao, K. L. Wu, K. A. Sun, Y. Wang, D. S. Wang, J. Li and Y. D. Li. 2019. Three-dimensional open nano-netcage electrocatalysts for efficient pH-universal overall water splitting. *Nat. Commun.*, 1: 4875.

[28] Liu, T. T., L. S. Xie, J. H. Yang, R. M. Kong, G. Du, A. M. Asiri, X. P. Sun and L. Chen. 2017. Self-standing CoP nanosheets array: A three-dimensional bifunctional catalyst electrode for overall water splitting in both neutral and alkaline media. *Chem. Electro. Chem.*, 4: 1840–1845.

[29] Tan, Y. W., M. Luo, P. Liu, C. Cheng, J. H. Han, K. Watanabe and M. W. Chen. 2019. Three-dimensional nanoporous $Co_9S_4P_4$ pentlandite as a bifunctional electrocatalyst for overall neutral water splitting. *ACS Appl. Mater. Interfaces*, 11: 3880–3888.

[30] Zeng, L. Y., K. A. Sun, Y. J. Chen, Z. Liu, Y. J. Chen, Y. Pan, R. Y. Zhao, Y. Q. Liu and C. G. Liu. 2019. Neutral-pH overall water splitting catalyzed efficiently by a hollow and porous structured ternary nickel sulfoselenide electrocatalyst. *J. Mater. Chem. A*, 7: 16793–16802.

[31] Yang, Y., H. Q. Yao, Z. H. Yu, S. M. Islam, H. Y. He, M. W. Yuan, Y. H. Yue, K. Xu, W. C. Hao, G. B. Sun, H. F. Li, S. L. Ma, P. Zapol and M. G. Kanatzidis. 2019. Hierarchical nanoassembly of $MoS_2/Co_9S_8/Ni_3S_2/Ni$ as a highly efficient electrocatalyst for overall water splitting in a wide pH range. *J. Am. Chem. Soc.*, 141: 10417–10430.

[32] Li, K. D., J. F. Zhang, R. Wu, Y. F. Yu and B. Zhang. 2016. Anchoring CoO domains on $CoSe_2$ nanobelts as bifunctional electrocatalysts for overall water splitting in neutral media. *Adv. Sci.*, 3: 1500426.

[33] Cheng, P. F., C. Yuan, Q. W. Zhou, X. B. Hu, J. Li, X. Z. Lin, X. Wang, M. L. Jin, L. L. Shui, X. S. Gao, R. Nötzel, G. F. Zhou, Z. Zhang and J. M. Liu. 2019. Core-shell MoS_2@CoO electrocatalyst for water splitting in neural and alkaline solutions. *J. Phys. Chem. C*, 123: 5833–5839.

[34] Liu, J. L., D. D. Zhu, T. Ling, A. Vasileff and S. Z. Qiao. 2017. S-$NiFe_2O_4$ ultra-small nanoparticle built nanosheets for efficient water splitting in alkaline and neutral pH. *Nano Energy*, 40: 264–273.

[35] Zhang, L., B. R. Liu, N. Zhang and M. M. Ma. 2017. Electrosynthesis of Co_3O_4 and $Co(OH)_2$ ultrathin nanosheet arrays for efficient electrocatalytic water splitting in alkaline and neutral media. *Nano Res.*, 11: 323–333.

[36] Li, R. Q., P. F. Hu, M. Miao, Y. L. Li, X. F. Jiang, Q. Wu, Z. Meng, Z. Hu, Y. Bando and X. B. Wang. 2018. CoO-modified Co_2N as a heterostructured electrocatalyst for highly efficient overall water splitting in neutral media. *J. Mater. Chem. A*, 6: 24767–24772.

[37]   Ohyama, J., T. Sato, Y. Yamamoto, S. Arai and A. Satsuma. 2013. Size specifically high activity of Ru nanoparticles for hydrogen oxidation reaction in alkaline electrolyte. *J. Am. Chem. Soc.*, 135: 8016–8021.

[38]   Yuan, C. Z., Y. F. Jiang, Z. W. Zhao, S. J. Zhao, X. Zhou, T. Y. Cheang and A. W. Xu. 2018. Molecule-assisted synthesis of highly dispersed ultrasmall RuO_2 nanoparticles on nitrogen-doped carbon matrix as ultraefficient bifunctional electrocatalysts for overall water splitting. *ACS Sustainable Chem. Eng.*, 6: 11529–11535.

[39]   Zhang, H., Y. Y. Liu, H. J. Wu, W. Zhou, Z. K. Kou, S. J. Pennycook, J. P. Xie, C. Guan and J. Wang. 2018. Open hollow Co-Pt clusters embedded in carbon nanoflake arrays for highly efficient alkaline water splitting. *J. Mater. Chem. A*, 6: 20214–20223.

[40]   Li, C. J., Y. Xu, S. L. Liu, S. L. Yin, H. J. Yu, Z. Q. Wang, X. N. Li, L. Wang and H. J. Wang. 2019. Facile construction of IrRh nanosheet assemblies as efficient and robust bifunctional electrocatalysts for overall water splitting. *ACS Sustainable Chem. Eng.*, 7: 15747–15754.

[41]   Zhu, Y. L., H. A. Tahini, Y. Wang, Q. Lin, Y. Liang, C. M. Doherty, Y. Liu, X. Y. Li, S. C. Smith, C. Selomulya, X. W. Zhang, Z. P. Shao and H. T. Wang. 2019. Pyrite-type ruthenium disulfide with tunable disorder and defects enables ultra-efficient overall water splitting. *J. Mater. Chem. A*, 7: 14222–14232.

[42]   Peng, Z. K., J. M. Liu, B. Hu, Y. P. Yang, Y. Q. Guo, B. J. Li, L. Li, Z. H. Zhang, B. B. Cui, L. H. He and M. Du. 2020. Surface engineering on nickel-ruthenium nanoalloys attached defective carbon sites as superior bifunctional electrocatalysts for overall water splitting. *ACS Appl. Mater. Interfaces*, 12: 13842–13851.

[43]   Yu, T. Q., Q. G. Xu, G. F. Qian, J. L. Chen, H. Zhang, L. Luo and S. B. Yin. 2020. Amorphous CoO_x-decorated crystalline RuO_2 nanosheets as bifunctional catalysts for boosting overall water splitting at large current density. *ACS Sustainable Chem. Eng.*, 8: 17520–17526.

[44]   Fan, R. L., Q. Q. Mu, Z. H. Wei, Y. Peng and M. G. Shen. 2020. Atomic Ir-doped NiCo layered double hydroxide as a bifunctional electrocatalyst for highly efficient and durable water splitting. *J. Mater. Chem. A*, 8: 9871–9881.

[45]   Suen, N. T., S. F. Hung, Q. Quan, N. Zhang, Y. J. Xu and H. M. Chen. 2017. Electrocatalysis for the oxygen evolution reaction: recent development and future perspectives. *Chem. Soc. Rev.*, 46: 337–365.

[46]   Wang, J., Y. Gao, H. Kong, J. Kim, S. Choi, F. Ciucci, Y. Hao, S. H. Yang, Z. P. Shao and J. Lim. 2020. Non-precious-metal catalysts for alkaline water electrolysis: Operando characterizations, theoretical calculations, and recent advances. *Chem. Soc. Rev.*, 49: 9154–9196.

[47]   Zhu, Y. L., W. Zhou, Y. J. Zhong, Y. F. Bu, X. Y. Chen, Q. Zhong, M. L. Liu and Z. P. Shao. 2017. A perovskite nanorod as bifunctional electrocatalyst for overall water splitting. *Adv. Energy Mater.*, 7: 1602122.

[48]   Tang, L., Z. Chen, F. Zuo, B. Hua, H. Zhou, M. Li, J. H. Li and Y. F. Sun. 2020. Enhancing perovskite electrocatalysis through synergistic functionalization of B-site cation for efficient water splitting. *Chem. Eng. J.*, 401: 126082.

[49]   He, B. B., K. Tan, Y. S. Gong, R. Wang, H. W. Wang and L. Zhao. 2020. Coupling amorphous cobalt hydroxide nanoflakes on $Sr_2Fe_{1.5}Mo_{0.5}O_{5+\delta}$ perovskite nanofibers to induce bifunctionality for water splitting. *Nanoscale*, 12: 9048–9057.

[50]   Zhang, Y. Q., H. B. Tao, Z. Chen, M. Li, Y. F. Sun and J. L. Luo. 2019. *In situ* grown cobalt phosphide (CoP) on perovskite nanofibers as an optimized trifunctional electrocatalyst for Zn-air batteries and overall water splitting. *J. Mater. Chem. A*, 7: 26607–26617.

[51]   Wu, X. H., G. M. Yang, H. Liu, W. Zhou and Z. P. Shao. 2018. Perovskite oxide/carbon nanotube hybrid bifunctional electrocatalysts for overall water splitting. *Electrochim. Acta*, 286: 47–54.

[52]   Yang, Y. Q., K. Zhang, H. L. Li, H. C. Chan, L. C. Yang and Q. S. Gao. 2017. MoS_2-Ni_3S_2 heteronanorods as efficient and stable bifunctional electrocatalysts for overall water splitting. *ACS Catal.*, 7: 2357–2366.

[53]   Czioska, S., J. Y. Wang, X. Teng and Z. F. Chen. 2018. Hierarchically structured $CuCo_2S_4$ nanowire arrays as efficient bifunctional electrocatalyst for overall water splitting. *ACS Sustainable Chem. Eng.*, 6: 11877–11883.

[54] Liu, C. C., D. B. Jia, Q. Y. Hao, X. R. Zheng, Y. Li, C. C. Tang, H. Liu, J. Zhang and X. L. Zheng. 2019. P-doped iron-nickel sulfide nanosheet arrays for highly efficient overall water splitting. *ACS Appl. Mater. Interfaces*, 11: 27667–276276.

[55] Zhang, Q., C. Ye, X. L. Li, Y. H. Deng, B. X. Tao, W. Xiao, L. J. Li, N. B. Li and H. Q. Luo. 2018. Self-interconnected porous networks of NiCo disulfide as efficient bifunctional electrocatalysts for overall water splitting. *ACS Appl. Mater. Interfaces*, 10: 27723–27733.

[56] Hu, P., Z. Y. Jia, H. Che, W. Y. Zhou, N. Liu, F. Li and J. S. Wang. 2019. Engineering hybrid $CoMoS_4/Ni_3S_2$ nanostructures as efficient bifunctional electrocatalyst for overall water splitting. *J. Power Sources*, 416: 95–103.

[57] Sun, C. C., Q. C. Dong, J. Yang, Z. Y. Dai, J. J. Lin, P. Chen, W. Huang and X. C. Dong. 2016. Metal-organic framework derived $CoSe_2$ nanoparticles anchored on carbon fibers as bifunctional electrocatalysts for efficient overall water splitting. *Nano Res.*, 9: 2234–2243.

[58] Chen, T. and Y. W. Tan. 2018. Hierarchical $CoNiSe_2$ nano-architecture as a high-performance electrocatalyst for water splitting. *Nano Res.*, 11: 1331–1344.

[59] Mushtaq, N., C. Qiao, H. Tabassum, M. Naveed, M. Tahir, Y. Q. Zhu, M. Naeem, W. Younas and C. B. Cao. 2020. Preparation of a bifunctional ultrathin nickel phosphide nanosheet electrocatalyst for full water splitting. *Sustain. Energy Fuels*, 4: 5294–5300.

[60] Jiang, Y. Y., Y. Z. Lu, J. Y. Lin, X. Wang and Z. X. Shen. 2018. A hierarchical MoP nanoflake array supported on Ni foam: A bifunctional electrocatalyst for overall water splitting. *Small Methods*, 2: 1700369.

[61] Riyajuddin, S., S. K. Tarik Aziz, S. Kumar, P. G. Nessim and P. K. Ghosh. 2020. 3D-Graphene decorated with g-C_3N_4/Cu_3P composite: A noble metal-free bifunctional electrocatalyst for overall water splitting. *Chem. Cat. Chem.*, 12: 1394–1402.

[62] Cao, S., N. You, L. Wei, C. Huang, X. M. Fan, K. Shi, Z. H. Yang and W. X. Zhang. 2020. CoP microscale prism-like superstructure arrays on Ni foam as an efficient bifunctional electrocatalyst for overall water splitting. *ACS Inorg. Chem.*, 59: 8522–8531.

[63] Li, Y., F. M. Li, Y. Zhao, S. N. Li, J. H. Zeng, H. C. Yao and Y. Chen. 2019. Iron doped cobalt phosphide ultrathin nanosheets on nickel foam for overall water splitting. *J. Mater. Chem. A*, 7: 20658–20666.

[64] Wang, Y., S. L. Zhao, Y. L. Zhu, R. S. Qiu, T. Gengenbach, Y. Liu, L. H. Zu, H. Y. Mao, H. T. Wang, J. Tang, D. Y. Zhao and C. Selomulya. 2020. Three-dimensional hierarchical porous nanotubes derived from metal-organic frameworks for highly efficient overall water splitting. *iScience*, 23: 100761.

[65] Suo, N., C. Chen, X. Q. Han, X. Q. He, Z. Y. Dou, Z. H. Lin, L. L. Cui and J. B. Xiang. 2020. The construction of hydrangea-like vanadium-doped iron nickel phosphide as an enhanced bifunctional electrocatalyst for overall water splitting. *ACS Appl. Energy Mater.*, 3: 9449–9458.

[66] Lin, J. H., Y. T. Yan, C. Li, X. Q. Si, H. H. Wang, J. L. Qi, J. Cao, Z. X. Zhong, W. D. Fei and J. C. Feng. 2019. Bifunctional electrocatalysts based on Mo-doped NiCoP nanosheet arrays for overall water splitting. *Nano-Micro Lett.*, 11: 55.

[67] Wu, Y. T., H. Wang, S. Ji, B. G. Pollet, X. Y. Wang and R. F. Wang. 2020. Engineered porous Ni_2P-nanoparticle/Ni_2P-nanosheet arrays via the Kirkendall effect and Ostwald ripening towards efficient overall water splitting. *Nano Res.*, 13: 2098–2105.

[68] Zhao, H. Y., Y. W. Wang, L. Fang, W. W. Fu, X. H. Yang, S. L. You, H. J. Zhang and Y. Wang. 2019. Cation-tunable flower-like $(Ni_xFe_{1-x})_2P$@graphitized carbon films as ultra-stable electrocatalysts for overall water splitting in alkaline media. *J. Mater. Chem. A*, 7: 20357–20368.

[69] Li, S. S., X. W, M. Li, J. Liu, C. R. Li, H. P. Wu, D. Y. Guo, F. M. Ye, S. L. Wang, L. Cheng and A. P. Liu. 2019. Self-supported ternary $(Ni_xFe_y)_2P$ nanoplates arrays as an efficient bifunctional electrocatalyst for overall water splitting. *Electrochim. Acta*, 319: 561–568.

[70] Wang, L. Y., C. D. Gu, X. Ge, J. L. Zhang, H. Y. Zhu and J. P. Tu. 2017. Anchoring Ni_2P sheets on $NiCo_2O_4$ nanocone arrays as optimized bifunctional electrocatalyst for water splitting. *Adv. Mater. Interfaces*, 4: 1700481.

[71] Zhang, F. S., J. W. Wang, J. Luo, R. R. Liu, Z. M. Zhang, C. T. He and T. B. Liu. 2018. Extraction of nickel from NiFe-LDH into Ni_2P@NiFe hydroxide as a bifunctional electrocatalyst for efficient overall water splitting. *Chem. Sci.*, 9: 1375–1384.

[72] Gao, Z., F. Q. Liu, L. Wang and F. Luo. 2019. Hierarchical Ni$_2$P@NiFeAlO$_x$ nanosheet arrays as bifunctional catalysts for superior overall water splitting. *Inorg. Chem.*, 58: 3247–3255.

[73] You, B., N. Jiang, M. L. Sheng and M. W. Bhushan. 2016. Hierarchically porous urchin-like Ni$_2$P superstructures supported on nickel foam as efficient bifunctional electrocatalysts for overall water splitting. *ACS Catal.*, 6: 714–721.

[74] Li, Y. J., H. C. Zhang, M. Jiang, Y. Kuang, X. M. Sun and X. Duan. 2016. Ternary NiCoP nanosheet arrays: An excellent bifunctional catalyst for alkaline overall water splitting. *Nano Res.*, 9: 2251–2259.

[75] Wang, P., J. Q. Chen, C. Li, W. P. Li, T. H. Wang and C. H. Liang. 2020. Three-dimensional heterostructured NiCoP@NiMn-layered double hydroxide arrays supported on Ni foam as a bifunctional electrocatalyst for overall water splitting. *ACS Appl. Mater. Interfaces*, 12: 4385–4395.

[76] Sahasrabudhe, A., H. Dixit, R. Majee and S. Bhattacharyya. 2018. Value added transformation of ubiquitous substrates into highly efficient and flexible electrodes for water splitting. *Nat. Commun.*, 9: 2014.

[77] Kang, Q. L., M. Y. Li, J. W. Shi, Q. Y. Liu and F. Gao. 2020. A universal strategy for carbon-supported transition metal phosphides as high-performance bifunctional electrocatalysts towards efficient overall water splitting. *ACS Appl. Mater. Interfaces*, 12: 19447–19456.

[78] Tang, W. K., X. F. Liu, Y. Li, Y. H. Lu, Z. M. Song, Q. Wang, R. H. Yu and J. L. Shui. 2020. Boosting electrocatalytic water splitting via metal-metalloid combined modulation in quaternary Ni-Fe-P-B amorphous compound. *Nano Res.*, 13: 447–454.

[79] Zhang, G. H., H. Y. Ge, L. X. Zhao, J. Y. Liu, F. B. Wang, S. W. Fan and G. D. Li. 2021. NiMn$_{1.5}$PO$_4$ thin layer supported on Ni foam as a highly efficient bifunctional electrocatalyst for overall water splitting. *Electrochim. Acta*, 367: 137567.

[80] Tian, G. Q., S. R. Wei, Z. T. Guo, S. W. Wu, Z. L. Chen, F. M. Xu, Y. Cao, Z. Liu, J. Q. Wang, L. Ding, J. C. Tu and H. Zeng. 2021. Hierarchical NiMoP$_2$-Ni$_2$P with amorphous interface as superior bifunctional electrocatalysts for overall water splitting. *J. Mater. Sci. Technol.*, 77: 108–116.

[81] Li, J. Y., J. Li, X. M. Zhou, Z. M. Xia, W. Gao, Y. Y. Ma and Y. Q. Qu. 2016. Highly efficient and robust nickel phosphides as bifunctional electrocatalysts for overall water-splitting. *ACS Appl. Mater. Interfaces*, 8: 10826–10834.

[82] Jiang, D. L., W. X. Ma, R. Yang, B. Quan, D. Li, S. C. Meng and M. Chen. 2020. Nickel-manganese bimetallic phosphides porous nanosheet arrays as highly active bifunctional hydrogen and oxygen evolution electrocatalysts for overall water splitting. *Electrochim. Acta*, 329: 135121.

[83] Lian, Y. B., H. Sun, X. B. Wang, P. W. Qi, Q. Q. Mu, Y. J. Chen, J. Ye, X. H. Zhao, Z. Deng and Y. Peng. 2019. Carved nanoframes of cobalt–iron bimetal phosphide as a bifunctional electrocatalyst for efficient overall water splitting. *Chem. Sci.*, 10: 464–474.

[84] Li, X., X. L Wang, J. Zhou, L. Han, C. Sun, Q. Q. Wang and Z. M. Su. 2018. Ternary hybrids as efficient bifunctional electrocatalysts derived from bimetallic metal-organic-frameworks for overall water splitting. *J. Mater. Chem. A*, 6: 5789–5796.

[85] Zhang, G. W., B. Wang, J. L. Bi, D. Q. Fang and S. C. Yang. 2019. Constructing ultrathin CoP nanomeshes by Er-doping for highly efficient bifunctional electrocatalysts for overall water splitting. *J. Mater. Chem. A*, 7: 5769–5778.

[86] Tian, L. H., X. D. Yan and X. B. Chen. 2016. Electrochemical activity of iron phosphide nanoparticles in hydrogen evolution reaction. *ACS Catal.*, 6: 5441–5448.

[87] Huang, X. K., L. Gong, H. Xu, J. H. Qin, P. Ma, M. Yang, K. Z. Wang, L. Ma, X. Mu and R. Li. 2020. Hierarchical iron-doped CoP heterostructures self-assembled on copper foam as a bifunctional electrocatalyst for efficient overall water splitting. *J. Colloid Interface Sci.*, 569: 140–149.

[88] Wang, X. X., G. W. She, L. X. Mu and W. S. Shi. 2020. Amorphous Co-Mo-P-O bifunctional electrocatalyst via facile electrodeposition for overall water splitting. *ACS Sustainable. Chem. Eng.*, 8: 2835–2842.

[89] Wu, K. L., Z. Chen, W. C. Cheong, S. J. Liu, W. Zhu, X. Cao, K. A. Sun, Y. Lin, L. R. Zheng, W. S. Yan, Y. Pan, D. S. Wang, Q. Peng, C. Chen and Y. D. Li. 2018. Toward bifunctional overall water splitting electrocatalyst: General preparation of transition metal phosphide nanoparticles decorated N-doped porous carbon spheres. *ACS Appl. Mater. Interfaces*, 10: 44201–44208.

[90]   Selvam, N. C. S., J. Lee, G. H. Choi, M. J. Oh, S. Y. Xu, B. Lim and P. J. Yoo. 2019. MXene supported Co_xA_y (A=OH, P, Se) electrocatalysts for overall water splitting: Unveiling the role of anions in intrinsic activity and stability. *J. Mater. Chem. A*, 7: 27383–2739.

[91]   Ji, L. L., J. Y. Wang, X. Teng, T. J. Meyer and Z. F. Chen. 2020. CoP nanoframes as bifunctional electrocatalysts for efficient overall water splitting. *ACS Catal.*, 10: 412–419.

[92]   Wang, Y. Q., L. Zhao, X. L. Sui, D. M. Gu and Z. B. Wang. 2019. Hierarchical CoP_3/NiMoO$_4$ heterostructures on Ni foam as an efficient bifunctional electrocatalyst for overall water splitting. *Ceram. Int.* 45: 17128–17136.

[93]   Hu, G. J., J. X. Xiang, J. Li, P. Liu, R. N. Ali and B. Xiang. 2019. Urchin-like ternary cobalt phosphosulfide as high-efficiency and stable bifunctional electrocatalyst for overall water splitting. *J. Catal.*, 371: 126–134.

[94]   Du, C. C., M. X. Shang, J. X. Mao and W. B. Song. 2017. Hierarchical MoP/Ni$_2$P heterostructures on nickel foam for efficient water splitting. *J. Mater. Chem. A*, 5: 15940–15949.

[95]   Han, H., F. Yi, S. Choi, J. Kim, J. Kwon, K. Park and T. Song. 2020. Self-supported vanadium-incorporated cobalt phosphide as a highly efficient bifunctional electrocatalyst for water splitting. *J. Alloys Compd.*, 846: 156350.

[96]   Tong, J. H., C. Y. Li, L. L. Bo, X. L. Guan, Y. X. Wang, D. Y. Kong, H. Wang, W. P. Shi and Y. N. Zhang. 2021. Bimetallic Fe-Co chalcogenophosphates as highly efficient bifunctional electrocatalysts for overall water splitting. *Int. J. Hydrog. Energy*, 46: 3354–3364.

[97]   Yu, F., H. Q. Zhou, Y. F. Huang, J. Y. Sun, F. Qin, J. M. Bao, W. A. Goddard, S. Chen and Z. F. Ren. 2018. High-performance bifunctional porous non-noble metal phosphide catalyst for overall water splitting. *Nat. Commun.*, 9: 2551.

[98]   Zhang, X. L., C. Liang, X. Y. Qu, Y. F. Ren, J. J. Yin, W. J. Wang, M. S. Yang, W. Huang and X. C. Dong. 2020. Sandwich-structured Fe-Ni$_2$P/MoS$_x$/NF bifunctional electrocatalyst for overall water splitting. *Adv. Mater. Interfaces*, 7: 1901926.

[99]   Guo, M. J., S. Y. Song, S. S. Zhang, Y. Yan, K. Zhan, J. H. Yang and B. Zhao. 2020. Fe-Doped Ni-Co phosphide nanoplates with planar defects as an efficient bifunctional electrocatalyst for overall water splitting. *ACS Sustainable Chem. Eng.*, 8: 7436–7444.

[100]   Xie, Y. Y., M. Q. Chen, M. K. Cai, J. Teng, H. F. Huang, Y. N. Fan, M. Barboiu, D. W. Wang and C. Y. Su. 2019. Hollow cobalt phosphide with N-doped carbon skeleton as bifunctional electrocatalyst for overall water splitting. *Inorg. Chem.*, 58: 14652–14659.

[101]   Lu, M. J., L. Li, D. Chen, J. Z. Li, N. I. Klyui and W. Han. 2020. MOF-derived nitrogen-doped CoO@CoP arrays as bifunctional electrocatalysts for efficient overall water splitting. *Electrochim. Acta*, 330: 135210.

3

Metal-organic Frameworks Derived Electrocatalysts for Energy Conversion Reaction

Xijun Wei,[1] Yingze Song,[1,] Qiang Zhang,[2] Bingbing Hu,[3] Yuan Cen,[4] Juan Gong,[5] Xiao Liu[5] and Qi Wan[1]*

3.1 Introduction

With the economic growth and rapidly rising population in recent decades, the rate of consumption of unsustainable energy sources is increasing, including natural gas, petroleum and coal, also causing serious environmental pollution.[1,2] At present, it is becoming increasingly urgent to develop a technological revolution in developing renewable and clean energy to reduce, recycle and preferably replace the consumption of fossil fuels, because of the combined crisis of both environment and energy. Among various energy conversion and storage devices, the electrochemical oxidation and reduction are the most basic reactions.[3] Recently, the redox reactions between hydrogen, oxygen and titanium dioxide, such as hydrogen evolution reaction (HER),[4] oxygen evolution reaction (OER),[5] oxygen reduction reaction (ORR)[6] and carbon dioxide reduction reaction (CO$_2$RR)[7] have gained increased attention, for

[1] State Key Laboratory of Environment-Friendly Energy Materials, School of Materials Science and Engineering, Graphene Joint Innovation Centre, Southwest University of Science and Technology, Mianyang, 621010, China.
[2] School of Chemistry & Chemical Engineering, Chongqing University of Technology, Chongqing 400054, China.
[3] College of Materials Science and Engineering, Chongqing Jiaotong University, Chongqing 400074, China.
[4] College of Chemistry and Chemical Engineering, Chongqing University, Chongqing 400044, China.
[5] College of Chemistry and Materials Science, Sichuan Normal University, Chengdu, 610066, P. R. China.
* Corresponding author: yzsong@swust.edu.cn

their critical roles in many energy technologies, including water splitting, metal-air batteries and fuel cells, etc. Designing and preparing efficient electrocatalysts to reduce the reaction energy barriers of ORR, OER, HER and CO_2RR to improve energy efficiency is an important way to promote the practical application of these energy technology.[8,9]

The design principles of advanced electrocatalysis for the above systems should be able to provide abundant tri-phase regions for surface reaction and ensure fast mass and electron transport, which possess excellent electrical conductivity, high electrochemical activities, more reactive sites and cycling stability.[10] Precious metals and their oxides showed excellent electrocatalytic performance, such as Pt for HER and ORR, IrO_2 and RuO_2 for OER.[11,12] However, due to the high cost and limited reserves of precious metal, the development of non-precious metal-based electrocatalysts as alternatives is becoming more and more urgent, such as carbon-based materials, transition metal compounds, alloys and single atom electrocatalysts.[13,14] Although some progress has been made in the development of electrocatalysts based on non-precious metals, the design of more advanced electrocatalysts, further exploration of catalytic mechanisms and unsatisfactory cyclic stability have been promoting the further progress of this field.

Metal-organic framework (MOFs), also known as porous coordination polymers (PCPs), are emerging as a kind of promising porous materials consisting of multidentate organic ligands and metal ions or clusters to form the structure of various dimensions. In the past decade, more than 20000 different MOFs have been studied and reported with adjustable constituents type, geometry and functionality.[15] The morphology, structure, composition, pore size distribution and functionality of MOFs also can be tuned by changing the ligands and metal nodes.[16] Therefore, MOFs exhibit versatile architectures, larger surface areas, tunable pore sizes and component, which receives much attention for using in applications such as gas storage and separation, heterogeneous catalysis, drug delivery, proton conduction, sensing and luminescence.[17]

However, due to the poor electrical conductivity of MOFs that make them difficult to use as electrocatalysts directly, there have been a few reports of electrically conductive MOFs applied as active materials, which are obtained by complex synthesis and high cost method that can not be used on a large scale.[18] The MOFs with adjustable structure and composition exhibits huge advantages by using as precursors to prepare various functional materials. To be specific, the organic linkers in MOFs are ideal source for various carbon materials, while the metal ions provide the source for various metal compounds. In 2008, the first example for preparation of porous carbons by using MOFs as templates was reported;[19] from then on, the pyrolysis of MOFs has been widely adopted as an effective way to obtain carbon-based hybrid nanomaterials with various compositions and morphologies.

The MOFs-derived carbon-based hybrid nanomaterials have many advantages for application in $OER/HER/ORR/CO_2RR$ electrocatalysts, as following: (1) there is a large tunable variety of MOFs in both composition and structure. Due to the easy regulation of the composition of MOFs, the precursor with corresponding components can be designed according to the application requirements, and then the corresponding derivative electrocatalysts can be obtained through appropriate preparation methods; (2) the morphologies of MOFs are heritable. MOFs has a variety of morphologies, such as nanocubes, nanospheres, nanorods and nanowires, and the MOFs-derived electrocatalyst can usually retain the morphology of these MOFs' precursors by a controlled preparation process; (3) the metal ions and organic ligands in the MOFs are arranged periodically, which plays an important role in preventing the aggregation of metal compounds, metal nanoparticles and other composite nanostructures during pyrolysis, and the homogeneous dispersed derivative electrocatalysts can be obtained with improved utilization efficiency; (4) MOFs with a very high internal pore volume allows additional precursors to enter and polymerize within the pores, which can prepare derived electrocatalysts with complex structures and compositions, including single-atom metal sites and the heteroatom species coordinating with them; (5) the MOFs-derived carbon-based hybrid nanomaterials exhibit good electrical conductivity because of the carbon residue, which can accelerate the ion transfer and electron conduction rate, resulting in good electrocatalytic performance.

In recent years, the preparation of electrocatalysts by using MOFs as templates with different approaches has attracted much attention, and it has been reported and summarized in previous research papers and reviews.[20] In this monograph, a comprehensive overview on MOFs-derived carbon-based hybrid nanomaterials as electrocatalysts for OER/HER/ORR/CO_2RR will be given, which can provide an important reference for the development of MOFs-derived electrocatalyst in the future. The MOFs-derived electrocatalysts are becoming a kind of important electrode materials in the field of energy conversion. As an extensively on-going hot topic with unique structure and performance, MOFs-derived electrocatalysts will show application prospect for other electrochemical energy storage and conversion technologies in the near future.

3.2 MOFs-derived carbon-based hybrid nanomaterials for hydrogen evolution reaction

Electrochemical hydrogen evolution reaction (HER) is of great importance for high-efficiency and low-cost production of hydrogen.[21] HER primarily involves two routes: the Volmer-Tafel pathway and the Volmer-Heyrovsky pathway.[22] There are three major processes during HER, including electron transfer, proton diffusion and bubble release, which largely dictate the HER performance. Therefore, the fabrication of an effective HER electrocatalysts should satisfy three principles. Firstly, the electrocatalysts used should possess a high intrinsic electrocatalytic activity, which can facilitate electron transfer between protons and electrode. Secondly, the electrode materials should have a highly porous structure, which can provide sufficient channels for proton diffusion and bubble transport inside the electrode. Thirdly, the electrode surface should promote the release of hydrogen bubbles, since bubbles can block ion diffusion and reduce the effective surface area of the electrode.[23–25]

The HER efficiency is significantly dependent on the selection of electrocatalysts. Currently, platinum and ruthenium oxide-based materials are the electrocatalysts of choice for HER in both acid and base media with high cost, limited availability and natural scarcity, which have hindered their wide-spread application.[26,27] Thus, significant efforts have been devoted to the development of high-efficiency HER electrocatalysts based on earth-abundant, cost-effective materials. Among these, MOFs are emerging as a class of crystalline porous materials. MOFs-based materials possess unique advantages, such as high specific surface area, crystalline porous structure, diverse and tunable chemical components, which offer attractive functionalities in catalyzing processes, by lowering reaction potentials, and speeding up reaction rates.[28,29]

Unfortunately, original MOFs show poor stability in strong acid or strong base owing to the unstable active sites and improper protection mechanisms.[30] The application of pristine MOF in electrocatalytic water splitting is limited for their low conductivity and their relatively inert activity for HER due to the lack of proper active component.[31]

Recently, more attentions are focused on the utilization of MOFs as precursors to produce a wide range of carbon-based electrocatalysts with specific designed chemical components and structures. Moreover, MOFs-derived transition metals and transition metals carbonaceous compounds (TM/TMCs@C) nanocomposites, featuring tunable compositions, controllable morphologies, and changeable structures, hold great promise as the advanced HER electrocatalysts.[32] The synthesis of MOFs-derived TM/TMCs@C nanocomposites generally consist of two main procedures-preparation of the MOFs precursor and subsequent thermal treatment under controlled temperature and atmosphere conditions.[33] The compositions of MOFs-derived TMs/TMCs@C nanocomposites mainly involve two types of chemical species (inexpensive transition metal element and nonmetal element). Because each species shows a certain affinity for the H* intermediate, altering the catalyst components is thus an effective method for optimizing the HER activity.[34] In view of this, the following chapter will present the advances of several kinds of TMs/TMCs@C-based HER electrocatalysts based on MOFs derivatives, such as bimetal, alloys, transition metal chalcogenides, phosphides, nitrides, carbides as well as single-atom electrocatalysts.

3.2.1 MOFs-derived carbon-based metal bimetallic and alloy electrocatalysts for HER

Transition metals (TM) nanoparticles derived from MOFs precursors play two pivotal roles in modifying the HER activity.[35] One is that the electrons located in the 3d orbitals of TM atom will interact with the outer-shell electrons in the carbon atom, tuning the adsorption ability of the active sites for H*; the other is increasing the electrical conductivity of the catalytic system and accelerating the electron transfer. Bimetallic-based electrocatalysts can be easily prepared by a straightforward pyrolysis of MOFs precursors. Specifically, ZIF-67 dodecahedra were first grown *in situ* on the $Cu(OH)_2$ nanowires as the precursors through a coprecipitation method, and this material was then converted to a hybrid composite composed of Cu and Co bimetal nanoparticles embedded within the N-doped mesoporous carbon framework (CuCo@NC) via thermal annealing at 800°C under flowing Ar.[36] The CuCo@NC electrocatalyst required 145 mV of overpotential at a current density of 10 mA cm^{-2} in 0.5 M H_2SO_4, which was much lower than those of Cu@NC (> 450 mV) and

Co@NC (380 mV). In addition, the catalyst showed a Tafel slope as low as 79 mV dec^{-1} and a negligible current decay after 8 h in 0.5 M H$_2$SO$_4$.

Unlike the multiple metal strategy, in which metal nanoparticles are isolated and embedded within carbon matrix, alloying is a unique strategy in which the metal species can be mixed to afford an atomic-scale solid solution, which can generate a synergistic effect to boost the HER activity. Yang et al. prepared a composite consisting of ternary FeCoNi alloys encapsulated within graphitic carbon (FeCoNi@C) by thermal annealing an Fe-Co-Ni-PB precursor at 600°C under flowing N$_2$ (Fig. 3.1a–d).[37]

When investigated as an HER electrocatalyst, the as-prepared composite afforded a current density of 10 mA cm^{-2} at the required overpotential of 149 mV in 1 M KOH, showing much higher electrocatalytic activity than its single metal counterpart (Fig. 3.1e). DFT calculations further unraveled that upon increasing the

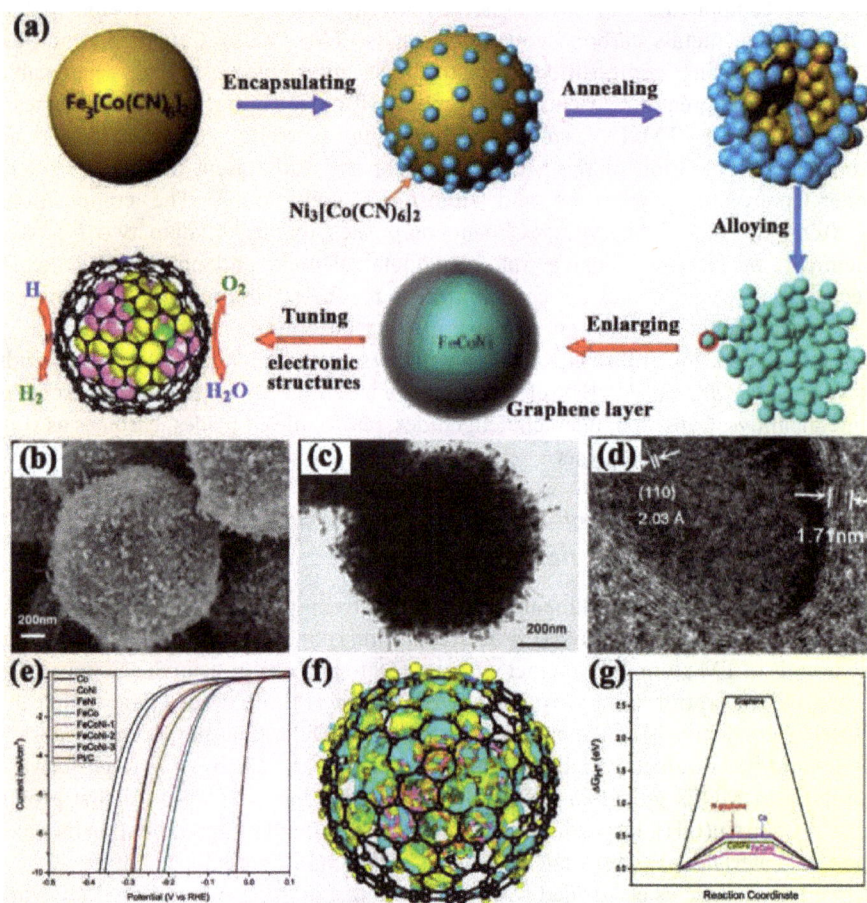

Figure 3.1. Fe-Co-Ni-PB-derived FeCoNi@C nanocomposite: (a) schematic illustration of the synthetic process, (b) FESEM image, (c, d) TEM images, (e) LSV curve in 1.0 M KOH, (f) charge density distribution, and (g) ΔG_{H^*} value. Reproduced with permission from Ref. 37. Copyright 2016, American Chemical Society.

degrees of freedom of the alloys or altering the metal proportions in the FeCoNi ternary alloys, the electronic structures of the nanocomposites could be deliberately tuned by varying the number of electrons transferred between the alloys and the graphitic carbon, thus modifying the HER activity (Fig. 3.1f and g).

3.2.2 MOF-derived carbon-based metal chalcogenides electrocatalysts for HER

Among transition metal chalcogenides, most transition metal sulfides are semiconductors, exhibiting inferior electrical conductivities compared to those of their selenide and telluride counterparts, which have lower charge transfer abilities during the HER process.[38] Nevertheless, this issue could be effectively alleviated by constructing transition metal sulfides@C nanocomposites through the sulfidation of MOFs precursors. Tian et al. reported an advanced NiS_2@C electrocatalyst synthesized by the sulfidation of a Ni-BTC precursor.[39] The HER activity of the NiS_2@C electrocatalyst outperforms most nickel sulfide-based electrocatalysts, with an overpotential of only 219 mV to achieve a current density of 10 mA cm^{-2} in 1 M KOH. Compared to metal sulfides, the corresponding metal selenides usually exhibit a better HER activity owing to their higher intrinsic electrical conductivity, Huang et al. prepared a nanocomposite consisting of coral-like NiSe encapsulated within a N-doped carbon framework by reacting a Ni-MOF with Se vapor at 600°C under flowing Ar.[40] It only required an overpotential of 123 mV in 0.5 M H_2SO_4 to afford a current density of 10 mA cm^{-2}. Additionally, this catalyst also possessed a Tafel slope as low as 53 mV dec^{-1}. Transition metals alloyed with Te have been demonstrated to have improved tolerance to electrolytes with a broad range of pH values. Wang et al. synthesized and compared the HER activity of several cobalt tellurides with different stoichiometric ratios by reacting the ZIF-67 with Te vapor at controlled temperatures (Fig. 3.2a–c).[41] It was found that the HER activity increased in the order of CoTe@C < CoTe$_2$@C< CoTe$_{1.1}$@C in 1.0 M KOH (Fig. 3.2d and 2e). After a continuous reaction for 20 h or cycling for 1000 cycles, CoTe$_{1.1}$@C also showed a negligible activity decay. EIS analysis suggested that the CoTe$_{1.1}$@C possessed the strongest electron transfer ability (Fig. 3.2f).

3.2.3 MOF-derived carbon-based metal phosphides and nitrides electrocatalysts for HER

For the application of transition metal phosphides in HER, P atoms can serve as a base and trap positively charged protons during HER due to their high electronegativity and the negative charge feature, and both metal and P atoms within transition metal phosphides can function as active sites, greatly boosting the HER activity.[42] Wang et al. synthesized porous Ni_2P@C nanosheets by the phosphorization of NiO nanosheet/MOF-74 precursors (Fig. 3.3a–d).[43] The as-synthesized Ni_2P@C nanosheet exhibited an excellent electrocatalytic performance towards HER in 1 M KOH, with a low overpotential of 168 mV at a current density of 10 mA cm^{-2} (Fig. 3.3e), a small Tafel slope of 63 mV dec^{-1} (Fig. 3.3f), as well as a negligible current decay after 10 h (Fig. 3.3g).

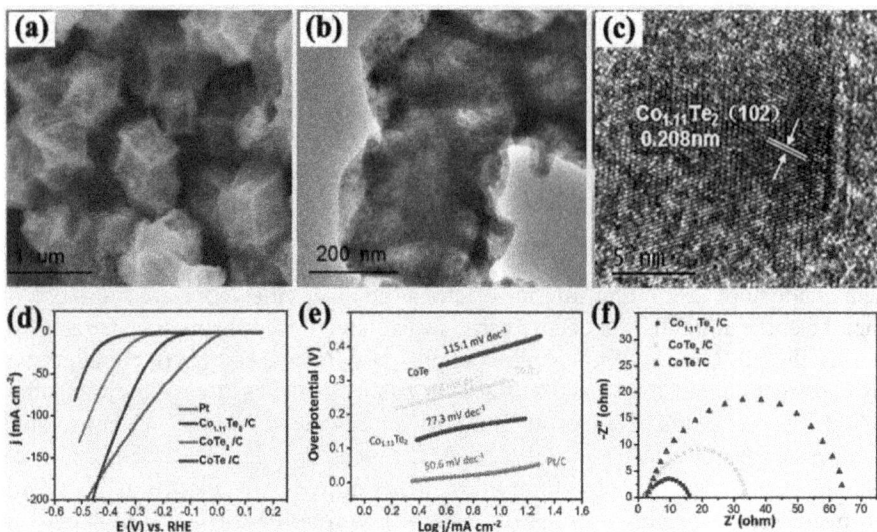

Figure 3.2. ZIF-67 derived CoTe$_{1.1}$@C nanocomposites: (a) FESEM image, (b, c) TEM images, (d) LSV curve, (e) Tafel slope, (f) EIS spectrum in 1.0 M KOH, respectively. Reproduced with permission from Ref. 41. Copyright 2018, Elsevier.

For the transition metal nitrides, the formation of metal–nitrogen bonds in the density of states of the metal d-band usually generates a smaller deficiency, endowing the metal nitrides with electron donating character, which leads to higher catalytic activities.[44] Chen et al. synthesized hybrid composites consisting of Co$_{5.47}$N nanoparticles encapsulated within porous carbon through the nitridation of ZIF-67 polyhedra at 700°C under flowing NH$_3$ (Fig. 3.4a–g).[45] During the nitridation, the N and Co species tended to combine and form Co-N compounds. The resulting hybrid composites exhibited an outstanding HER performance in 1 M KOH, affording a current density of 10 mA cm^{-2} at a low overpotential of 149 mV (Fig. 3.4h) and a small Tafel slope of 51 mV dec^{-1} (Fig. 3.4i). Moreover, the cobalt nitrides are more stable than that of bare Co during the cycle durability tests (Fig. 3.4j).

3.2.4 *MOF-derived carbon-based carbides and single-atom electrocatalysts for HER*

Transition metal carbides are of particular interest as HER electrocatalysts because of their electronic structures, which are similar to that of Pt, and exceptional chemical stability resistance to alkaline and acidic corrosion, which enables efficient HER activity over a wide range of pH values.[46,47] Among the investigated metal carbides electrocatalysts, a Mo-based carbide was demonstrated to exhibit the closest HER activity to that of Pt by both experimental and theoretical analyses. Wu et al. employed a cage-confinement pyrolysis for synthesizing the nanostructured octahedral MoC$_x$ nanoparticles (Fig. 3.5a–d); a Cu-based MOF [HKUST-1, Cu$_3$(BTC)$_2$(H$_2$O)$_3$] was used as a host, and then Mo-based Keggin-type POMs (H$_3$PMo$_{12}$O$_{40}$) were periodically inserted in the largest pores of the host, resulting in the formation of the

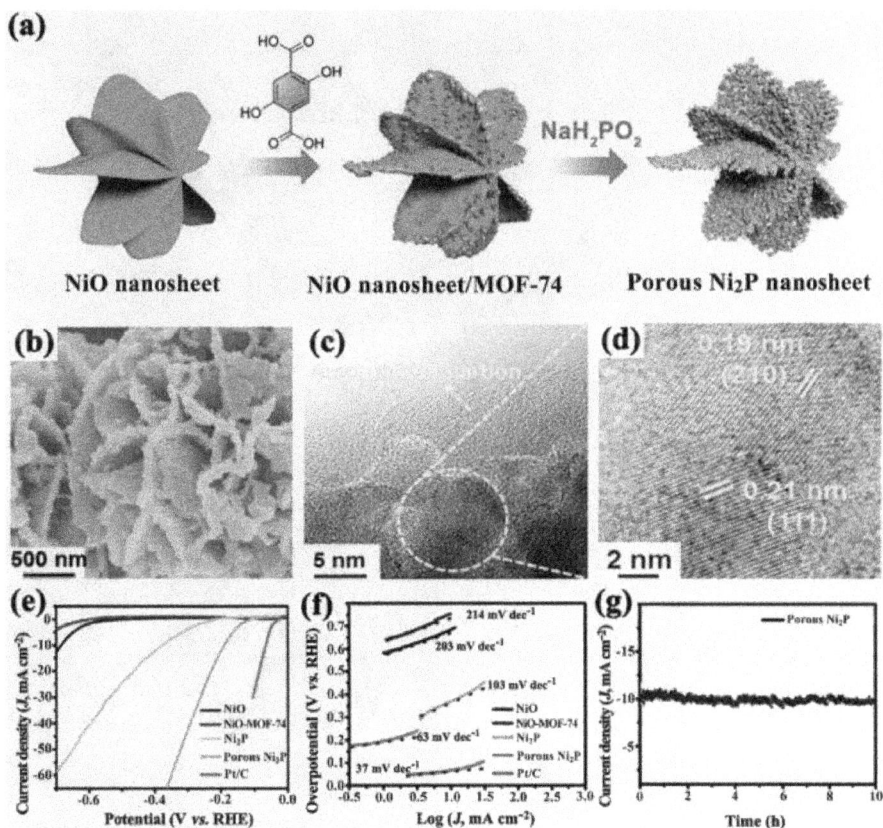

Figure 3.3. Ni-MOF-74-derived porous $Ni_2P@C$ nanocomposite: (a) schematic illustration of synthetic process, (b) FESEM image, (c, d) TEM images, (e) LSV curve, (f) Tafel slope, and (g) cycle durability in 1.0 M KOH, respectively. Reproduced with permission from Ref. 43. Copyright 2018, Royal Society of Chemistry.

Cu-MOF@Mo-POMs precursor.[48] The as-prepared MoC_x-based HER electrocatalyst afforded a current density of 10 mA cm^{-2} at overpotentials of only 142 and 151 mV in 0.5 M H_2SO_4 (Fig. 3.5e) and 1.0 M KOH (Fig. 3.5f), respectively.

Theoretically, isolated single-atom electrocatalysts (SACs) could offer remarkably better HER activities as compared to the corresponding bulk and nanosized electrocatalysts owing to their much higher atom utilization and unique low coordination and unsaturation.[49] Recent studies have shown that MOFs could function as ideal templates to prepare carbon-supported single-metal atoms, showing exceptional catalytic activities. Fan et al. achieved the first atomically isolated Ni species anchored on graphitic carbon (A-Ni-C) by employing a Ni-MOF as a precursor.[50] As shown in Fig. 3.6a–d, three main steps were involved in preparation of the A-Ni-C electrocatalysts. When the A-Ni-C hybrid was used as the HER electrocatalyst, it could drive a current density of 10 mA cm^{-2} in 0.5 M H_2SO_4 at an overpotential of only 34 mV (Fig. 3.6e) and achieved a Tafel slope of 41 mV dec^{-1} (Fig. 3.6f), much lower than those of the Ni@C electrocatalyst.

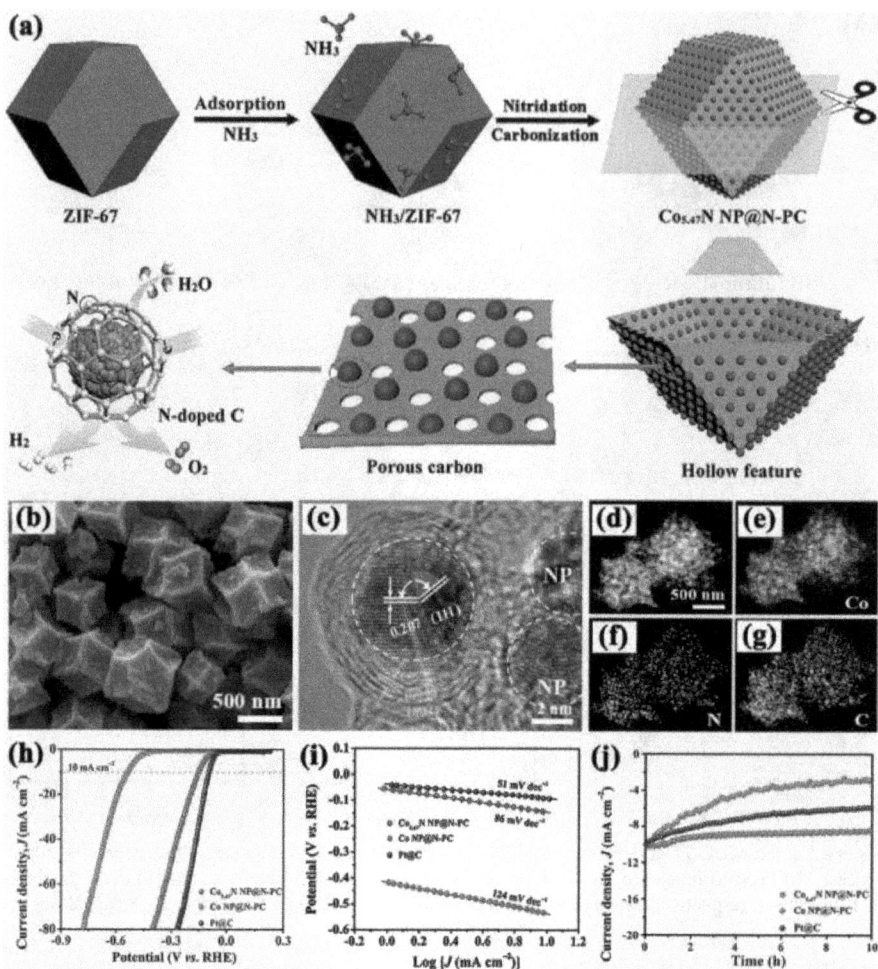

Figure 3.4. ZIF-67-derived $Co_{5.47}N$ NP@N-PC nanocomposite: (a) schematic diagram illustration of synthetic process, (b) FESEM image, (c) TEM image, (d) HAADF image, corresponding (e) Co, (f) N and (g) C elemental mappings, (h) LSV curve, (i) Tafel slope, and (j) cycle durability in 1.0 M KOH, respectively. Reproduced with permission from Ref. 45. Copyright 2018, American Chemical Society.

Electrode materials derived from MOFs have great potential as electrocatalysts, but there also have been some problems for HER. To improve the performance of MOFs derived materials as HER electrocatalyst, there should be an emphasis on the control of particle size during synthesis method that could help to increase the exposed active sites. Very important factor for HER electrocatalysis is high conductivity, which can be enhanced by incorporating organic linkers with a high number of conjugate bonds.

Thanks to the well-developed crystalline porous structure, vast combinations of chemical components and tailorable structures, MOFs-derived materials are a family of ideal HER electrocatalysts. This chapter highlights the recent developments and latest strategies of MOFs derivatives (including in bimetal, alloys, transition metal

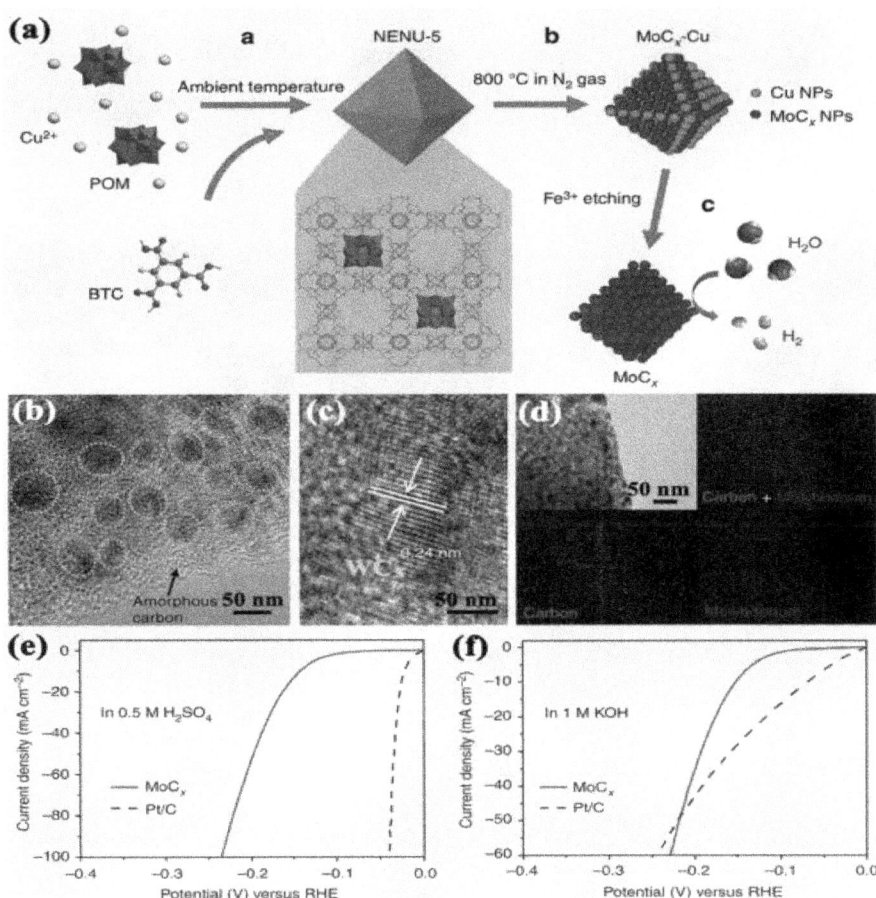

Figure 3.5. (a) Schematic diagram for the synthesis of octahedral MoC$_x$ nanoparticle and its (b, c) HRTEM images, (d) corresponding elemental mappings, LSV curves in (e) 0.5 M H$_2$SO$_4$ and (f) 1.0 M KOH, respectively. Reproduced with permission from Ref. 48. Copyright 2015, Nature Publishing Group.

chalcogenides, phosphides, nitrides, carbides and single-atom electrocatalysts) for implication in electrocatalysis for HER. The synthesis and modification methods of various MOFs-based electrocatalysts always aim to introduce catalytic active species or interfaces to the systems and expose uniformly dispersed catalytically active sites, further promoting synergistic functions between catalytically active species or between compounds and supporting frameworks. In any event, MOFs-derived electrocatalysts have drawn a bright picture for the development of HER.

3.3 MOFs-derived carbon-based hybrid nanomaterials for oxygen evolution reaction

Due to the four-electron transfer involved, the overpotential of the oxygen evolution reaction (OER) is relatively large as well as the formation of O-O double bond requires higher activation energy, so the rate of obtaining oxygen is slow.[51,52] In order to further

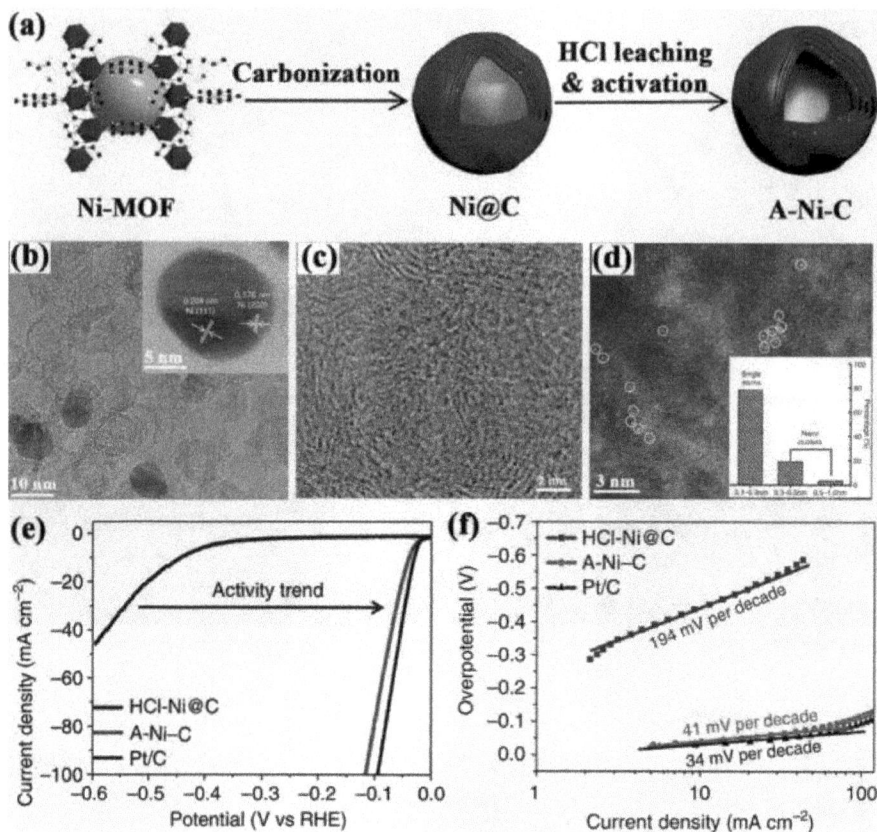

Figure 3.6. Ni-MOF-derived porous A-Ni-C nanocomposites: (a) schematic illustration of the synthetic process, (b) TEM image, (c) HRTEM image, (d) high-angle annular dark-field scanning transmission electron microscopy (HAADF) image, (e) LSV curve and (f) Tafel slope in 0.5 M H_2SO_4. Reproduced with permission from Ref. 50. Copyright 2016, Nature Publishing Group.

improve the reaction efficiency and reduce the energy consumption required for the reaction, we need to find highly efficient electrocatalysts. The ideal electrocatalysts should have the following characteristics: (1) high-efficiency catalytic activity, low cost and abundant resources; (2) high electrical conductivity and good stability; (3) abundant active sites, high exposure, and dispersed catalytic centers. The high cost and low abundance of traditional precious metal ruthenium-based and iridium-based materials limit their further development.[53] Recently, low-cost transition metals, and transition metal compounds such as transition metal oxides/phosphides have attracted great attention as a substitute for traditional precious metal electrocatalysts. Nevertheless, transition metal compounds have lower specific surface area and tend to agglomerate together, causing the active sites to be masked and the catalytic activity to decrease.[54,55] The MOFs possess a large specific surface area, uniform pore structure and adjustable pore size, which are composed of different metal ions and organic ligands to form nanocubes, nanorods, nanospheres and other structures with various morphology. However, the application of poorly conductive MOFs directly to OER electrocatalysts is limited by the slow electron migration rate. When MOFs are annealed in different atmospheres, the original structure of MOFs will be preserved well, forming various and controllable metals or metal compounds and other derivatives which are widely used in OER electrocatalysts. On the one hand, MOFs derivatives retain the original ordered structure and periodic active sites. On the other hand, the combination of the derivatives with highly conductive carbon materials can accelerate the electron transport rate. MOFs derivatives are mainly divided into the following categories: MOFs-generated metal and heteroatom-doped carbon materials, oxides and alloy compounds, carbides and nitrides, phosphides, chalcogenides and other derivatives.[56,57] The following chapter will be divided into the above categories to introduce the application of MOFs derivatives in OER, as well as the current shortcomings and trends.

3.3.1 MOF-derived carbon-based metal and heteroatom-doped materials electrocatalysts for OER

The MOFs with heteroatom-containing organic ligands *in situ* form heteroatom-doped conductive carbon materials with metal particles by carbonization have gained more attention for OER. The nitrogen-containing Co-based zeolite imidazole ester skeleton ZIF-67 has been widely used as precursors to prepare OER electrocatalysts. For example, Li et al. synthesized cobalt nanoparticles (NPs) uniformly embedded in nitrogen-rich carbon materials (PNC/Co) via facile carbonization process.[58] *In situ* oxidation of cobalt nanoparticles became the active site of OER. Not only PNC/Co maintained the original rhombic dodecahedron but also the conductivity of the material has been improved benefiting from the rich defects of heteroatom (N) doped carbon-based materials. In addition, Liu et al. applied the cobalt boron imidazole skeleton rich in B and N heteroatoms,[59] in which the heteroatom-containing organic ligands would restitute the metal nodes during carbonization in an inert atmosphere (Fig. 3.7a). Metal Co particles and a small amount of Co oxide were encapsulated in a B and N double heteroatom co-doped graphitic carbon shell. Carbonization temperature was one of the factors affecting OER activity; when the carbonization temperature was 900°C, it was beneficial to the formation of surface oxides and the crystallization of Co particles, and the catalytic activity was optimal (Fig. 3.7b and 3.7c). Surface oxidized materials have suitable adsorption effect on OH*, O* and OOH*. Conductive carbon layer via B, N double-doped MOFs-derived carbon can promote electron transfer and improve conductivity (Fig. 3.7d).

Most MOFs derivatives are powder materials, which is easy to accumulate and not conducive to the exposure and utilization of active sites. MOFs derivatives grow directly on porous conductive substrates which have been proven to be a feasible way to increase OER activity. The reason is that the conductive substrates are in excellent contact with the material, which can reduce the charge transfer resistance and facilitate the efficient transmission of electrons. Among them, foamed nickel (NF) has been widely used owing to its large surface area and 3D porous structure. Recently, Yuan et al. prepared a Co-MOFs directly on foam nickel, forming Co-MOFs//NF via a simple hydrothermal method (Fig. 3.8a).[60] The reason why nitrogen-doped carbon nanotubes growed directly on NF was that dicyandiamide

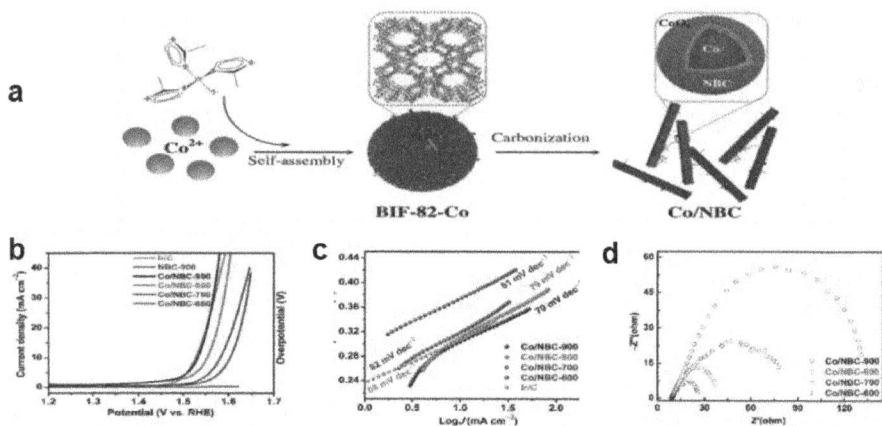

Figure 3.7. (a) Illustration of the synthesis of Co/NBC, (b) polarization curves and (c) the corresponding Tafel plots of the Co/NBC samples, NBC-900, and commercial Ir/C, (d) Nyquist plots of the Co/NBC samples. Reproduced with permission from Ref. 59. Copyright 2018, Wiley-VCH.

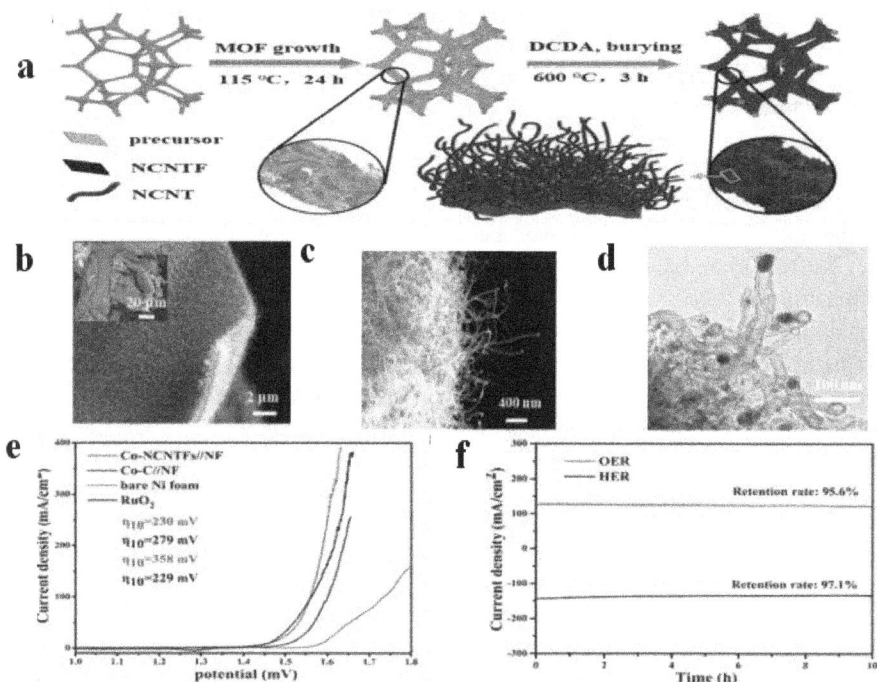

Figure 3.8. (a) Synthesis scheme of Co-MOF//NF and Co-NCNTFs//NF, (b) the NCNT framework, the inset of (b) is the NCNT framework grown on Ni foam, (c) and the interlaced NCNTs, (d) TEM image of Co-NCNTFs powder scraped from Co-NCNTFs//NF, (e) OER polarization curves of samples with 85% iR compensation in a 1 M KOH solution at a scan rate of 1 mV s^{-1}, (f) stability test of Co-NCNTFs//NF for OER and HER at a current density of around 10 mA cm^{-2}. Reproduced with permission from Ref. 60. Copyright 2020, American Chemical Society.

(DCDA) added during carbonization provided a nitrogen source and a secondary carbon source. The structural characterization diagram indicated that the Co-MOF on the NF became a rough wrinkled CNT framework, forming high-speed electronic transmission pipeline cross-linked with each other (Fig. 3.8b–d). The synthesized Co-NCNTF exhibited excellent catalytic activity. In 1 M KOH solution, it had an ultra-low overpotential of 230 mV at a current density of 10 mA cm^{-2}, which is far superior to the OER activity of most nickel-based materials (Fig. 3.8e), and maintained excellent stability for 10 h at a current density of 130 mA cm^{-2} (Fig. 3.8f). In order to construct independent electrodes without adhesives, in addition to the aforementioned conductive substrates, carbon nanotubes and graphene with remarkable electrical conductivity, mechanical strength and high surface area are also used to enhance catalytic activity frequently. Yan et al. combined hollow microspherical Ni-MOF with 2D graphene film, under the action of a nitrogen-containing carbon source-dicyandiamide (DCDA),[61] where the Ni metal particles were wrapped by MOFs-derived nitrogen-doped carbon nanotubes. The carbon nanotubes were connected with the 2D graphene film to form an integrated three-dimensional self-supporting heterogeneous film. The unique 3D structure of the materials prevented the aggregation of the 2D graphene film effectively. The Ni core was wrapped by defect-rich nitrogen-doped carbon nanotubes, which can effectively prevent metal particles from agglomerating. The highly conductive graphene and nitrogen-containing carbon nanotubes were contacted with the nickel core directly. In addition, the opening 3D heterostructure was conducive to the rapid transmission of electrons. The material was provided with the multifunctional catalytic activity of OER, HER and ORR, and its OER performance can reach a current density of 10 mA cm^{-2} at a low overpotential of 260 mV.

3.3.2 MOF-derived carbon-based metal oxides electrocatalysts for OER

As we all know, non-precious metal oxides are low in toxicity, rich in resources, and easy to prepare. Many metal oxides derived from MOFs have been widely used to catalyze OER reactions because of their remarkable electrochemical activity and favorable stability in alkaline solutions. For instance, Zhu et al. prepared a hollow rod-shaped Co_3O_4 derived from a MOFs template with well structural advantage.[62] The hollow structure was conducive to mass transmission and exhibited excellent OER performance. The synergistic effect of bimetals or trimetals with different valences can modify the coordination environment, adjust the electronic structure, and obtain the best capacity to bind to oxygen, sequentially further optimizing OER performance. Lu et al. selected Co and Fe with excellent catalytic activity to synthesize MOF-74-Co/Fe-derived spinel-like $CoFe_2O_4$/C porous hybrid nanorods.[63] In the process of synthesis, polyaniline was used to capture Co^{2+} and Fe^{3+} to promote the uniform growth and nucleation of MOFs on the NF, forming NF@$CoFe_2O_4$/C after carbonization. Spinel bimetallic compounds can be synthesized by bimetallic MOF, which regulates the electronic environment between different active metal sites to enhance the activity of OER. In addition, the bimetallic oxides are synthesized by ionic exchange with metal ions to construct heterogeneous interfaces and change the electronic structure of metals.

The synthesis of bimetal and polymetal MOFs are usually prepared by one-pot hydrothermal method, but it has disadvantages such as single structure, poor stability and few active sites. The synthesis of polymetallic MOFs by an ionic exchange method is easy to form hollow, core-shell and other layered structures, and thus optimize the activity of OER. For instance, Wang et al. synthesized (Ru-Co)O$_x$ with a hollow structure on carbon cloth by exchanging Co-MOF in Ru^{3+}solution.[64] As is shown in Fig. 3.9a, the heterostructures of Ru-doped CoO$_x$ and Co-doped RuO$_x$ uniformly adhere to the surface of the carbon cloth, forming hollow nanosheets (Fig. 3.9b–e). The heterostructures can optimize the electronic structure of the material surface, because Ru has the abundant d-orbital electrons and suitable oxygen adsorption energy. From the XPS, Co 2p$_{3/2}$ and Ru 3p$_{3/2}$ are all positively offset, which reduces the electron density of the metal and enhances the catalytic activity (Fig. 3.9f–h). Hence, the dual functional catalytic activity of HER and OER can be achieved. For the OER performance, the current density is as low as 171.2 mV at the current density of 10 mA cm^{-2} (Fig. 3.9i). In addition, the synergistic effect of the three metals is also beneficial to enhance the catalytic activity. For example, Tuo et al. synthesized a multilayer structure of Fe-MIL-101-NH$_2$ by ionic exchange.[65] The three-dimensional MOFs octahedron is wrapped by the two-dimensional CoNiFe-MOF, forming a core-shell CoNiFeO$_x$-NC after carbonization. The core-shell structure effectively prevents the collapse of the structure during carbonization. During the ionic exchange process, Ni^{2+} substitutes part of Fe in the Fe-MIL-101-NH$_2$, while the etching of Co^{2+} converted the original MOF into a concave octahedral structure, which provided an environment for the formation of two-dimensional CoNiFe-MOF. From the charge density distribution of NiFe$_2$O$_4$ and Co-NiFe$_2$O$_4$, the transition of charge from Fe to Co to Ni occurs due to the strong electronegativity of

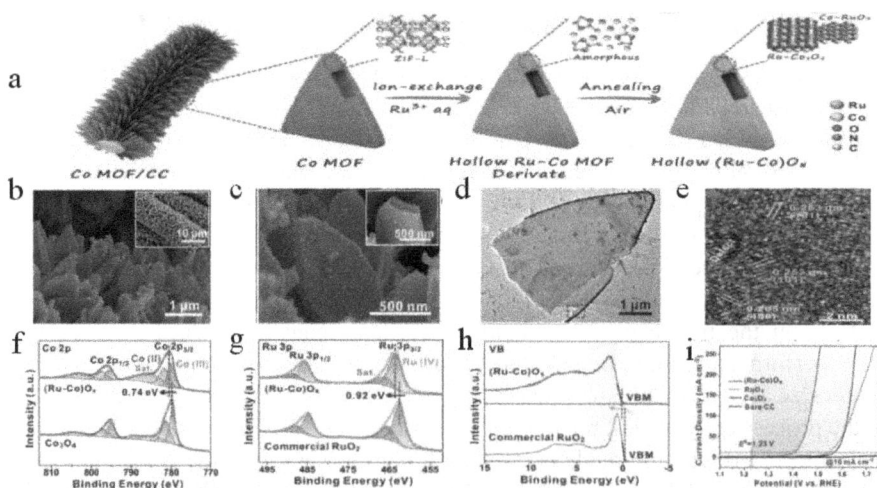

Figure 3.9. (a) Schematic illustration of the fabrication process of hollow (Ru-Co)O$_x$ nanosheet arrays on carbon cloth. SEM images of (Ru-Co)O$_x$ nanoarrays inset in (b) is a low-magnification image and inset in (c) shows a broken hollow nanosheet. (d) TEM and (e) HRTEM images of (Ru-Co)O$_x$. (f) High-resolution Co 2p XPS spectra of (Ru-Co)O$_x$ and Co$_3$O$_4$. (g) High-resolution Ru 3p and (h) valence band XPS spectra of (Ru-Co)O$_x$ and commercial RuO$_2$ (i) OER polarization curves. Reproduced with permission from Ref. 64. Copyright 2020, Wiley-VCH.

Ni. After adjusting the electronic environment, the oxygen binding strength at OH site of the Ni-Co coordination will decrease, and its OER performance in alkaline solution only exhibits a low potential of 265 mV at 50 mA cm^{-2}.

3.3.3 MOF-derived carbon-based metal carbides and nitrides electrocatalysts for OER

Compared with oxides, metal nitrides have brilliant conductivity and suitable adsorption energy. Negative nitrogen can accept protons, hence it has excellent OER activity and has been widely used in catalysis and energy storage fields.[66–67] Prussian blue analogues (PBAs) with an open distribution framework and uniform metal active centers belong to the category of MOFs. Among them, PBAs-derived metal nitrides can retain their excellent structural characteristics. Kang et al. synthesized mesoporous $Co_3N@AN$-CNC using the Prussian blue analogue $Co_3[Co(CN)_6]_2$ nanocubes as the precursor.[68] $Co_3N@AN$-CNC with a porous cubic structure was coated with a highly conductive carbon of *in situ* grown nitrogen-doped carbon to further enhance conductivity. In the CV result, it can be seen that $Co_3N@AN$-CNC possesses more abundant OER active sites than Co_3O_4. Due to the above advantages, $Co_3N@AN$-CNC exhibits an ultra-low overpotential of 280 mV at 10 mA cm^{-2} and excellent cycling stability. On the other hand, metal carbides also exhibited excellent catalytic activity, while its OER activity is limited because of the agglomeration and uneven carbonization during high-temperature carbonization. The ordered arrangement of the metal nodes in the bimetallic carbide derived from MOFs can effectively restrain agglomeration. The 3d metal atom synthesizes the bimetallic carbide with other metals, which can effectively improve the adsorption energy of the carbide to the intermediate and activate its OER performance. Chen et al. exchanged and etched ZIF-L with cobalt ions and tungsten ions, forming $Co_6W_6C@NC$ vertical porous nanosheets on flexible carbon cloth.[69] The metal Co was converted into catalytically active CoO_x and $Co(OH)_2$ in the OER process, and tungsten was mixed into CoO_x and $Co(OH)_2$ to adjust the electronic environment and enhance the OER activity. The vertical nitrogen-containing porous carbon nanosheets derived from ZIF-L can efficiently transport electrons.

3.3.4 MOF-derived carbon-based metal phosphides electrocatalysts for OER

It has been demonstrated that MOF-derived metal phosphides have eminent OER performance. Yu et al. synthesized carbon-containing Ni_5P_4/Ni_2P derivatives by using Ni-Ni-PBA as precursor which showed excellent catalytic activity for OER.[70] He et al. synthesized MOF-derived Ni-Co mixed metal phosphide/carbon nanoboxes via a surfactant-mediated method.[71] In addition, other high-efficiency OER mixed metal phosphides have also been developed. Hong et al. synthesized $Na(OH)_2$-mediated FeCo-MOF in aqueous solution, forming a sort of $(Fe_xCo_y$-P/C) compound after carbonization and phosphorus doping.[72] The synergy of phosphorus and iron doping reduced resistance and strengthened performance. The measure of optimizing the electronic structure by doping metal has been widely used in brine electrolysis. Lin et al. successfully doped Ru^{3+} into the two-dimensional MOF MIL-53 (NiFe) on

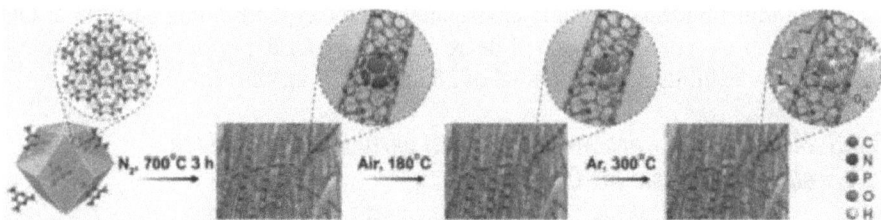

Figure 3.10. Synthesis and morphological characterization of Part-Ph Co@Co–P@NPCNTs. Reproduced with permission from Ref. 74. Copyright 2020, Wiley-VCH.

NF by a hydrothermal method at 120°C for 3 hours.[73] Ru particles were uniformly grown on the surface of the two-dimensional MOF, and then phosphatized to obtain Ru-doped nickel iron phosphide (Ru-NiFeP/NF) loaded on NF. Two-dimensional MOFS derived uniformly arranged ultra-thin nanosheet arrays and fully exposed active sites on 2D nanosheets, and the electronic structure by Ru doping in bimetallic phosphides was changed to improve OER performance. Jiao et al. partially phosphatized Co@Co-P nanoparticles with a core-shell structure into N and P co-doped carbon nanotubes, which effectively inhibited the surface oxidation of Co and the reunion of metal nanotube particles (Fig. 3.10).[74] It avoids corrosion in strong alkaline solutions and effectively improves conductivity at the same time.

3.3.5 MOF-derived carbon-based metal chalcogenides electrocatalysts for OER

Owing to their different structures, transition metal chalcogenides can be divided into layered transition metal chalcogenides and non-layered transition metal chalcogenides, abbreviated as MX_2. The catalytically active transition metal M is mainly composed of element (including Fe, Co, Ni, etc.), and X is composed of three elements (S, Se, and Te). The layered transition metal chalcogenide compound is similar to the structure of sandwich, and the upper and lower layers of sulfur atoms (X) and the middle metal atoms (M) are combined by a strong covalent bond. The layers are combined by weak van der waals forces, which are easy to fall off to obtain a single-layer or few-layer layered structure. Meanwhile, the unique 2D layered structure can provide a larger specific surface area and sufficient active sites. Non-layered transition metal chalcogenide compounds have a more stable three-dimensional structure. With regard to transition metal chalcogenide compounds, it's convenient for the chalcogen elements to transfer electrons on the transition metal and obtain better OER activity owing to their stronger electronegativity.[75,76] Nevertheless, the current application in catalysis is not satisfactory due to surface oxidation, poor structural stability, and easy accumulation of layered structures. Therefore, it is necessary to optimize the crystal structure and surface electronic states to further improve conductivity. Co and its sulfides with good catalytic activity are widely used in catalytic reactions such as OER and HER. Doping with MOFs-derived transition metal chalcogenides can optimize the surrounding electron density and improve OER performance. For example, Du et al. obtained $Cu-Co(CO_3)(OH)$ nanobelts by hydrothermal *in situ* growth on NF, and the Cu-Co-MOF was prepared by using

Cu-Co(CO$_3$)(OH) to reaction with H3BTC, then, the Cu-doped Co$_9$S$_8$ nanorod array was synthesized via the vulcanization process.[77] It has been confirmed that Cu-Co$_9$S$_8$ with abundant void structure can be obtained after vulcanization for 6 h, which is conducive to ion transport. The increased activity was analyzed by DFT calculation. The density of state (DOS) of Cu-Co$_9$S$_8$ is close to the Fermi level with a wider peak, indicating an efficient charge transfer rate. In addition, the intermediate activation energy was greatly reduced due to Cu doping, which increased the inherent catalytic activity of the material and promoted the OER process. Zhao et al. designed a two-dimensional Co-MOF[Co(BDC)$_2$(SPDP)$_2$(DMF)(H$_2$O)] by using SPDP and H$_2$BDC organic ligands.[78] When this Co-MOF was carbonized at a high temperature of 900°C, the obtained N, O, S triple-doped carbon coated Co$_9$S$_8$ derivatives exhibited good OER performance. Xuan et al. used Fe to optimize the characteristics of nickel's oxygen adsorption energy and electronic structure.[79] The Prussian blue analogue (NiFe-PBA) was used as sacrificial template by hydrothermal reaction with sulfur powder, obtaining the carbon-coated Fe$_5$Ni$_4$S$_8$/NiS mixed phase after carbonization process. The current density can reach 10 mA cm^{-2} with only an overpotential of 200 mV, and it also exhibited excellent cycling stability. The NiFe bimetallic selenide derived from the Prussian blue analogues also possesses excellent OER activity. Guo et al. obtained NiFe-Se/CFP by electrodepositing NiFe-PBA on carbon fiber paper (CFP) and then selenizing at 450°C for 30 min. During the OER process, the surface of materials changes from nickel-iron selenide to bimetallic oxide or hydroxide, which can provide a current density of 10 mA cm^{-2} at a low potential of 281 mV.[80]

The advantages of MOFs derivatives as OER electrocatalysts are as follows: the shape and structure of MOFs are controllable, and different organic ligands can be selected to synthesize MOFs in different morphology. The homogeneous structure of MOFs can also be used to design core shell, hollow shell, and other advantageous structures with excellent OER performance through ionic exchange, etching, or self-assembly after synthesis. In addition, the synergistic effect of polymetallic and ion doping is the key to improve the performance of OER by changing the electronic structure and adding active sites. However, the specific reaction mechanism of various MOFs derivatives in OER is still unclear. Further studies on the mechanism of MOFs will provide a theoretical basis for improving the performance of OER.

3.4 MOFs-derived carbon-based hybrid nanomaterials for oxygen reduction reaction

Similar to batteries, fuel cells convert the chemical energy of fuel and oxidant into electric energy. Unlike the batteries, they do not need to be recharged, as long as the fuel and oxidant supply is maintained. When hydrogen is used as fuel, the fuel cell only produces electricity, water, and some heat. Compared with thermal engines, fuel cells have the advantages of high efficiency, no environmental pollution, and unlimited sources of reactants.[81,82] Therefore, fuel cells are expected to have a wide range of commercial applications in transportation, stationary power generation, portable power generation and other fields, thereby helping to solve global energy supply and clean environmental issues. Among the existing fuel cells, the proton exchange membrane fuel cell (PEMFC) has been actively developed for use in automobiles, portable electronic devices, and cogeneration systems due to its simplicity, low operating temperature, high power density, and fast start-up. PEMFCs are especially well suited as the main power sources for automobiles and buses.[83] Fuel cell vehicles (FCV) have been considered as one of the ultimate solutions for the automotive business, and have far-reaching advantages over pure electric vehicles (EVs). Indeed, the first mass produced FCVs (Toyota Mirai), which have been commercially sold in Japan since 2014 and are going to be available in North America in 2015 at a price of ~ 57 000 dollars. One of the main reasons for the high sale price of the Mirai is the high Pt loading in the fuel cell stacks.[84]

At the anode of PEMFC, hydrogen is oxidized to produce electrons and protons, which are transferred to the cathode through the external circuit and the proton exchange membrane, respectively ($H_2 \rightarrow 2\ H^+ + 2\ e^-$). At the cathode, oxygen is reduced to water by reacting with protons and electrons ($1/2 O_2 + 2\ H^+ + 2\ e^- \rightarrow H_2O$). Both the anode and cathode electrodes are composed of Pt-based nanoparticles that are highly dispersed on carbon black to promote the reaction rates of the hydrogen oxidation reaction (HOR) and oxygen reduction reaction (ORR). Because HOR reacts very fast to Pt, the Pt loading on the anode is reduced to below 0.05 mg cm^{-2}. On the other hand, at the cathode, even on the best Pt-based electrocatalysts, the slow reaction kinetics of ORR requires a higher Pt loading (about 0.4 mg cm^{-2}) to achieve ideal fuel cell performance. Pt is a scarce and expensive metal as shown in Fig. 3.11.[85] Therefore, it would be advantageous to reduce its loading, or even completely replace it with a rich and cheap metal.

MOFs as the precursors of a new type of high-porosity carbon-based electrocatalyst have received extensive attention. Their unique internal structure provides an opportunity to obtain a higher surface area than other porous materials, such as zeolite and activated carbon, etc.[86] Different from the harsh operating conditions and high energy consumption of traditional synthesis processes, MOFs-derived carbon-based materials have many advantages as electrocatalysts: (1) the structure, composition and function of derivatives are flexible and tunable because MOFs' precursors can be modularly designed according to the target properties through the self-assembly of metal ions/clusters and the bridging of organic ligands;[87] (2) heteroatom-doped

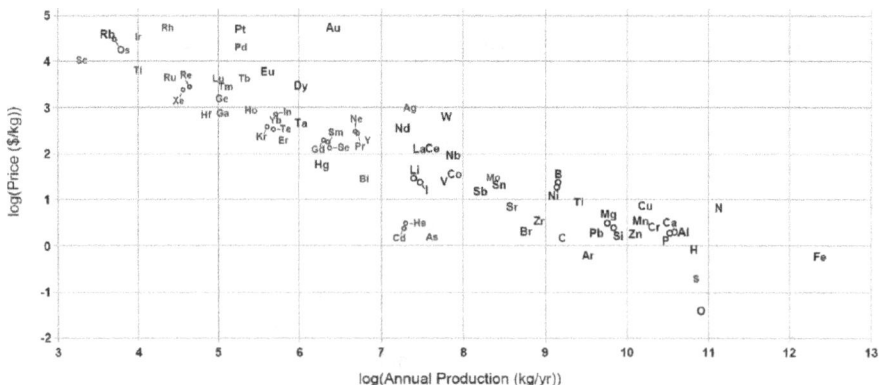

Figure 3.11. Price of the elements (in $/kg) versus their annual production (in kg/yr). Reproduced with permission from Ref. 85. Copyright 2016, American Chemical Society.

carbon materials of different non-metals or metal elements are easy to achieve, which is attributed to the ultra-high specific surface area, different pore size distribution, and ordered porous structure of MOFs, which are easy to adsorb organic molecules.[88] To improve the electron transport characteristics and the catalytic activity of ORR, single/dual heteroatom-doped carbon-based electrocatalysts are widely fabricated and divided into two forms, namely nonmetal-doped porous carbon materials (N, S, B, P, et al.) and metal-doped porous carbon materials (Fe, Ni, Co, Cu, et al.), which show more active sites and enhance electrocatalytic activity for ORR.[89]

3.4.1 Synthesis methods of MOF-derived porous carbon electrocatalysts

3.4.1.1 Carbonization conditions

The synthesis of MOF-derived porous carbon electrocatalysts mainly includes two steps: the design of precursors and the control of carbonization conditions. Porous carbon electrocatalysts derived from MOFs must undergo high-temperature pyrolysis. High temperature can increase the degree of graphitization and further increase the conductivity of the catalyst, thereby increasing the ORR activity. However, high temperature will reduce the amount of heteroatom doping (such as nitrogen), thereby reducing the number of active sites. Therefore, determining the appropriate temperature is a key factor for preparing high-efficiency porous carbon-based electrocatalysts for ORR by using various MOFs precursors.[90] The synthesis of metal-free porous carbon electrocatalysts usually by using zinc-containing MOFs (such as ZIF-8), and the pyrolysis temperature is set above 900°C, which is conducive to the complete removal of Zn metal by evaporation. And when use the other MOFs as precursors to prepare pure carbon electrocatalysts that need to be cleaned with acid after carbonization process.[91] For the synthesis of transition metal-doped carbon electrocatalysts, the carbonization temperature usually covers a wide range from 700–1000°C. The metal doping not only provides a highly active synthesis electrocatalysts, but also increases the graphitization of the material degree, because transition metals are good graphitization electrocatalysts.[92] In order to maintain the MOFs structure during the carbonization process in the presence of metals, the mesoporous silicon protective calcination strategy or MOFs composite strategy is very effective.

3.4.1.2 Structure design of MOFs precursors

In addition to the carbonization conditions, the design of the precursors is also a key factor in the preparation of high-efficiency electrocatalysts. The preparation of MOFs generally undergoes coordination and self-assembly of metal ions and organic ligands to form a cyclic framework porous structure. The synthesis methods of MOFs include *in situ* hydrothermal synthesis, room temperature stirring, microwave synthesis, seed crystal growth and so on, which provide layout for the design and preparation of MOFs with different morphologies (3D mesh, 2D film, et al.).[93] In order to realize the versatility of MOFs precursors, MOFs composite precursors are also designed. By introducing oxygen-containing functional groups or charged templates, *in situ* growth on carbon nanotubes (CNT), graphene oxide (GO), lactate dehydrogenase (LDH), cobalt hydroxide, cobalt oxalate and other design templates were used to prepare MOFs precursors. In addition, post-synthesis treatment is also used, that is, through solid-phase or liquid-phase mixing, a secondary heteroatom source is added to the prepared MOFs precursor.[94]

3.4.1.3 Synthesis strategy of porous carbon electrocatalysts from different MOFs precursors

In the above two steps, we divide MOFs-derived carbon electrocatalysts into three categories according to the difference of MOFs precursors. The simplest method is to use MOFs as the precursor, and use the direct carbonization method to prepare the porous carbon electrocatalysts. The second method is to use MOFs and secondary heteroatom sources as precursors to prepare porous carbon electrocatalysts. The third method is to use MOF composite materials (such as MOFs/CNT, MOFs/rGO, MOFs/ LDH, et al.) as precursors to prepare porous carbon electrocatalysts (MOF-based composite carbonization method). A detailed description is shown in Fig. 3.12.[95]

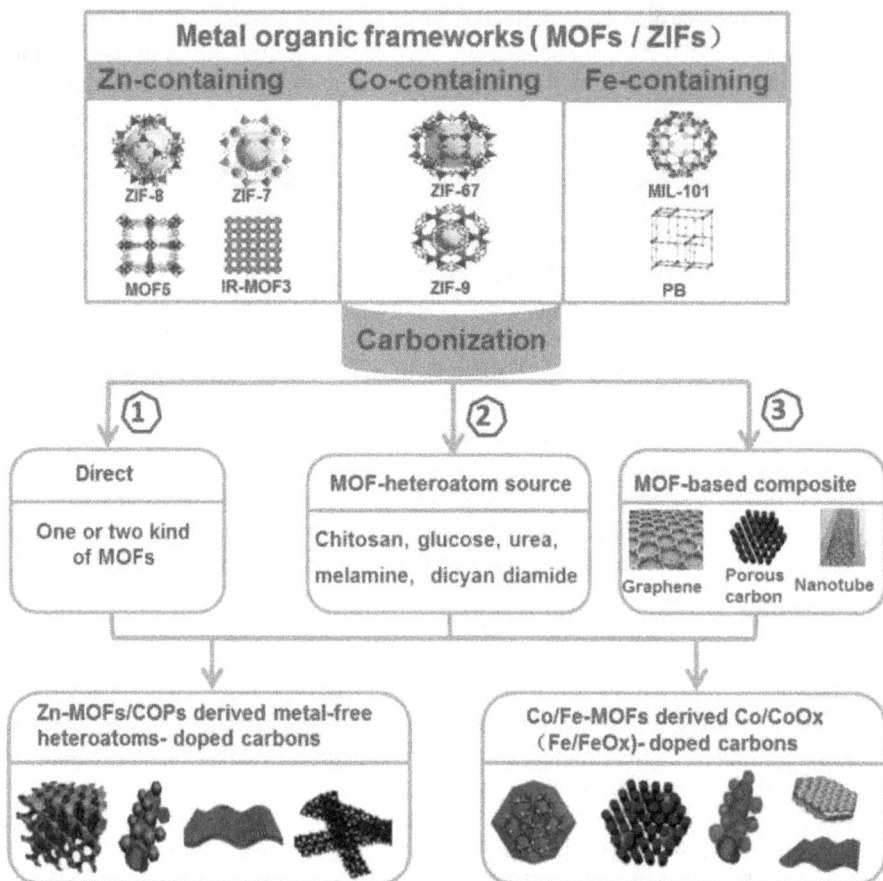

Figure 3.12. Schematic illustration of MOF-derived porous carbon electrocatalysts. Generally, we can adopt three methods to obtain MOF-derived porous carbon electrocatalysts, (1) the direct carbonization of MOFs, (2) the carbonization of MOF and heteroatom source mixture, (3) the carbonization of MOF-based composites. As a result, MOF-derived, metal-free, or transition-metal-doped porous carbon electrocatalysts can be obtained by rationally designing the MOFs precursors. Reproduced with permission from Ref. 95. Copyright 2018, Wiley-VCH.

3.4.2 MOF-derived metal-free heteroatom-doped porous carbon electrocatalysts

3.4.2.1 N-Doped carbon electrocatalysts

The synthesis method of MOFs-derived N-doped carbon electrocatalysts depends on the nature, existence form and type of doping element and the prospect of oriented electrocatalyst. The basic characteristics of MOF precursors, such as size, topology, surface characteristics, pore volume and pore structure are closely related to the microstructure, spatial distribution of pores and catalytic activity of the MOF-derived carbon composite electrocatalyst. The comprehensive strategies reported in the past can be divided into two categories: (1) direct carbonization of nitrogen-containing MOFs precursors (such as ZIF-67) is called "*in situ*" nitrogen doping;[96] (2) the post-synthesis modification strategy of the carbonization process, that is, the use of nitrogen-containing guest molecules to introduce nitrogen into the catalyst.[97] MOFs can be used as a direct precursor to construct heteroatom-doped porous carbon electrocatalysts without any additional secondary carbon source or pore former, which can be attributed to its clear porous structure and organic components with rich carbon content. If the organic ligands of MOFs contain a variety of heteroatoms, the heterogeneous atom doping is easy to achieve and the cost is low, and can be used as an ideal precursor to prepare carbon-based electrocatalysts with uniformly distributed catalytic centers and high-density active sites.

In the process of direct carbonization growth, MOFs are usually used as carbon and nitrogen sources as self-sacrificing templates, while nitrogen-doping occurs during the conversion of nitrogen-containing MOFs precursors into carbon nanostructures. For example, Hong and co-workers presented a highly porous zeolite-type MOFs (ZIF-8) as precursor, which becomes a chemically and thermally stable electrocatalyst. The aromatic methyl imidazole ligand of ZIF-8 is oxygen-free, and the abundant nitrogen is directly incorporated into the aromatic ring to promote the consolidation of highly enriched nitrogen-containing active sites to the carbon matrix. In their research, nitrogen-doped carbon was obtained by a highly graphitic carbon framework through a high temperature (1000°C) process, during which ZIF-8 reduces metal zinc and zinc ion by evaporation (boiling point 908°C) and possible residual is removed after acid treatment (Fig. 3.13a and b).[98] The synthesized N-doped graphite porous carbon has a uniform rhombohedral dodecahedron, with a relatively uniform N distribution, hierarchical porous structure, high density of active sites, and uniform distribution of active sites (as shown in Fig. 3.13c–g).[98] The optimized electrocatalyst (NGPC-1000-10) shows superior electrocatalytic activity, and cycling stability for ORR process and brilliant methanol tolerance in alkaline medium, which can be attributed to the synergy of high surface area and porosity, enriching active sites being derived from the high graphitic-N part, and a high degree of graphitization.

3.4.2.2 Binary or ternary nonmetallic heteroatom-doped carbon-based electrocatalysts

Heteroatom-doped carbon-based electrocatalysts have a large asymmetric spin and optimized carbon skeleton charge density, which will produce potential synergistic effects between different heteroatoms. The composite heteroatom doped carbon-

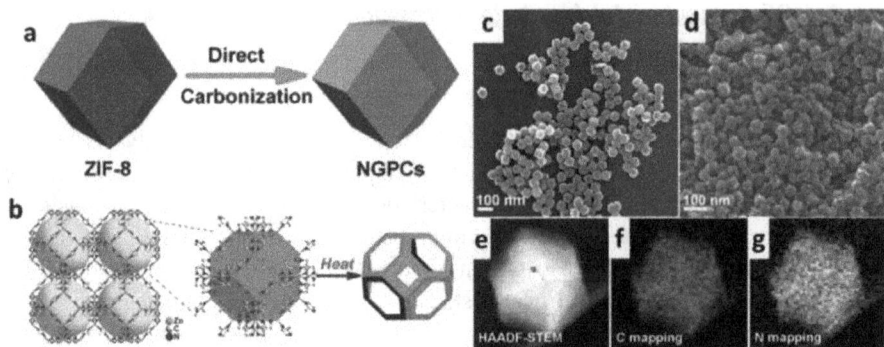

Figure 3.13. (a, b) Schematic illustration of ZIF-8-driven template synthesis of highly graphitized nitrogen-doped porous carbon. (c) SEM images of monodispersed ZIF-8 and (d) NGPC-1000-10. (e–g) HAADF–STEM images of a single carbon polyhedron and the corresponding C-, and N-elemental mappings. Reproduced with permission from Ref. 98. Copyright 2014, Royal Society of Chemistry.

based catalyst can further improve the electrocatalytic activity for ORR. A large number of studies have proved that heteroatom (such as N, S, B, P) doped carbon electrocatalysts have high-efficiency ORR catalytic activity; the reason is that it has obvious electronegativity, and can significantly change its electronic and chemical properties. Among heteroatoms, S and N co-doped electrocatalysts have a significant synergistic effect and potential energy to replace C atoms. This is mainly due to the combination of sulfur-related active sites and hierarchical porous structure, which promotes the rapid diffusion of oxygen molecules and electrolyte ions to the catalytic sites and releases products from the catalytic sites. Compared with traditional liquid organic CVD technology, S-containing functions can be fused into sp^2 hybrids of carbon atoms through the direct pyrolysis process of host MOFs and S-containing guest molecules; the latter has better uniformity and higher S content. For example, Gao et al. reported using liquid S-hexane mixed vapor to form S-doped graphene/Cu at 950°C, resulting in the presence of S atoms in linear nanodomains with a low content instead of the hetero S atom being doped.[95] Dai et al. developed the MOF-5 precursor by embedding urea and DMSO (as N and S precursors, respectively) in the pores of MOF and carbonizing in ultrapure N_2 atmosphere (Fig. 3.14).[90] The characterization results show that N and S atoms have been introduced into the porous carbon framework, the optimized electrocatalyst. Due to the synergistic effect of N and S, the final nitrogen-sulfur co-doped porous carbon electrocatalyst shows a more positive starting potential, long-term stability and excellent methanol tolerance for ORR in 0.1 M KOH, which can even be compared with commercial Pt/C electrocatalysts.

Moreover, MOF-5 was used as template, and the ternary co-doped porous carbon electrocatalyst (NPS-C-MOF-5) of nitrogen (N), phosphorus (P) and sulfur (S) was prepared. Dicyandiamid (DCDA), triar-ylphosphine (TPP), and dimethyl sulfoxide (DMSO) were introduced as N, P, and S precursors, which achieved uniform doping of N, P, and S in the synthesized carbon material (Fig. 3.15).[99] When heteroatoms (N, P, S) are doped into the carbon framework, the pore structure of the

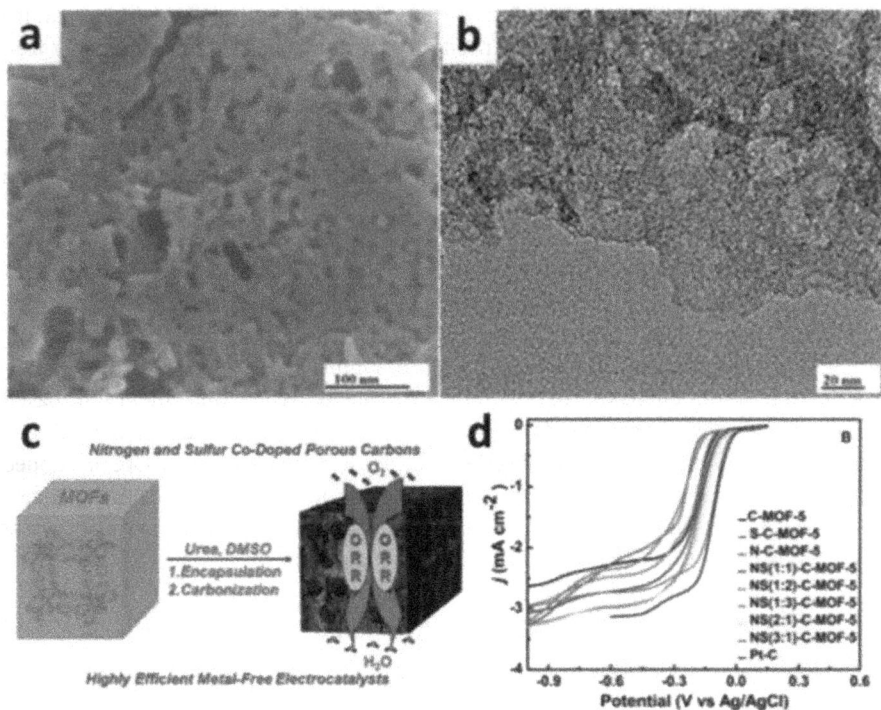

Figure 3.14. (a) SEM and (b) HRTEM images of the typical NS (3:1)-C-MOF-5 electrocatalyst; (c) synthesis procedure for MOF-5 templated N and S co-doped porous carbon as metal-free electrocatalysts for ORR; (d) LSVs of different samples (different atomic ratio of N and S) at 1600 rpm. Reproduced with permission from Ref. 90. Copyright 2014, Royal Society of Chemistry.

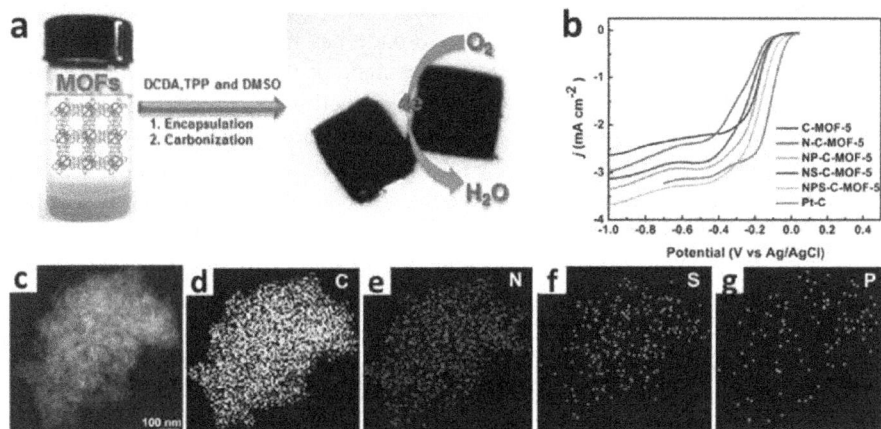

Figure 3.15. (a) Schematic illustration of the synthesis of MOF-templated NPS-C-MOF-5 as a metal-free electrocatalyst for the ORR; (b) LSVs of different samples at a rotation rate of 1600 rpm; (c) STEM images of NPS-C-MOF-5; (d–g) the corresponding C-, O-, N-, P-, and S-elemental mappings, respectively. Reproduced with permission from Ref. 99. Copyright 2014, Nature Publishing Group.

materials was changed. It can be inferred that some of the micropores of heteroatom-doped materials are sintered, and the ratio of micropores increased, which may lead to different activity. On account of the synergistic effects of N, P, and S ternary-doped carbon, the NPS-C-MOF-5 electrocatalyst displayed a more positive onset potential (about −0.006 V at 1600 rpm), and demonstrated a superior electrocatalytic activity, excellent methanol tolerance and outstanding long-term stability for ORR in 0.1 M KOH.[96]

In this part, the importance for application in ORR of design method of MOFs precursor, and the synthesis strategy of MOFs-derived heteroatom-doped porous carbon catalyst have been emphasized, because the MOFs precursors often determine the microstructure of the derived porous carbon electrocatalysts, which in turn affects its ORR performance. Generally speaking, to synthesize high-performance metal-free carbon electrocatalysts, the MOF-heteroatom carbonization and MOF-based composite carbonization methods are more effective, because these two methods can introduce more elements into the obtained samples to achieve non-metal atoms doped in the carbon matrix. Therefore, in order to obtain a highly efficient metal-doped porous carbon catalyst, the design of the MOFs template becomes very important because the catalyst can be synthesized by direct carbonization.

3.5 MOFs-derived carbon-based hybrid nanomaterials for CO_2 reduction reaction

Carbon dioxide electroreduction reaction (CO_2RR) has received more and more research attention in recent years due to its excellent energy storage performance.[100–102] The intermittent energy (such as wind energy, tidal energy, etc.) could be efficiently converted into chemicals (such as formic acid, CO and other compounds) by CO_2RR, which was also conducive to the recycling of the greenhouse gas CO_2, providing new solutions method for efficient storage and conversion of intermittent energy.[103–105]

Although electroreduction of carbon dioxide has many advantages, there are still many problems. (1) High reaction overpotential: CO_2 molecules have strong chemical stability, and the chemical adsorption activation process needs a high energy barrier to easily cause electrochemical polarization, leading to overpotential and energy loss.[106–108] (2) Low Faraday efficiency: In aqueous solution, CO_2RR and HER occur at the same time, and there are many kinds of CO_2RR products, resulting in low Faraday efficiency of the target product.[109,110] Therefore, how to design and prepare a catalyst with a single product and high Faraday efficiency of CO_2RR is a difficult point.

MOFs with unique morphology and electronic structure possess excellent CO_2 adsorption capacity, which are potential electrocatalysts for CO_2RR.[111,112] However, due to the discrete metal centers and insulating organic ligands of MOFs materials, the electrical conductivity is too poor, and high-efficiency conversion cannot be achieved. In order to solve this problem, MOFs-derived electrocatalysts with high electron transport capabilities are obtained by MOFs modified materials to enhance the conversion of CO_2RR. MOF-derived electrocatalysts mainly include heterojunction-type electrocatalysts grown on highly conductive supports (such as graphene, metals, etc.)[113-116] and porous carbon materials derived from MOFs carbonization.[117,118] Compared with other types of electrocatalysts, MOFs derivative electrocatalysts with a rich spatial structure and orderly arranged active sites can easily enhance the chemisorption of CO_2 to obtain low overpotential and high Faraday efficiency of CO_2RR. At present, the research on MOFs derivative electrocatalysts mainly focuses on the influence of morphology, active sites, electron density and other factors on the reaction performance. Morphology, active site structure, and local electron density of MOFs-derived carbon-based hybrid nanomaterials are comprehensively summarized. Furthermore, the challenges and prospects for future research into MOFs derivative electrocatalysts for CO_2RR are included.

3.5.1 The influence of the morphology of MOFs derivatives on the performance of CO_2RR

The CO_2 gas phase enters the electrolyte through dissolution, and then is adsorbed to the active sites in the pores of the MOFs derivative to react. The different morphology and structure of MOF derivatives have an impact on the chemisorption performance of CO_2RR.[117,119] The morphology influences of one-dimensional stick-like, two-dimensional plane and three-dimensional on the performance of CO_2RR are summarized (Fig. 3.16).

The one-dimensional rod-shaped MOFs-derived electrocatalysts possess the advantages of a porous structure similar to carbon nanotubes and a large specific surface area. The unique morphological structure formed between the metal and the one-dimensional carbon material greatly enhances its conductivity and CO_2 adsorption and conversion capacity. Wang et al. introduced active metals into ZIF-8 through Ni^{2+} ion exchange, and then pyrolyzed and evaporated Zn in N_2 atmosphere under 950°C to obtain one-dimensional carbon nanotube materials.[120] The electrocatalysts exhibit excellent CO_2RR performance, which is attributed to the special adsorption capacity of carbon nanostructures doped with nano-metal particles and nitrogen for

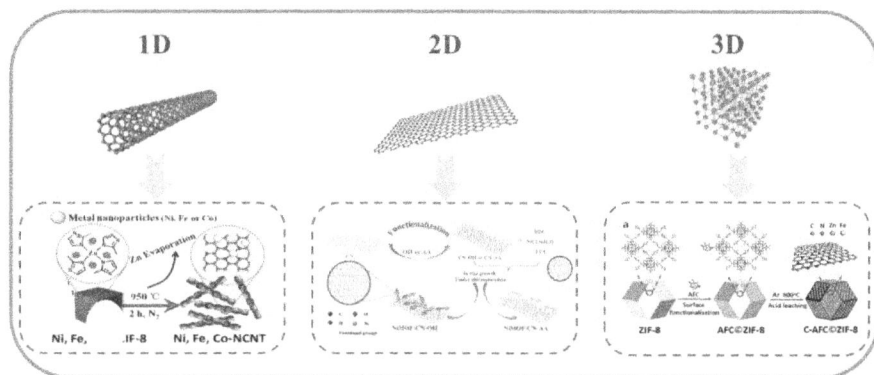

Figure 3.16. Different shapes of MOFs-derived carbon-based hybrid nanomaterials MOF-derived electrocatalysts. Reproduced with permission from Ref. 120. Copyright 2018, American Chemical Society. Reproduced with permission from Ref. 113. Copyright 2020, the Royal Society of Chemistry. Reproduced with permission from Ref. 135. Copyright 2017, Elsevier.

CO_2. Moreover, one-dimensional MOF derivative materials can also be prepared by one-dimensional MOF pyrolysis derivatization. Deng et al. used pyrolysis of BiBTC to prepare a Bi-based catalyst confined by carbon nanotubes.[121] The experimental results show that the synergistic effect of Bi_2O_3 nanoparticles and carbon nanorods in Bi_2O_3@C promotes the rapid reduction and oriented conversion of CO_2. Specifically, Bi_2O_3 nanoparticles help to improve reaction kinetics and formic acid selectivity, and carbon nanorods help improve the selectivity of formic acid and the total current density.

The two-dimensional MOFs material with the unique ultra-thin planar structure possess many exposed active sites and open pore structure, which facilitates the adsorption of CO_2 at the active sites and the diffusion of products.[122,123] Hod et al. used electrophoresis to prepare two-dimensional functionalized iron porphyrin materials as electrocatalysts with high electrochemical surface (1015 cm^2/mg) and Faraday efficiency of synthesis gas (100%).[124] Zhong et al. prepared a layered bimetallic two-dimensional conjugated MOFs ($PcCu-O_8-Zn$) with copper phthalocyanine (CuN_4) and zinc bis(dihydroxy) complex (ZnO_4), which has high CO selectivity (88%) and switching frequency 0.39 s^{-1}, and high stability (> 10 h).[125] In addition, the unique planar structure of the two-dimensional MOF can also be used to modify the electrode surface to enhance the adsorption performance of the electrode surface for CO_2 and reaction intermediates. For example, De Luna et al. reported that MOF films were grown *in situ* on gold nanostructured microelectrodes (AuNMEs) to inhibit CO production, and showed high selectivity and current density of C_2H_4 and CH_4 in the CO_2RR reaction.[107]

Three-dimensional MOFs materials with a rich morphology and structure, including spherical,[126] cubic,[127–129] octahedron,[130] dodecahedron,[131,132] etc., can support higher metal active sites. However, the pore size of common three-dimensional MOF materials is micropores (< 2 nm), which may not be able to effectively use the active sites located inside the three-dimensional structure.[111] To solve this problem, researchers have proposed many strategies. For example, expanding the pore size (mesopores or macropores) of carbon materials facilitates the diffusion of CO_2 into the internal active sites of the MOFs.[133] Ye et al. used Zn-MOF-74 and melamine as precursors to be calcinated to obtained nitrogen-doped mesoporous carbon (NPC), which can effectively promote the active sites and electrolyte in the NPC to improve the utilization of active sites.[134] Wang et al. reported that ferric ammonium citrate was selectively restricted to the surface of ZIF-8 nanoparticles to prepare C-AFC@ZIF-8 by a post-synthesis modification strategy.[135] The result showed that the high-exposure iron-nitrogen active sites possessed higher reactivity for electro-reduction of carbon dioxide.

3.5.2 The effect of the active site structure of MOFs derivatives on the performance of CO_2RR

The generation of $*CO_2^-$ by CO_2 obtaining electrons on the active site is recognized as the rate-determining step of the CO_2 activation process, requiring a high thermodynamic energy barrier to be crossed.[136] Therefore, the research for improving the efficiency of CO_2 activation must be focused on how to design the active site structure of MOF derivatives.

MOFs derivatives generally are nitrogen-doped carbon materials with metal-nitrogen species considered as active sites of the CO_2 activation,[137] such as pyridine nitrogen, pyrrole nitrogen and graphite nitrogen. The MOF material with different metals and ligands possess different coordination number and reactivity. For example, Wang et al. reported that the current density of Ni-N/C (> 50 mA/cm^2) was higher than Fe-N/C (~ 6 mA/cm^2), which could be attributed to the synergistic effects of high content nitrogen doping and unsaturated coordination (Ni-N).[135,138] Therefore, suitable active metals and active centers with different saturations have a greater impact on the activation performance of CO_2. In addition, the activation performance of CO_2 is related to the spatial position of the active metal atoms, which can be tuned by the type of organic ligands. For example, the current density and the TOF of CO, respectively, was 25.1 mA/cm^2 and 4932 h^{-1} of CO_2RR (0.73V vs RHE) on CoPP@CNT (Co-N$_4$), but the current density and the TOF of CO was only 4.5 mA/cm^2 and 480.2 h^{-1} on Co-N$_5$/HNPCSs (Co-N$_5$).[139,140]

In addition, MOF-derived materials with defect structures possess high electron cloud density at the defect site, which can help to improve the efficiency of CO_2 activation.[141] For example, Cu-MOF with a large number of defects on the foamed copper substrate was synthesized by controlling the growth kinetics of MOF.[142] The surface of Cu-MOF with Cu^{2+} defects is beneficial to form an O-Cu-O-C structure between Cu^{2+} and CO_2. In ionic liquid-acetonitrile electrolyte, the current density of CO_2RR of is 65.8 mA/cm^2, and the Faraday efficiency of formate reaches 90.5% on Cu-MOF/Cu electrode under -1.8 V vs Ag/Ag$^+$. Wang et al. also reported that a catalyst with intrinsic carbon defects was synthesized by heating the nitrogen-free MOFs precursor.[143] The CO Faraday efficiency of CO_2RR on this carbon defects material was 94.5%. The results show that the performance of CO_2RR is positively correlated with the concentration of intrinsic carbon defects. The local charge density of carbon atoms of MOF-derivatives carbon was changed by introduction of intrinsic carbon defects, helping to activate CO_2 molecules to form reactive intermediate COOH*, and then forming CO.

3.5.3 The influence of the local electron density of MOF derivatives on the performance of CO_2RR

CO_2 is converted into a variety of intermediates on the catalyst surface by obtaining electrons and H$^+$. These reaction intermediates continue to be adsorbed on the active sites on the catalyst surface under static electricity. Therefore, the adsorption type of intermediates can be controlled by changing the local electron density to the selectivity of the target product.

The adsorption of reaction intermediates on the catalyst surface is divided into C coordination (*COOH, *CO, etc.) and O coordination (*OCHO, etc.), depending on the type of product. For example, the main product of *COOH is CO, and the main product of *OCHO is HCOOH. The high local electron density of the catalyst surface is conducive to the formation of carbon coordination products. Conversely, the low

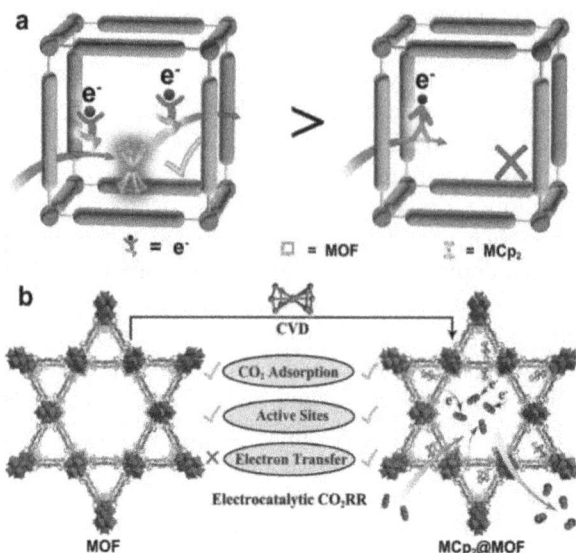

Figure 3.17. The schematic presentation for the advantages of MCp_2@MOF in electrocatalytic CO_2RR. Reproduced with permission from Ref. 146. Copyright 2020, Elsevier.

local electron density is more conducive to the generation of oxygen coordination products. However, in general, the local electron density and donating ability of MOFs are low due to the discrete active sites, which greatly limit the efficiency of CO_2 adsorption activity. Wang et al. reported that some electron-rich units (such as ε-$PMo_8{}^V$-$Mo_4{}^{VI}$-$O_{35}Zn_4$ clusters) were inserted into MOFs to improve the local electron density of MOF materials and the electron transfer efficiency of CO_2RR.[144] This innovative work provides a new method for improving the electronic power of MOF materials in CO_2RR. Using a similar method, 1,10-phenanthroline with a strong electron-donating was doped into ZIF-8, which showed that the CO Faraday efficiency of CO_2RR was as high as 90.6% at $-1.1V$ vs. RHE.[145] The Sp^2 hybridized carbon atom of the methylimidazole ligand in ZIF-8 was the catalytically active site for CO_2 reduction. Due to the doping of the electron-donating ligand, electron transfer was induced to improve the electrons density of the original catalytically active site (3.702 to 4.884).

In addition to electron-rich groups, the local electron density of the MOF active sites can also be changed by doping metal clusters (Fig. 3.17). $CoCp_2$-MOF-545-Co was synthesized by introducing metallocene ($CoCp_2$) into MOF-545-Co. The CO Faraday efficiency of CO_2RR on $CoCp_2$@MOF-545-Co electrode was as high as 97%, while the CO Faraday efficiency on MOF-545-Co electrode was only 54.6%.[146] The DFT calculation results showed that the binding energy of Co and CO_2 on a single porphyrin structure was only $-0.24eV$, and the binding energy was $-0.92eV$ after adding $CoCp_2$ due to C-Co bond being formed between $CoCp_2$ and Co-TCPP to enhance the electron donating ability of the site.

3.6 Summary and outlook

In summary, MOFs-derived carbon-based hybrid nanomaterials have demonstrated great application prospect in the field of HER/OER/ORR/CO_2RR, due to their controllable morphology and structure, unique physical and chemical properties, versatility in the tuning of the composition and structures in multiple dimensions, and excellent electrochemical performance. As summarized from the above chapters in this monograph, MOFs-derived carbon-based metal oxides/sulfides/phosphides/ selenides and single atom electrocatalysts were mostly prepared by pyrolysis of MOFs precursors under various atmosphere or hydrothermal process. The performance of the MOFs-derived electrocatalyst can be optimized by adjusting the technological parameters of preparation.

The development of MOFs-derived electrocatalysts for HER/OER/ORR/CO_2RR has acquired significant progress. Nevertheless, there are still several challenges to realize practical applications of these electrocatalysts. Firstly, in order to understand how different organic linkers, metals source, guest species and morphology affect the structure and properties of MOF-derived materials, a systematic study of different types of MOFs is required. Secondly, a systematic investigation of various dimensions (0 D, 1 D, 2 D and 3 D) of MOFs-derived materials is required, which has important influence on catalytic performance. Thirdly, the role of pyrolysis temperature, atmosphere, and the presence of other species/additives in the conversion of MOFs to various carbon, nanohybrids, and nanocomposites needs to be properly understood. Fourthly, in-depth theoretical calculations, *in situ* spectroscopic measurements, electrochemical modeling, and high-resolution imaging are urgently needed to investigate the catalysis mechanisms of MOFs-derived materials.

Acknowledgements

This work was supported by Project of Southwest University of Science and Technology (No. 20zx7142), Project of State Key Laboratory of Environment-Friendly Energy Materials (SWUST, Grant Nos. 19FKSY16 and 18ZD320304), Scientific Research Foundation of Chongqing University of Technology (2020ZDZ022), and Scientific Research Foundation Project of Chongqing Jiaotong University (2020020086). The authors acknowledge the support from State Key Laboratory of Environment-Friendly Energy Materials (Mianyang, China).

References

[1]   Montoya, J. H., L. C. Seitz, P. Chakthranont, A. Vojvodic, T. F. Jaramillo and J. K. Nørskov. 2016. Materials for Solar Fuels and Chemicals. *Nat. Mater.*, 16: 70–81.

[2]   Chu, S. and A. Majumdar. 2012. Opportunities and challenges for a sustainable energy future. *Nature*, 488: 294–303.

[3]   Wang, H. F., L. Y. Chen, H. Pang, S. Kaskel and Q. Xu. 2020. MOF–derived electrocatalysts for oxygen reduction, oxygen evolution and hydrogen evolution reactions. *Chem. Soc. Rev.*, 49: 1414–1448.

[4]   Zhu, J., L. S. Hu, P. X. Zhao, L. Y. Lee and K. Y. Wong. 2020. Recent advances in electrocatalytic hydrogen evolution using nanoparticles. *Chem. Rev.*, 120(2): 851–918.

[5] Sun, H. M., Z. H. Yan, F. M. Liu, W. C. Xu, F. Y. Cheng and J. Chen. 2020. Self-supported transition-metal-based electrocatalysts for hydrogen and oxygen evolution. *Adv. Mater.*, 32: 1806326.

[6] He, Y. H., S. W. Liu, C. Priest, Q. R. Shi and G. Wu. 2020. Atomically dispersed metal-nitrogen-carbon catalysts for fuel cells: Advances in catalyst design, electrode performance, and durability improvement. *Chem. Soc. Rev.*, 49: 3484–3524.

[7] Goyal, A., G. Marcandalli, V. A. Mints and M. T. M. Koper. 2020. Competition between CO_2 reduction and hydrogen evolution on a gold electrode under well-defined mass transport conditions. *J. Am. Chem. Soc.*, 142: 4154–4161.

[8] Wei, C., R. R. Rao, J. Y. Peng, B. T. Huang, I. E. L. Stephens, M. Risch, Z. C. J. Xu and Y. S. Horn. 2019. Recommended practices and benchmark activity for hydrogen and oxygen electrocatalysis in water splitting and fuel cells. *Adv. Mater.*, 31: 1806296.

[9] Yang, Z. K., C. M. Zhao, Y. T. Qu, H. Zhou, F. Y. Zhou, J. Wang, Y. Wu and Y. D. Li. 2019. Trifunctional self-supporting cobalt-embedded carbon nanotube films for ORR, OER, and HER triggered by solid diffusion from bulk metal. *Adv. Mater.*, 31: 1808043.

[10] Tang, C., H. F. Wang and Q. Zhang. 2018. Multiscale principles to boost reactivity in gas-involving energy electrocatalysis. *Acc. Chem. Res.*, 51: 881–889.

[11] Guo, H. L., Q. C. Feng, J. X. Zhu, J. S. Xu, Q. Q. Li, S. L. Liu, K. W. Xu, C. Zhang and T. X. Liu. 2019. Cobalt nanoparticle-embedded nitrogen-doped carbon/carbon nanotube frameworks derived from a metal-organic framework for tri–functional ORR, OER and HER electrocatalysis. *J. Mater. Chem. A*, 7: 3664–3672.

[12] Qin, Q., H. Jang, P. Li, B. Yuan, X. Liu and J. Cho. 2019. A tannic acid-derived N–, P–, Co–doped carbon–supported iron–based nanocomposite as an advanced trifunctional electrocatalyst for the overall water splitting cells and zinc-air batteries. *Adv. Energy Mater.* 9: 1803312.

[13] Jiang, H., J. X. Gu, X. S. Zheng, M. Liu, X. Q. Qiu, L. B. Wang, W. Z. Li, Z. F. Chen, X. B. Ji and J. Li. 2019. Defect-rich and ultrathin N doped carbon nanosheets as advanced trifunctional metal-free electrocatalysts for the ORR, OER and HER. *Energy Environ. Sci.*, 12: 322–333.

[14] Guo, Y. Y., P. F. Yuan, J. N. Zhang, H. C. Xia, F. Y. Cheng, M. F. Zhou, J. Li, Y. Y. Qiao, S. C. Mu and Q. Xu. 2018. Co_2P–CoN double active centers confined in N-doped carbon nanotube: Heterostructural engineering for trifunctional catalysis toward HER, ORR, OER, and Zn-Air batteries driven water splitting. *Adv. Funct. Mater.*, 28: 1805641.

[15] Furukawa, H., K. E. Cordova, M. O'Keeffe and O. M. Yaghi. 2013. The chemistry and applications of metal–organic frameworks. *Science*, 341: 1230444.

[16] Stassen, I., N. Burtch, A. Talin, P. Falcaro, M. Allendorf and R. Ameloot. 2017. An updated roadmap for the integration of metal-organic frameworks with electronic devices and chemical sensors. *Chem. Soc. Rev.*, 46: 3185–3241.

[17] Zhu, Q. L. and Q. Xu. 2014. Metal–organic framework composites. *Chem. Soc. Rev.*, 43: 5468–5512.

[18] Sheberla, D., J. C. Bachman, J. S. Elias, C. J. Sun, Y. S. Horn and M. Dinca. 2017. Conductive MOF electrodes for stable supercapacitors with high areal capacitance. *Nature Mater.*, 16: 220–224.

[19] Liu, B., H. Shioyama, T. Akita and Q. Xu. 2008. Metal-organic framework as a template for porous carbon synthesis. *J. Am. Chem. Soc.* 130: 5390–5391.

[20] Zhang, H., X. M. Liu, Y. Wu, C. Guan, A. K. Cheetham and J. Wang. 2018. MOF-derived nanohybrids for electrocatalysis and energy storage: Current status and perspectives. *Chem. Commun.*, 54: 5268–5288.

[21] Davis, S. J., N. S. Lewis, M. Shaner, S. Aggarwal, D. Arent and I. L. Azevedo. 2018. Net-zero emissions energy systems. *Science*, 360: 9793.

[22] Ojha, K., S. Saha, P. Dagar and A. K. Ganguli. 2018. Nanocatalysts for hydrogen evolution reactions. *Phys. Chem. Chem. Phys.*, 20: 6777–6799.

[23] Zhao, G., K. Rui, S. X. Dou and W. Sun. 2018. Heterostructures for electrochemical hydrogen evolution reaction: A review. *Adv. Funct. Mater.*, 28: 1803291.

[24] Meyer, T. J., M. V. Sheridan and B. D. Sherman. 2017. Mechanisms of molecular water oxidation in solution and on oxide surfaces. *Chem. Soc. Rev.*, 46: 6148–6169.

[25] Liu, Y. F., K. Zhong, K. Luo, M. X. Gao, H. G. Pan and Q. D. Wang. 2009. Size–dependent kinetic enhancement in hydrogen absorption and desorption of the Li–Mg–N–H system. *J. Am. Chem. Soc.*, 131: 1862–1870.

[26]   Murthy, A. P., J. Madhavan and K. Murugan. 2018. Recent advances in hydrogen evolution reaction catalysts on carbon/carbon–based supports in acid media. *J. Power Sources*, 398: 9–26.

[27]   Schalenbach, M., O. Kasian, M. Ledendecker, F. D. Speck, A. M. Mingers and K. J. J. Mayrhofer. 2018. The electrochemical dissolution of noble metals in alkaline media. *Electrocatalysis*, 9: 153–161.

[28]   Xu, Y., Q. Li, H. Xue and H. Pang. 2018. Metal-organic frameworks for direct electrochemical applications. *Coord. Chem. Rev.*, 376: 292–318.

[29]   LiS. L. and Q. Xu. 2013. Metal-organic frameworks as platforms for clean energy. *Energy Environ. Sci.*, 6: 1656–1683.

[30]   Downes, C. A. and S. C. Marinescu. 2017. Electrocatalytic metale organic frameworks for energy applications. *ChemSusChem.*, 10: 4374–4392.

[31]   Wang, L., L. Ren, X. Wang, X. Feng, J. Zhou and B. Wang. 2018. Multivariate MOF-templated pomegranate-like Ni/C as efficient bifunctional electrocatalyst for hydrogen evolution and urea oxidation. *ACS Appl. Mater. Interfaces*, 10: 4750–4756.

[32]   Liu, J., D. Zhu, C. Guo, V. Aasileff and S. Z. Qiao. 2017. Design strategies toward advanced MOF-derived electrocatalysts for energy-conversion reactions. *Adv. Energy Mater.*, 7: 1700518.

[33]   Humagain, G., K. MacDougal, J. MacInnis, J. M. Lowe, R. H. Coridan and S. MacQuarrie. 2018. Highly efficient, biochar–derived molybdenum carbide hydrogen evolution electrocatalyst. *Adv. Energy Mater.*, 8: 1801461.

[34]   Xue, Y., S. Zheng, H. Xue and H. Pang. 2019. Metal-organic framework composites and their electrochemical applications. J. Mater. Chem. A, 7: 7301–7327.

[35]   Kuang, M., Q. Wang, P. Han and G. Zheng. 2017. Cu, Co-embedded N-enriched mesoporous carbon for efficient oxygen reduction and hydrogen evolution reactions. *Adv. Energy Mater.* 7: 1700193.

[36]   Xu, Y., W. Tu, B. Zhang, S. Yin, Y. Huang and M. Kraft. 2017. Nickel nanoparticles encapsulated in few-layer nitrogen-doped graphene derived from metal-organic frameworks as efficient bifunctional electrocatalysts for overall water splitting. *Adv. Mater.*, 29: 1605957.

[37]   Yang, Y., Z. Lin, S. Gao, J. Su, Z. Lun and G. Xia. 2016. Tuning electronic structures of nonprecious ternary alloys encapsulated in graphene layers for optimizing overall water splitting activity. *ACS Catal.*, 7: 469–479.

[38]   Roger, I., M. A. Shipman and M. D. Symes. 2017. Earth–abundant catalysts for electrochemical and photoelectrochemical water splitting. *Nature Rev. Chem.*, 1: 306–327.

[39]   Tian, T., L. Huang, L. Ai and J. Jiang. 2017. Surface anion-rich NiS_2 hollow microspheres derived from metal-organic frameworks as a robust electrocatalyst for the hydrogen evolution reaction. *J. Mater. Chem. A*, 5: 20985–20992.

[40]   Huang, Z., J. Liu, Z. Xiao, H. Fu, W. Fan and B. Xu. 2018. A MOF-derived coral-like NiSe@ NC nanohybrid: An efficient electrocatalyst for the hydrogen evolution reaction at all pH values. *Nanoscale*, 10: 22758–22765.

[41]   Wang, H., Y. Wang, L. Tan, L. Fang, X. Yang and Z. Huang. 2019. Component-controllable cobalt telluride nanoparticles encapsulated in nitrogen-doped carbon frameworks for efficient hydrogen evolution in alkaline conditions. *Appl. Catal. B. Environ.*, 244: 568–575.

[42]   Buss, J. A., M. Hirahara, Y. Ueda and T. Agapie. 2018. Molecular mimics of heterogeneous metal phosphides: Thermochemistry, hydrid-proton isomerism, and HER reactivity. *Angew. Chem. Int. Ed.*, 57: 16329–16333.

[43]   Wang, Q., Z. Liu, H. Zhao, H. Huang, H. Jiao and Y. Du. 2018. MOF-derived porous Ni_2P nanosheets as novel bifunctional electrocatalysts for the hydrogen and oxygen evolution reactions. *J. Mater. Chem. A.*, 6: 18720–18727.

[44]   Balogun, M. S., Y. Huang, W. Qiu, H. Yang, H. Ji and Y. Tong. 2017. Updates on the development of nanostructured transition metal nitrides for electrochemical energy storage and water splitting. Mater. *Today*, 20: 425–451.

[45]   Chen, Z., Y. Ha, Y. Liu, H. Wang, H. Yang and H. Xu. 2018. *In situ* formation of cobalt nitrides/graphitic carbon composites as efficient bifunctional electrocatalysts for overall water splitting. ACS Appl. Mater. *Interfaces*, 10: 7134–7144.

[46]   Gao, Q., W. Zhang, Z. Shi, L. Yang and Y. Tang. 2019. Structural design and electronic modulation of transition metal-carbide electrocatalysts toward efficient hydrogen evolution. *Adv. Mater.*, 31: 1802880.

[47]   Liang, Q., H. Jin, Z. Wang, Y. Xiong, S. Yuan and X. Zeng. 2019. Metal-organic frameworks derived reverse-encapsulation Co-NC@Mo$_2$C complex for efficient overall water splitting. *Nano Energy*, 57: 746–752.

[48]   Wu, H. B., B. Y. Xia, L. Yu, X. Y. Yu and X.W. Lou. 2015. Porous molybdenum carbide nano–octahedrons synthesized via confined carburization in metal-organic frameworks for efficient hydrogen production. *Nat. Commun.*, 6: 6512.

[49]   Sun, T., L. Xu, D. Wang and Y. Li. 2019. Metal organic frameworks derived single atom catalysts for electrocatalytic energy conversion. *Nano. Res.*, 12: 2067–2080.

[50]   Fan, L., P. F. Liu, X. Yan, L. Gu, Z. Z. Yang and H. G. Yang. 2016. Atomically isolated nickel species anchored on graphitized carbon for efficient hydrogen evolution electrocatalysis. *Nat. Commun.*, 7: 10667.

[51]   Wei, C., S. Sun, D. Mandler, X. Wang, S. Z. Qiao and Z. J. Xu. 2019 .Approaches for measuring the surface areas of metal oxide electrocatalysts for determining their intrinsic electrocatalytic activity. *Chem. Soc. Rev.*, 48: 2518–2534.

[52]   Yang, M., Y. Zhou, Y. Cao, Z. Tong, B. Dong and Y. Chai. 2020. Advances and challenges of Fe-MOFs based materials as electrocatalysts for water splitting. *Applied Materials Today*, 20: 100692.

[53]   Antolini, E. 2014. Iridium as catalyst and cocatalyst for oxygen evolution/reduction in acidic polymer electrolyte membrane electrolyzers and fuel cells. *ACS Catal.*, 4: 1426–1440.

[54]   Lu, F., M. Zhou, Y. Zhou and X. Zeng. 2017. First-row transition metal based catalysts for the oxygen evolution reaction under alkaline conditions: Basic principles and recent advances. *Small*, 13: 1701931.

[55]   Lyu, F., Q. Wang, S. M. Choi and Y. Yin. 2019. Noble-metal-free electrocatalysts for oxygen evolution. *Small*, 15: 1804201.

[56]   Wang, H., Q. Zhu, R. Zou and Q. Xu. 2017. Metal-organic frameworks for energy applications. *Chem.*, 2: 52–80.

[57]   Du, J., F. Li and L. Sun. 2021. Metal-organic frameworks and their derivatives as electrocatalysts for the oxygen evolution reaction. *Chem. Soc. Rev.*, 50: 2663–2695.

[58]   Li, X., Z. Niu, J. Jiang and L. Ai. 2016. Cobalt nanoparticles embedded in porous N-rich carbon as an efficient bifunctional electrocatalyst for water splitting. *J. Mater. Chem. A*, 4: 3204–3209.

[59]   Liu, M. R., Q. L. Hong, Q. H. Li, Y. Du, H. X. Zhang, S. Chen, T. Zhou and J. Zhang. 2018. Cobalt boron imidazolate framework derived cobalt nanoparticles encapsulated in B/N codoped nanocarbon as efficient bifunctional electrocatalysts for overall water splitting. *Adv. Funct. Mater.*, 28: 1801136.

[60]   Yuan, Q., Y. Yu, Y. Gong and X. Bi. 2020. Three-dimensional N-doped carbon nanotube frameworks on Ni foam derived from a metal-organic framework as a bifunctional electrocatalyst for overall water splitting. *ACS Appl. Mater. Inter.*, 12: 3592–3602.

[61]   Yan, L., Y. Xu, P. Chen, S. Zhang, H. Jiang, L. Yang, Y. Wang, L. Zhang, J. Shen and X. Zhao. 2020. A freestanding 3D heterostructure film stitched by MOF-derived carbon nanotube microsphere superstructure and reduced graphene oxide sheets: A superior multifunctional electrode for overall water splitting and Zn–Air Batteries. *Adv. Mater.*, 32: 2003313.

[62]   Zhu, Y. P., T. Y. Ma, M. Jaroniec and S. Z. Qiao. 2017. Self-templating synthesis of hollow Co$_3$O$_4$ microtube arrays for highly efficient water electrolysis. *Angew. Chemie. Int. Ed.*, 56: 1324–1328.

[63]   Lu, X. F., L. F. Gu, J. W. Wang, J. X. Wu, P. Q. Liao and G. R. Li. 2017. Bimetal-organic framework derived CoFe$_2$O$_4$/C porous hybrid nanorod arrays as high-performance electrocatalysts for oxygen evolution reaction. *Adv. Mater.*, 29: 1604437.

[64]   Wang, C. and L. Qi. 2020. Heterostructured inter-doped ruthenium-cobalt oxide hollow nanosheet arrays for highly efficient overall water splitting. *Angew. Chemie. Int. Ed.*, 59: 17219−17224.

[65]   Chen, C., Y. Tuo, Q. Lu, H. Lu, S. Zhang, Y. Zhou, J. Zhang, Z. Liu, Z. Kang and X. Feng. 2021. Hierarchical trimetallic Co-Ni-Fe oxides derived from core-shell structured metal-organic frameworks for highly efficient oxygen evolution reaction. *Appl. Catal. B Environ.*, 287: 119953.

[66]   Gao, X., X. Liu, W. Zang, H. Dong, Y. Pang, Z. Kou, P. Wang, Z. Pan, S. Wei and S. Mu. 2020. Synergizing in-grown Ni₃N/Ni heterostructured core and ultrathin Ni₃N surface shell enables self-adaptive surface reconfiguration and efficient oxygen evolution reaction. *Nano Energy*, 78: 105355.

[67]   Zhu, C., A. L. Wang, W. Xiao, D. Chao, X. Zhang, N. H. Tiep, S. Chen, J. Kang, X. Wang and J. Ding. 2018. *In situ* grown epitaxial heterojunction exhibits high-performance electrocatalytic water splitting. *Adv. Mater.*, 30: 1705516.

[68]   Kang, B. K., S. Y. Im, J. Lee, S. H. Kwag, S. B. Kwon, S. Tiruneh, M. Kim, J. H. Kim, W. S. Yang and B. Lim. 2019. *In-situ* formation of MOF derived mesoporous Co₃N/amorphous N-doped carbon nanocubes as an efficient electrocatalytic oxygen evolution reaction. *Nano Res.*, 12: 1605–1611.

[69]   Chen, J., B. Ren, H. Cui and C. Wang. 2020. Constructing pure phase tungsten-based bimetallic carbide nanosheet as an efficient bifunctional electrocatalyst for overall water splitting. *Small*, 16: 1907556.

[70]   Yu, X., Y. Feng, B. Guan, X. W. D. Lou and U. Paik. 2016. Carbon coated porous nickel phosphides nanoplates for highly efficient oxygen evolution reaction. *Energ. Environ. Sci.*, 9: 1246–1250.

[71]   He, P., X. Y. Yu and X. W. Lou. 2017. Carbon-incorporated nickel-cobalt mixed metal phosphide nanoboxes with enhanced electrocatalytic activity for oxygen evolution. *Angewandte Chemie International Edition*, 56: 3897–3900.

[72]   Hong, W., M. Kitta and Q. Xu. 2018. Bimetallic MOF-derived FeCo-P/C nanocomposites as efficient catalysts for oxygen evolution reaction. *Small Methods*, 2: 1800214.

[73]   Lin, Y., M. Zhang, L. Zhao, L. Wang, D. Cao and Y. Gong. 2021. Ru doped bimetallic phosphide derived from 2D metal organic framework as active and robust electrocatalyst for water splitting. *Appl. Surf. Sci.*, 536: 147952.

[74]   Jiao, J., W. Yang, Y. Pan, C. Zhang, S. Liu, C. Chen and D. Wang. 2020. Interface engineering of partially phosphidated Co@Co–P@ NPCNTs for highly enhanced electrochemical overall water splitting. *Small*, 16: 2002124.

[75]   Peng, X., Y. Yan, X. Jin, C. Huang, W. Jin, B. Gao and P. K. Chu. 2020. Recent advance and prospectives of electrocatalysts based on transition metal selenides for efficient water splitting. *Nano Energy*, 105234.

[76]   Lee, W. S. V., T. Xiong, X. Wang and J. Xue. 2021. Unraveling MoS₂ and transition metal dichalcogenides as functional zinc-ion battery cathode: A perspective. *Small Methods*, 5: 2000815.

[77]   Du, X., H. Su and X. Zhang. 2019. Metal-organic framework-derived Cu-doped Co9S8 nanorod array with less low-valence Co sites as highly efficient bifunctional electrodes for overall water splitting, ACS Sustain. *Chem. Eng.*, 7: 16917–16926.

[78]   Zhao, J., R. Wang, S. Wang, Y. Lv, H. Xu and S. Zang. 2019. Metal-organic framework-derived Co₉S₈ embedded in N, O and S-tridoped carbon nanomaterials as an efficient oxygen bifunctional electrocatalyst. *J. Mater. Chem. A*, 7: 7389–7395.

[79]   Xuan, C., W. Lei, J. Wang, T. Zhao, C. Lai, Y. Zhu, Y. Sun and D. Wang. 2019. Sea urchin-like Ni-Fe sulfide architectures as efficient electrocatalysts for the oxygen evolution reaction. *J. Mater. Chem. A*, 7: 12350−12357.

[80]   Guo, Y., C. Zhang, J. Zhang, K. Dastafkan, K. Wang, C. Zhao and Z. Shi. 2021. Metal organic framework-derived bimetallic NiFe selenide electrocatalysts with multiple phases for efficient oxygen evolution reaction. *ACS Sustain. Chem. Eng.*, 9: 2047–2056.

[81]   Chandran, P., A. Ghosh and S. Ramaprabhu. 2018. High-performance Platinum-free oxygen reduction reaction and hydrogen oxidation reaction catalyst in polymer electrolyte membrane fuel cell. *Sci. Re.*, 8: 3591–3511.

[82]   Jaouen, F., M. Lefèvre, J.-P. Dodelet and M. Cai. 2006. Heat-treated Fe/N/C Catalysts for O₂ electroreduction: Are active sites hosted in micropores? *J. Phys. Chem. B*, 110: 5553–5558.

[83]   Xiang, Z., Y. Xue, D. Cao, L. Huang, J. F. Chen and L. Dai. 2014 Highly efficient electrocatalysts for oxygen reduction based on 2D covalent organic polymers complexed with non-precious metals. *Angew. Chem. Int. Ed.*, 53: 2433–2437.

[84]   Debe, M. K. 2012. Electrocatalyst approaches and challenges for automotive fuel cells. *Nature*, 486: 43–51.

[85]   Shao, M. H., Q. W. Chang, J. Dodelet, and R. Chenitz. 2016. Recent advances in electrocatalysts for oxygen reduction reaction. *Chem. Rev.*, 116: 3594–3657.

[86]   Furukawa, H., N. Ko, Y. B. Go, N. Aratani, S. B. Choi, E. Choi, A. Ö. Yazaydin, R. Q. Snurr, M. O'Keeffe, J. Kim and O. M. Yaghi. 2010. Ultrahigh porosity in metal-organic frameworks. *Science*, 329: 424–428.

[87]   Hu, M., J. Reboul, S. Furukawa, N. L. Torad, Q. Ji, P. Srinivasu, K. Ariga, S. Kitagawa and Y. Yamauchi. 2012. Direct carbonization of al-based porous coordination polymer for synthesis of nanoporous carbon. *J. Am. Chem. Soc.*, 134: 2864–2867.

[88]   Liu, B., H. Shioyama, T. Akita and Q. Xu. 2008. Metal-organic framework as a template for porous carbon synthesis. *J. Am. Chem. Soc.*, 130: 5390–5391.

[89]   Cai, G. and H. L. Jiang. 2017. A modulator-induced defect-formation strategy to hierarchically porous metal-organic frameworks with high stability. *Angew. Chem. Int. Ed.*, 56: 563–567.

[90]   Li, J. S., Y. Y. Chen, Y. J. Tang, S. L. Li, H. Q. Dong, K. Li, M. Han, Y. Q. Lan, J. C. Bao and Z. H. Dai. 2014. Metal-organic framework templated nitrogen and sulfur co-doped porous carbons as highly efficient metal-free electrocatalysts for oxygen reduction reactions. *J. Mater. Chem. A*, 2: 6316–6319.

[91]   Armel, V., J. Hannauer and F. Jaouen. 2015. Effect of ZIF-8 crystal size on the O2 electro-reduction performance of pyrolyzed Fe–N–C catalysts. *Catalysts*, 5: 1333–1351.

[92]   Ahn, S. H. and A. Manthiram. 2017. Self-templated synthesis of Co- and N-doped carbon microtubes composed of hollow nanospheres and nanotubes for efficient oxygen reduction reaction. *Small*, 13: 1603437.

[93]   Zhang, H., H. Osgood, X. Xie, Y. Shao and G. Wu. 2017. Engineering nanostructures of PGM-free oxygen-reduction catalysts using metal-organic frameworks. *Nano Energy*, 31: 331–350.

[94]   Qiao, L., A. Zhu, Y. Liu, Y. Bian, R. Dong, D. Zhong, H. Wu and J. Pan. 2018. Metal-organic framework-driven copper/carbon polyhedron: Synthesis, characterization and the role of copper in electrochemistry properties. *J. Mater. Sci.*, 53: 7755–7766.

[95]   Yang, L., X. Zeng, W. Wang and D. Cao. 2018. Recent progress in MOF-derived, heteroatom–doped porous carbons as highly efficient electrocatalysts for oxygen reduction reaction in fuel cells. *Adv. Funct. Mater.*, 28: 1704537.

[96]   Zeng, M., Y. Liu, F. Zhao, K. Nie, N. Han, X. Wang, W. Huang, X. Song, J. Zhong and Y. Li. 2016. Zinc–Air batteries: Metallic cobalt nanoparticles encapsulated in nitrogen-enriched graphene shells: its bifunctional electrocatalysis and application in zinc–air batteries. *Adv. Funct. Mater.*, 26: 4234–4234.

[97]   Morozan, A., M. T. Sougrati, V. Goellner, D. Jones, L. Stievano and F. Jaouen. 2013. Effect of furfuryl alcohol on metal organic framework-based Fe/N/C electrocatalysts for polymer electrolyte membrane fuel cells. *Electrochim. Acta*, 119: 192–205.

[98]   Zhang, L. J., Z. X. Su, F. L. Jiang, L. L. Yang, J. J. Qian, Y. F. Zhou, W. M. Li and M. C. Hong. 2014. Highly graphitized nitrogen-doped porous carbon nanopolyhedra derived from ZIF-8 nanocrystals as efficient electrocatalysts for oxygen reduction reactions. *Nanoscale*, 6: 6590–6602.

[99]   Li, J. S., S. L. Li, Y. J. Tang, K. Li, L. Zhou, N. Kong, Y. Q. Lan, J. C. Bao and Z. H. Dai. 2014. Heteroatoms ternary-doped porous carbons derived from MOFs as metal-free electrocatalysts for oxygen reduction reaction. *Sci. Rep.*, 4: 5130.

[100] Song, R. B., W. Zhu, J. Fu, Y. Chen, L. Liu, J. R. Zhang, Y. Lin and J. J. Zhu. 2019. Electrode materials engineering in electrocatalytic CO_2 reduction: Energy input and conversion efficiency. *Adv. Mater.*, 32: 1903796.

[101] Hu, L., Y. W. Hu, R. Liu, Y. C. Mao, M. S. Balogun and Y. X. Tong. 2019. Co-based MOF-derived $Co/CoN/Co_2P$ ternary composite embedded in N- and P-doped carbon as bifunctional nanocatalysts for efficient overall water splitting. *Int. J. Hydrogen Energy*, 44: 11402.

[102] Gu, J., C. S. Hsu, L. Bai, H. M. Chen and X. Hu. 2019. Atomically dispersed Fe^{3+} sites catalyze efficient CO_2 electroreduction to CO. *Science*, 364: 1091.

[103] Zhang, H. B., J. W. Nai, L. Yu and X. W. Lou. 2017. Metal-organic-framework-based materials as platforms for renewable energy and environmental applications. *Joule*, 1: 77.

[104] Jiao, L. and H. L. Jiang. 2019. Metal-organic-framework-based single-atom catalysts for energy applications. *Chem. Rev.*, 5: 786.

[105] Gong, Y. N., L. Jiao, Y. Y. Qian, C. Y. Pan, L. Zheng, X. C. Cai, B. Liu, S. Yu and H. L. Jiang. 2020. Regulating the coordination environment of MOF-emplated single atom nickel electrocatalysts for boosting CO_2 reduction. *Angew. Chem.*, 59: 2705.

[106] Ross, M. B., P. De Luna, Y. F. Li, C. T. Dinh, D. Kim, P. Yang and E. H. Sargent. 2019. Designing materials for electrochemical carbon dioxide recycling. *Nat. Catal.*, 2: 648.

[107] Rosen, B. A., A. Salehi-Khojin, M. R. Thorson, W. Zhu, D. T. Whipple, P. J. Kenis and R. I. Masel. 2011. Ionic liquid-mediated selective conversion of CO_2 to CO at low overpotentials. *Science*, 334: 643.

[108] Gao, S., Y. Lin, X. Jiao, Y. Sun, Q. Luo, W. Zhang, D. Li, J. Yang and Y. Xie. 2016. Partially oxidized atomic cobalt layers for carbon dioxide electroreduction to liquid fuel. *Nature*, 529: 68.

[109] Nitopi, S., E. Bertheussen, S. B. Scott, X. Liu, A. K. Engstfeld, S. Horch, B. Seger, I. E. L. Stephens, K. Chan, C. Hahn, J. K. Norskov, T. F. Jaramillo and I. Chorkendorff. 2019. Progress and perspectives of electrochemical CO_2 reduction on copper in aqueous electrolyte. *Chem. Rev.*, 119: 7610.

[110] Fan, L., C. Xia, P. Zhu, Y. Lu and H. Wang. 2020. Electrochemical CO_2 reduction to high-concentration pure formic acid solutions in an all-solid-state reactor. *Nat. Commun.*, 11: 3633.

[111] Aubrey, M. L., B. M. Wiers, S. C. Andrews, T. Sakurai, S. E. Reyes-Lillo, S. M. Hamed, C. J. Yu, L. E. Darago, J. A. Mason, J. O. Baeg, F. Grandjean, G. J. Long, S. Seki, J. B. Neaton, P. Yang and J. R. Long. 2018. Electron delocalization and charge mobility as a function of reduction in a metal-organic framework. *Nat. Mater.*, 17: 625.

[112] Shao, P., L. C. Yi, S. M. Chen, T. H. Zhou and J. Zhang. 2020. Metal-organic frameworks for electrochemical reduction of carbon dioxide: The role of metal centers. *J. Energy Chem.*, 40: 156.

[113] Zhao, L., Z. Zhao, Y. Li, X. Chu, Z. Li, Y. Qu, L. Bai and L. Jing. 2020. The synthesis of interface-modulated ultrathin Ni(ii) MOF/g-C₃N₄ heterojunctions as efficient photocatalysts for CO_2 reduction. *Nanoscale*, 12: 10010.

[114] Kang, X., B. Wang, K. Hu, K. Lyu, X. Han, B. F. Spencer, M. D. Frogley, F. Tuna, E. J. L. McInnes, R. A. W. Dryfe, B. Han, S. Yang and M. Schroder. 2020. Quantitative electro-reduction of CO_2 to liquid fuel over electro-synthesized metal-organic frameworks. *J. Am. Chem. Soc.*, 142: 17384.

[115] De Luna, P., W. Liang, A. Mallick, O. Shekhah, F. P. Garcia de Arquer, A. H. Proppe, P. Todorovic, S. O. Kelley, E. H. Sargent and M. Eddaoudi. 2018. Metal-organic framework thin films on high-curvature nanostructures toward tandem electrocatalysis. *ACS Appl. Mater. Inter.*, 10: 31225.

[116] Mu, Q. Q., W. Zhu, X. Li, C. F. Zhang, Y. H. Su, Y. B. Lian, P. W. Qi, Z. Deng, D. Zhang, S. Wang, X. Zhu and Y. Peng. Electrostatic charge transfer for boosting the photocatalytic CO_2 reduction on metal centers of 2D MOF/rGO heterostructure. *Appl. Catal. B-Environ.*, 262: 118144.

[117] Tang, J. and Y. Yamauchi. 2016. Carbon materials: MOF morphologies in control. *Nat. Chem.*, 8: 638.

[118] Wang, Q. and D. Astruc. 2020. State of the art and prospects in metal-organic framework (MOF)-based and MOF-derived nanocatalysis. *Chem. Rev.*, 120: 1438.

[119] Kang, Y. S., Y. Lu, K. Chen, Y. Zhao, P. Wang and W. Y. Sun. 2019. Metal-organic frameworks with catalytic centers: From synthesis to catalytic application. *Coord. Chem. Rev.*, 378: 262.

[120] Zhang, S., Q. Wu, L. Tang, Y. Hu, M. Wang, J. Zhao, M. Li, J. Han, X. Liu and H. Wang. 2018. Individual high-quality N-doped carbon nanotubes embedded with nonprecious metal nanoparticles toward electrochemical reaction. *ACS Appl. Mater. Inter.*, 10: 39757.

[121] Deng, P., F. Yang, Z. Wang, S. Chen, Y. Zhou, S. Zaman and B. Y. Xia. 2020. Metal-organic framework–derived carbon nanorods encapsulating bismuth oxides for rapid and selective CO_2 electroreduction to formate. *Angew. Chem. Int. Ed.*, 59: 10807.

[122] Wang, M., K. Torbensen, D. Salvatore, S. Ren, D. Joulie, F. Dumoulin, D. Mendoza, B. Lassalle-Kaiser, U. Isci, C. P. Berlinguette and M. Robert. 2019. CO_2 electrochemical catalytic reduction with a highly active cobalt phthalocyanine. *Nat. Commun.*, 10: 3602.

[123] Wu, J. X., S. Z. Hou, X. D. Zhang, M. Xu, H. F. Yang, P. S. Cao and Z. Y. Gu. 2019. Cathodized copper porphyrin metal-organic framework nanosheets for selective formate and acetate production from CO_2 electroreduction. *Chem. Sci.*, 10: 2199.

[124] Hod, I., M. D. Sampson, P. Deria, C. P. Kubiak, O. K. Farha and J. T. Hupp. 2015. Fe-porphyrin-based metal-organic framework films as high-surface concentration, heterogeneous catalysts for electrochemical reduction of CO_2. *ACS Catal.*, 5: 6302.

[125] Zhong, H., M. Ghorbani-Asl, K. H. Ly, J. Zhang, J. Ge, M. Wang, Z. Liao, D. Makarov, E. Zschech, E. Brunner, I. M. Weidinger, J. Zhang, A. V. Krasheninnikov, S. Kaskel, R. Dong and X. Feng.

2020. Synergistic electroreduction of carbon dioxide to carbon monoxide on bimetallic layered conjugated metal-organic frameworks. *Nat. Commun.*, 11: 1409.

[126] Song, X., H. Zhang, Y. Yang, B. Zhang, M. Zuo, X. Cao, J. Sun, C. Lin, X. Li and Z. Jiang. 2018. Bifunctional nitrogen and cobalt codoped hollow carbon for electrochemical syngas production. *Adv. Sci.*, 5: 1800177.

[127] Wang, W., Y. Gang, Z. Hu, Z. Yan, W. Li, Y. Li, Q. F. Gu, Z. Wang, S. L. Chou, H. K. Liu and S. X. Dou. 2020. Reversible structural evolution of sodium–rich rhombohedral Prussian blue for sodium-ion batteries. *Nat. Commun.*, 11: 980.

[128] Hu, H., B. Y. Guan and X. W. Lou. 2016. Construction of complex CoS hollow structures with enhanced electrochemical properties for hybrid supercapacitors. *Chem*, 1: 102.

[129] Zhao, K. M., W. W. Zhu, S. Q. Liu, X. L. Wei, G. Y. Ye, Y. K. Su and Z. He. 2020. Two-dimensional metal-organic frameworks and their derivatives for electrochemical energy storage and electrocatalysis. *Nanoscale Adv.*, 2: 536.

[130] Lv. S., Y. Tang, K. Zhang, D. Tang. 2018. Wet NH$_3$-Triggered NH$_2$–MIL–125(Ti) structural switch for visible fluorescence immunoassay impregnated on paper. *Anal. Chem.*, 90: 14121.

[131] Liu, C., Y. L. Tong, X. Q. Yu, H. Shen, Z. Zhu, Q. Li and S. Chen. 2020. MOF-based photonic crystal film toward separation of organic dyes. *ACS Appl. Mater. Inter.*, 12: 2816.

[132] Zhang, P., B. Y. Guan, L. Yu and X. W. D. Lou. 2017. Formation of double-shelled zinc-cobalt sulfide dodecahedral cages from bimetallic zeolitic imidazolate frameworks for hybrid supercapacitors. *Angew. Chem. Int. Ed.*, 56: 7141.

[133] Feng, L., K. Y. Wang, X. L. Lv, T. H. Yan and H. C. Zhou. 2020. Hierarchically porous metal-organic frameworks: Synthetic strategies and applications. *Nat. Sci. Rev.*, 7: 1743.

[134] Ye, L., Y. Ying, D. Sun, Z. Zhang, L. Fei, Z. Wen, J. Qiao and H. Huang. 2020. Highly efficient porous carbon electrocatalyst with controllable N-species content for selective CO$_2$ reduction. *Angew. Chem. Int. Ed.*, 59: 3244.

[135] Ye, Y. F., F. Cai, H. B. Li, H. H. Wu, G. X. Wang, Y. S. Li, S. Miao, S. H. Xie, R. Si, J. Wang and X. H. Bao. 2017. Surface functionalization of ZIF-8 with ammonium ferric citrate toward high exposure of Fe-N active sites for efficient oxygen and carbon dioxide electroreduction. *Nano Energy*, 38: 281.

[136] Han, N., Y. Wang, L. Ma, J. G. Wen, J. Li, H. C. Zheng, K. Q. Nie, X. X. Wang, F. P. Zhao, Y. F. Li, J. Fan, J. Zhong, T. P. Wu, D. J. Miller, J. Lu, S. T. Lee and Y. G. Li. 2017. Supported cobalt polyphthalocyanine for high–performance electrocatalytic CO$_2$ reduction. *Chem*, 3: 652.

[137] Koshy, D. M., A. T. Landers, D. A. Cullen, A. V. Ievlev, H. M. Meyer, C. Hahn, Z. N. Bao and T. F. Jaramillo. 2020. Direct characterization of atomically dispersed catalysts: Nitrogen-coordinated ni sites in carbon–based materials for CO$_2$ Electroreduction. *Adv. Energy Mater.*, 10: 2001836.

[138] Yan, C., H. Li, Y. Ye, H. Wu, F. Cai, R. Si, J. Xiao, S. Miao, S. Xie, F. Yang, Y. Li, G. Wang and X. Bao. 2018. Coordinatively unsaturated nickel-nitrogen sites towards selective and high-rate CO$_2$ electroreduction. *Energy. Environ. Sci.*, 11: 1204.

[139] Zhu, M., J. Chen, L. Huang, R. Ye, J. Xu and Y. F. Han. 2019. Covalently grafting cobalt porphyrin onto carbon nanotubes for efficient CO$_2$ electroreduction. *Angew. Chem. Int. Ed.*, 58: 6595.

[140] Pan, Y., R. Lin, Y. Chen, S. Liu, W. Zhu, X. Cao, W. Chen, K. Wu, W. C. Cheong, Y. Wang, L. Zheng, J. Luo, Y. Lin, Y. Liu, C. Liu, J. Li, Q. Lu, X. Chen, D. Wang, Q. Peng, C. Chen and Y. Li. 2018. Design of single-Atom Co-N$_5$ catalytic site: A robust electrocatalyst for CO$_2$ reduction with nearly 100% CO selectivity and remarkable stability. *J. Am. Chem. Soc.*, 140: 4218.

[141] Kang, X., L. Li, A. Sheveleva, X. Han, J. Li, L. Liu, F. Tuna, E. J. L. McInnes, B. Han, S. Yang and M. Schroder. 2020. Electro-reduction of carbon dioxide at low over-potential at a metal-organic framework decorated cathode. *Nat. Commun.*, 11: 5464.

[142] Nam, D. H., O. S. Bushuyev, J. Li, P. De Luna, A. Seifitokaldani, C. T. Dinh, F. P. Garcia de Arquer, Y. Wang, Z. Liang, A. H. Proppe, C. S. Tan, P. Todorovic, O. Shekhah, C. M. Gabardo, J. W. Jo, J. Choi, M. J. Choi, S. W. Baek, J. Kim, D. Sinton, S. O. Kelley, M. Eddaoudi and E. H. Sargent. 2018. Metal-organic frameworks mediate Cu coordination for selective CO$_2$ electroreduction. *J. Am. Chem. Soc.*, 140: 11378.

[143] Wang, W., L. Shang, G. Chang, C. Yan, R. Shi, Y. Zhao, G. I. N. Waterhouse, D. Yang and T. Zhang. 2019. Intrinsic carbon-defect-driven electrocatalytic reduction of carbon dioxide. *Adv. Mater.*, 31: 1808276.

[144] Wang, Y. R., Q. Huang, C. T. He, Y. Chen, J. Liu, F. C. Shen and Y. Q. Lan. 2018. Oriented electron transmission in polyoxometalate-metalloporphyrin organic framework for highly selective electroreduction of CO_2. *Nat. Commun.*, 9: 4466.

[145] Dou, S., J. Song, S. Xi, Y. Du, J. Wang, Z. F. Huang, Z. J. Xu and X. Wang. 2019. Boosting electrochemical CO_2 reduction on metal-organic frameworks via ligand doping. *Angew. Chem. Int. Ed.*, 58: 4041.

[146] Xin, Z. F., Y. R. Wang, Y. F. Chen, W. L. Li, L. Z. Dong and Y. Q. Lan. 2020., Metallocene implanted metalloporphyrin organic framework for highly selective CO_2 electroreduction. *Nano Energy*, 67: 104233.

4

Electrocatalysts for Oxidation of Methanol and Urea

Jing Li, Zhao Liu, Jiawei Shi and *Weiwei Cai**

‖‖

4.1 Introduction

Along with the development of human society, energy consumption is constantly increasing. The energy crisis and the related environmental problems caused by the large amount of direct use of fossil energy have become the primary issues for mankind to be solved urgently. Fuel cell is a power generation device that directly converts chemical energy into electrical energy.[1,2] Fuel cell has attracted wide attention since this energy conversion process is not restricted by the Carnot cycle and its energy conversion efficiency can be much higher than that of an internal combustion engine. At the same time, fuel cells also present various advantages, such as low environmental pollution, safety and reliability, simple operation, and wide application fields.

Proton exchange membrane fuel cell (PEMFC) possesses the characteristics of high energy conversion rate, low temperature startup, no electrolyte leakage, etc.[3–5] It is hence recognized as one of the most promising energy supplies for aerospace, military, electric vehicles, portable electronic equipment and regional power stations. However, due to the imperfect hydrogen storage and transportation technology, the hydrogen source has become the most important obstacle in the commercialization of PEMFC. Therefore, researchers have turned their research interest to use small organic molecules as alternative fuels represented by methanol,[6,7] ethanol,[8] formic acid,[9] dimethyl ether,[10] ethylene glycol[11] and urea,[12] among which, direct methanol fuel cell (DMFC) with methanol oxidation reaction (MOR) in the anode has achieved the fastest progress and has exhibited application potential as portable power sources. Alternatively, urea oxidation reaction (UOR) did not only act as anode reaction in

Sustainable Energy Laboratory, Faculty of Materials Science and Chemistry, China University of Geosciences Wuhan, 430074, China.
* Corresponding author: caiww@cug.edu.cn.

direct fuel cells, but also showed the application prospect in hydrogen production and water-treatment fields. As a result, electrocatalysts for MOR and UOR are in-detail discussion in this chapter.

4.2 Electrocatalysts for oxidation of methanol

DMFC using liquid methanol as a fuel has the advantages of being cost-effective, easy to obtain and store, and high energy density.[6,7] Although the power density of DMFC is still an order of magnitude lower than that of pure hydrogen fueled PEMFCs, the use of liquid methanol can avoid the storage and transportation difficulties of hydrogen gas. Therefore, DMFC has been considered as suitable energy conversion device for portable applications. DMFC is based on the electrochemical reaction of methanol and oxygen on the anode and cathode, respectively. The reaction equations occurring in DMFC are as follows:

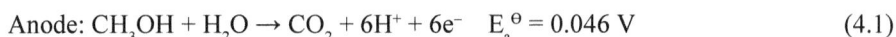

Anode: $CH_3OH + H_2O \rightarrow CO_2 + 6H^+ + 6e^-$ $\quad E_a^\ominus = 0.046$ V $\qquad$ (4.1)

Cathode: $3/2\ O_2 + 6H^+ + 6e^- \rightarrow 3H_2O$ $\qquad E_c^\ominus = 1.23$ V $\qquad$ (4.2)

Overall: $CH_3OH + 3/2\ O_2 \rightarrow CO_2 + 2H_2O$ $\quad E^\ominus = 1.18$ V $\qquad$ (4.3)

At 25°C, the thermodynamic potential of DMFC is ca. 1.20 V, and the conversion efficiency can reach 96.4%, which is far better than the internal combustion engine constrained by the Carnot cycle. Due to the multiple resistances generated during the reaction, polarization loss is caused, and the actual working potential of the DMFC is lower than the theoretical value.[13] The polarization loss of DMFC has been widely analyzed, and it is mainly caused by the slow kinetics of the electrochemical reactions, especially for the methanol oxidation reaction (MOR) that occurs on the anode. So far, at low temperature and in acidic electrolyte, the most active electrocatalyst in MOR research is still dominated by Pt and Pt alloys. Therefore, the reaction mechanism of methanol oxidation reported in the previous literature in most cases is carried out on Pt-based catalysts. The following are the primary reaction mechanism of MOR on Pt catalyst:

$CH_3OH + Pt \rightarrow Pt \cdot CH_2OH^* + H^+ + e^-$ $\qquad$ (4.4)

$Pt \cdot CH_2OH^* + Pt \rightarrow Pt_2 \cdot CHOH^* + H^+ + e^-$ $\qquad$ (4.5)

$Pt_2 \cdot CHOH^* + Pt \rightarrow Pt_3 \cdot COH^* + H^+ + e^-$ $\qquad$ (4.6)

$Pt_3 \cdot COH^* \rightarrow Pt \cdot CO^* + 2Pt + H^+ + e^-$ $\qquad$ (4.7)

$Pt + H_2O \rightarrow Pt \cdot OH^* + H^+ + e^-$ $\qquad$ (4.8)

$Pt \cdot OH^* + Pt \cdot CO^* \rightarrow 2Pt + CO_2 + H^+ + e^-$ $\qquad$ (4.9)

The above reaction process can be divided into two steps: first, CH_3OH gradually removes H on the active site of Pt, and finally adsorbs CO intermediate on Pt; second, hydrolysis occurs on the adjacent active site of Pt to generate OH*, OH* can oxidize the produced CO intermediate to generate CO_2, thereby releasing the active site of Pt. Developing CO-resistant Pt-based catalysts and Pt-free catalysts has therefore been carried out to promote the application pathway of DMFC.

4.2.1 Pt-based MOR catalysts

In MOR, methanol has a faster oxidation rate on the surface of Pt, but as the reaction proceeds, some intermediates will continue to accumulate on the surface of Pt because of their strong binding ability to Pt. As a result, the active sites of Pt are poisoned and the current density decays rapidly. Therefore, the catalytic reaction of methanol on the Pt catalyst is called a self-poisoning reaction. Experiments have proved that when the potential is lower than 0.45 V, intermediates such as CO will quickly accumulate on the active site of Pt, causing the Pt active site being poisoned. In addition, for Pt, only at a higher potential (> 0.6 V) hydrolysis and dissociation will occur to form Pt·OH*. Therefore, the generation of CO limits the use of Pt catalysts toward MOR, according to the abovementioned statements. Due to the slow MOR kinetic reaction rate, a large amount of Pt catalyst is still needed in the anode of DMFC. As a result, it is important to develop electrocatalysts toward MOR with low cost, strong redox activity and strong resistance to CO poisoning, which is also the key for large-scale commercialization of DMFC technology. Figure 4.1 shows the Pt-based catalysts that have been extensively studied in MOR so far. At present, most works are researching for an effective strategy to control the synthesis of Pt alloy catalysts with multi-dimensional nanostructures.[14–16] For example, it can be prepared into porous structures such as core-shell structure, hollow structure, nanotube (NT), nanowire (NW), etc. Studies have shown that the above porous structures can increase the active center of Pt while reducing the cost, and can also effectively prevent the dissolution and agglomeration of metal Pt nanoparticles, and effectively improve the stability of the Pt alloy catalyst and the catalytic performance of MOR. In addition, a large number of scientific research focuses on reducing the cost of the catalyst, introducing the transition metal elements rich in the earth (Fe, Co, Ni, Cu, etc.) into the Pt-based catalyst. By using the synergy between different components, geometric effect, ligand effect and electronic effect, the dependence on Pt can be reduced for MOR. As a result, catalytic performance of Pt-based alloy catalysts in DFMC can be ultimately improved.

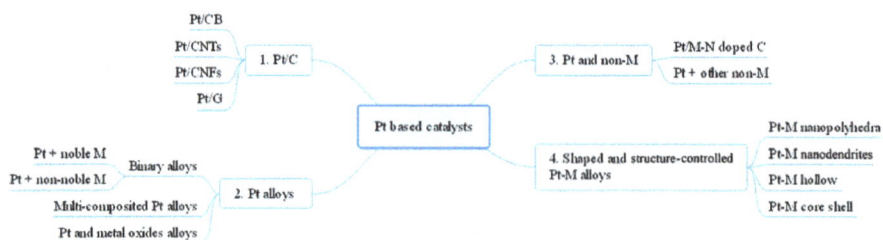

Figure 4.1. The representative Pt-based fuel cell catalyst.[16] Reproduced with permission from Ref. 16. Copyright 2020, the Royal Society of Chemistry.

Among them, Pt-M (Au, Ru, Co, Fe, Cu or Ni) catalysts have attracted attention. By reducing the number of invalid Pt atoms, more active sites could be exposed via synthesis strategies, such as in the core-shell structure. For Pt-based catalysts with a core-shell structure, the choice of the core material is extremely important. After a large number of experimental studies, it is generally believed that Pd is the most stable core material in acidic media. In addition, the size of Pd nanoparticles is also easy to be controlled and is hence considered as the most ideal core material. The core-shell structure is also helpful to avoid the agglomeration of the catalyst during the fuel cell operation. It can also effectively remove the intermediates from the Pt active sites to achieve great MOR stability in acidic condition. In order to reduce the cost of the catalyst, cost-effective 3d transition metals were also used as the core materials for the core-shell structure construction. Long et al.[17] used a hydrothermal synthesis method to synthesize a Cu-centered metal organic framework (Cu-MOF), which was directly carbonized at 600°C to form carbon-supported copper (Cu/C) nanoparticles. Partial replacement was subsequently carried out to get Cu@Pt/C catalyst with core-shell structure. The synthesized Cu@Pt/C catalyst has a Pt-rich PtCu alloy shell structure and the octahedral structure of Cu-MOF can also be achieved. This design is beneficial for reducing the use of Pt and exposing more Pt active sites. The electrochemical activity of Cu@Pt/C catalyst is much greater than that of commercial Pt/C catalyst. The performance of MOR was also improved by nearly 4 times, with less charge transfer resistance, and enhanced stability toward MOR. Sriphathoorat et al.[18] synthesized PtNiCo core-shell nanocrystals. The PtNiCo ternary metal nanocrystals are composed of a Pt-Ni-Co core and a dendritic Pt-Co bimetallic shell. Compared with commercial Pt/C, the best Pt-Ni and Pt-Co bimetallic nanoalloys, PtNiCo core-shell nanocrystals have better MOR catalytic activity.

PtRu alloy is always considered as the most effective MOR catalyst in acidic condition.[19] As we all know, the CO intermediate produced in MOR has a strong adsorption capacity on the surface of Pt, which causes the Pt active sites to be occupied and poisoned, and further MOR catalytic process cannot be taken on these Pt active sites. Fortunately, alloying with Ru can change the MOR process on Pt to a dual-functional mechanism and the poisoning effect of Pt active sites can be significantly reduced. Compared with Pt, Ru can dissociate water molecules at a lower potential to generate OH*, which can make CO_{ads} easy to be oxidized and removed. According to related literature,[20] when the atomic ratio of PtRu alloy is 1:1, this catalyst is the most active because it can minimize the dissociation potential. Huang et al.[21] synthesized a series of PtRu nanocrystals of different forms through a one-step wet chemical method (Fig. 4.2). The experimental results show that PtRu NWs has the best MOR activity and stability. The mass activity of PtRu NWs can reach 0.82 A·mg$_{Pt}^{-1}$ and the specific activity is 1.16 mA·cm^{-2}, which is better than other forms of PtRu catalysts and commercial Pt/C catalyst.

In order to make the Pt-based MOR catalyst more stable, the use of metal oxides, including TiO_2, WO_3, Fe_2O_3, SnO_2, CeO_2, etc., as support materials has also been widely researched. These oxides have strong mechanical strength and great stability under acidic conditions, and can effectively disperse Pt NPs. Among them, tungsten oxide (WO_3) is considered as a great support material for MOR catalysts. In the Pt/ WO_3-G catalyst prepared by Jin et al.,[22] the distribution of Pt nanoparticles is fine, with an average particle size of ca. 3.9 nm. Compared with commercial Pt/C, Pt/WO$_3$-G exhibited better MOR catalytic activity; due to this, metal oxide can effectively provide oxygen-containing substances to build an effective heterogeneous interface between Pt and metal oxides. CeO$_2$ is another representative metal oxide and has received widespread attention; due to this, CeO$_2$ has strong oxygen storage capacity and good stability in acidic environment. In addition, CeO$_2$ also has the advantages of multivalent states. Therefore, Chen et al.[23] studied the strong coupling effect of CeO$_2$ support, which can effectively inhibit the sintering of Pt nanoparticles and promote MOR activity (Fig. 4.3). Moreover, the supporting and anchoring capabilities of CeO$_2$ carriers with different morphologies to Pt nanoparticles are different, which directly affects the dispersion state of Pt nanoparticles and the interaction between Pt and the carrier.

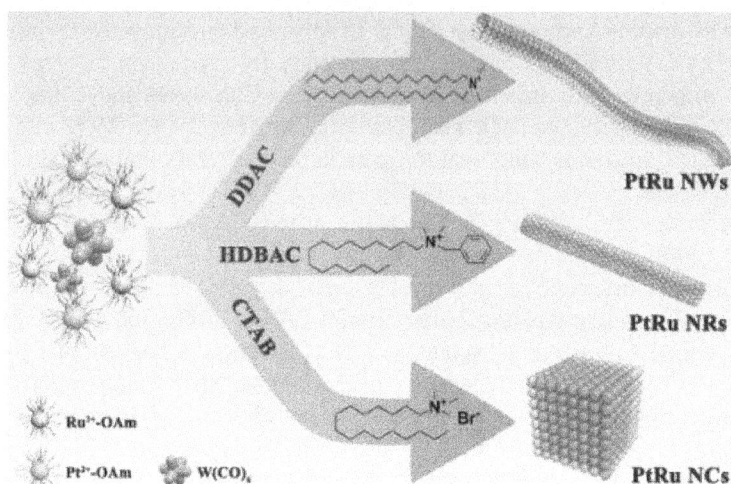

Figure 4.2. Synthetic routes of different forms of PtRu nanocrystals.[21] Reproduced with permission from Ref. 21. Copyright 2018, American Chemical Society.

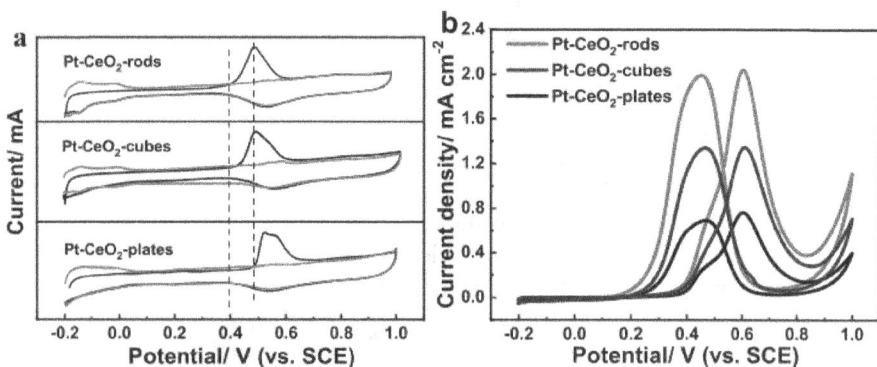

Figure 4.3. (a) CO-stripping curve test of Pt-CeO$_2$-rods, Pt-CeO$_2$-cubes, Pt-CeO$_2$-plates in 0.5 M H$_2$SO$_4$; (b) CV curves of the Pt-CeO$_2$-rods, Pt-CeO$_2$-cubes, Pt-CeO$_2$-plates catalysts in 0.5 M H$_2$SO$_4$ + 1.0 M CH3OH solution.[23] Reproduced with permission from Ref. 23. Copyright 2020, Elsevier.

4.2.1 Pt-free MOR catalysts

As the most important electrode material for MOR, Pt and Pt-based catalysts are still the most effective anode catalyst materials. The high price and low reserves of Pt seriously hinder the commercialization process of DMFC. The use of hetero-promoter metals to fabricate multi-element catalysts has improved catalyst activity and Pt utilization to a certain extent, while still cannot meet the requirements of DMFC commercialization. In order to reduce the cost of electrode catalysts in DMFC, researchers have studied the possibility of Pt-free catalysts toward MOR. Till now, Pt-free MOR catalysts mainly include: non-Pt metals and their alloys, metal carbides, perovskite-based oxides and MOF based catalyst.

Non-Pt metals, such as Pd, Ni, Cu, etc., have aroused the research interest of scholars due to their good catalytic activity toward methanol oxidation. Compared with Pt, reserve of Pd on Earth is relatively sufficient and it has therefore been considered as the candidate to replace Pt-based catalysts in DMFC. However, the catalytic activity of Pd and its alloys on methanol in acidic media is far lower than that of Pt. Therefore, this section mainly introduces the research on the catalytic performance of Pd on methanol in alkaline media.

Tan et al.[24] deposited highly dispersed Pd particles surrounded by CeO_2 on nitrogen-doped mesoporous carbon spheres with a core-shell structure to make a Pd-CeO_2/NMCS catalyst (Fig. 4.4), which was used to provide Pd sites containing adjacent precious metals. Oxygen species in CeO_2 support can promote the removal of CO. Great electrocatalytic activity can be hence achieved due to the enhanced resistance to CO poisoning in alkaline media. Compared with the commercial PtRu/C catalyst, the peak oxidation current density of the catalyst is about 6 times higher and the cycle stability is also much stronger. Through experimental and theoretical studies, strong electronic interaction between Pd and CeO_2 and N doped in the carrier reduces the adsorption energy of CO on the Pd surface. The enhanced resistance to CO poisoning can therefore improve the electrooxidation activity of methanol. Finally, in the fuel cell device, the Pd-CeO_2/NMCS catalyst showed better performance than the PtRu/C catalyst. The DMFC anode catalyst prepared with Pd metal as the substrate has greatly improved the resistance to CO toxicity, and the catalytic activity has also been greatly improved. Due to the advantages of Pd reserves, it can replace Pt as a new material for the preparation of DMFC catalysts.

Compared with the precious-metal Pd, non-precious metals like Ni and Cu have inherent advantages in price. As they also have methanol catalytic activity in alkaline media, it is possible to fundamentally solve the cost issue of DMFC. Therefore, there are more and more researches on Ni and Cu based MOR catalysts. Askari et al.[25] prepared a Ni/NiO catalyst with a peak current density of 15.94 mA·cm^{-2}. Nazal et al.[26] prepared a NiCu alloy catalyst and the peak MOR current density can reach 125.5 mA·cm^{-2} due to the addition of Cu metal. Compared with pristine Ni-based catalysts, the catalytic activity is greatly improved because the d-band holes of Ni are filled with Cu electrons. It can be concluded that Ni and Cu-based catalysts have good catalytic activity for the electrooxidation of methanol.

Among the metal carbide catalysts, WC has the properties of covalent compounds, ionic crystals and transition metal materialsThanks to these excellent

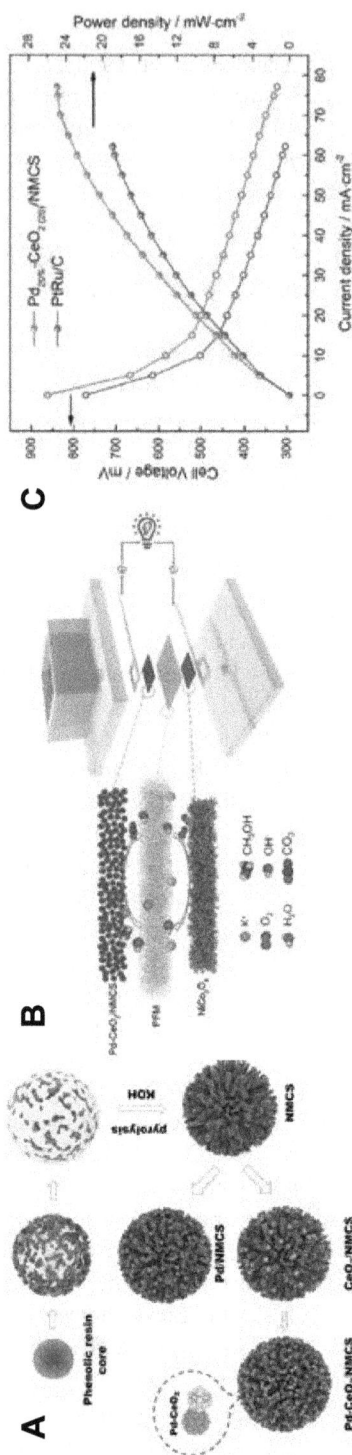

Figure 4.4. (a) Typical Synthetic Procedures of NMCS Supports and Pd-CeO$_2$/NMCS Catalysts. (b) Configuration of polymer fiber membrane-based membrane–electrode assembly and a DMFC design. (c) Polarization and power density plots of Pd-CeO$_2$/NMCS (5 mg·cm^{-2}) and PtRu/C (5 mg·cm^{-2}) catalysts at 20°C and ambient pressure. Reproduced with permission from Ref. 24. Copyright 2019, American Chemical Society.

physical properties of WC, it is an ideal co-catalyst to promote the electrooxidation of methanol in Pt-based catalysts.[27] Research on the MOR performance of WC has caused scholars' attention. The preparation process of WC is mainly based on high temperature reduction and carburizing of tungsten metal salt or mixture in a reducing atmosphere. WC is of good acid resistance and good catalytic activity for MOR. However, WC prepared by high-temperature carbonization can be easily coated by carbon shells and the catalytic activity was therefore reduced. Therefore, the carbon layers coated on the surface of the WC catalyst have to be removed at first to expose the active sites. Erich et al.[28] prepared WC and Pt modified WC catalysts as DMFC anode catalysts and evaluated their catalytic activity and stability. The temperature step-up desorption method and high-resolution electron energy loss spectroscopy studies showed that the WC surface is even more active than Pt and Ru for the dissociation of methanol and water. The desorption temperature of CO on WC is at least 100 K lower than that on Pt and Ru surface. In the process of catalytic MOR, WC has a high tolerance to CO poisoning. Electrochemical tests further confirmed the stability and activity of the WC film against methanol oxidation. Tian et al.[29] prepared a series of Pt-modified WC catalysts. DFT calculation results showed that the double-layer Pt-modified WC and Pt (111) catalysts have similar onset potentials for MOR, and the MOR activity is 2.4 times higher than that of pristine Pt. The synergistic effect of Pt and WC also proved that WC is a potential carrier for anode catalysts. Nie et al.[30] used batch microwave heating method and direct chemical reduction method to prepare WC to support Au, Pd and Pt as the anode catalyst for DMFC. AuPdPt@WC/C has greater hydrogen adsorption/desorption and platinum oxide generation/reduction currents than commercial Pt/C. Pt in AuPdPt@WC/C also has a larger specific surface area than the commercial Pt/C catalyst. The surface chemical and electrochemical test results showed that the catalytic activity and stability of AuPdPt@WC/C toward MOR are better than commercial Pt/C. The electronic conductivity, melting point and hardness of WC is similar to that of metals, which makes it unaffected by the environment such as the temperature in practical fuel cells. However, the oxidation mechanism of methanol on AuPdPt@WC/C and other WC catalysts or WC-supported catalyst need to be further studied. In addition, the effect of some reaction conditions, such as PH value, on the electrooxidation of methanol is still not very clear.

Because of the unique structural characteristics and good electrical conductivity, perovskite oxides are also studied as DMFC anode catalyst materials. Perovskite oxides are mainly divided into ABO_3 type (hexagonal system) and A_2BO_4 type (octahedral structure), where A metal is usually alkaline earth or rare earth metal ion and B metal is transition metal ion. In addition, perovskite-type rare earth oxides are rich in oxygen, and the active oxygen in the lattice can effectively oxidize species such as CO. Therefore, perovskite oxides have better resistance to CO poisoning than precious metals such as Pt. However, perovskite oxide as a catalyst is mostly used in alkaline conditions. For example, Singh et al.[31] prepared $La_{2-x}Sr_xNiO_4$ ($0 \leq x \leq 1$) perovskite oxide via a sol-gel method. At 0.55 V and 25°C, $La_{1.5}Sr_{0.5}NiO_4$ electrode provides a current density of more than 200 mA·cm^{-2} in 1 mol·L^{-1} KOH + 1 mol·L^{-1} CH_3OH. Balasubramanian et al.[32] studied the synergistic effect of perovskite oxide $LaNiO_3$ and Pt/C on MOR catalysis. $LaNiO_3$ powder combined with a small amount of Pt/C was measured in an acidic condition. Performance testing shows that the addition of $LaNiO_3$ provides more active sites for the catalyst. Due to the CO removal capacity of perovskite surface oxygen from the surface of Pt, $LaNiO_3$@Pt/C exhibited great catalytic activity. Although perovskite oxides, as a kind of non-noble metal oxides, have shown catalytic activity for the electrooxidation of methanol, their catalytic activity is still lower than that of Pt. By considering the CO oxidation ability and good electrical conductivity, perovskite oxides can act as a carrier for Pt metal. However, it should be noted that perovskite oxides have low acid corrosion resistance and are more suitable for alkaline medium DMFCs.

Metal organic framework (MOF) is a new type of porous crystalline material. In the past decade, scholars have carried out a lot of research on the application in catalysis because of its large specific surface area, high porosity, adjustable pores and abundant catalytic active sites. It has therefore become one of the most promising catalysts nowadays and is also studied for MOR catalysis in alkaline media. Lubna et al.[33] synthesized MOF of cobalt benzene tricarboxylic acid (Co BTC MOF) supported on reduced graphene oxide (rGO/CoBTCMOF) and studied the methanol oxidation reaction (MOR) in alkaline medium. Among all the prepared rGO/CoBTC composite materials, the peak MOR current density of 1 wt% rGO/Co BTCMOF in $1\ mol\cdot L^{-1}\ NaOH + 2\ mol\cdot L^{-1}\ CH_3OH$ solution can reach 130 mA$\cdot$cm^{-2}. Moreover, by selecting a suitable ratio of reduced graphene oxide and MOF, the synergistic effect between reduced graphene oxide and MOF can be achieved to promote the oxidation of CO on the electrode surface. Similarly, the NiO-MOF/rGO catalyst prepared by Tayyaba et al.[34] also showed good catalytic methanol electrooxidation activity in $1\ mol\cdot L^{-1}\ NaOH + 3\ mol\cdot L^{-1}\ CH_3OH$ solution, and the peak MOR current density of NiO-MOF/5wt % rGO can reach 275.85 mA$\cdot$cm^{-2}. It indicated that the MOF material has great activity for the catalytic MOR and therefore provides more choices for the anode catalyst of DMFC.

4.2.2 Conclusions

In general, the most effective MOR catalyst in acidic condition now is still PtRu alloy. It has rarely been reported that catalysts exhibit good catalytic activity without the participation of precious metals, which means that it does not have the possibility of application in the current DMFC system. Based on this, in the future MOR catalyst research, it is necessary to satisfy the needs of practical applications and carry out targeted catalyst composition and structure design.

4.3 Electrocatalysts for oxidation of urea

4.3.1 UOR mechanism

Energy crisis and environmental pollution have become two important problems in human society, which need to be solved urgently. The key to solve these problems is to develop green and sustainable hydrogen energy.[35,36] As a relatively mature hydrogen production technology, hydrogen production from electrolyzed water has attracted much attention due to its high efficiency and renewable energy sources such as solar energy and wind energy. Unfortunately, large-scale commercialization of hydrogen production from electrolyzed water is hindered by the strong dependence of oxygen evolution reaction (OER) and hydrogen evolution reaction (HER) on noble metal catalysis. Particularly, the problem of OER is more prominent because of the slow four electron transfer process and harsh operating conditions, resulting in insufficient catalytic activity and poor stability.[37-39] In order to avoid these problems, some thermodynamically more favorable electrooxidation reactions of small organic molecules, such as methanol,[40] ethanol[41] and urea,[42] are used to replace the anodic OER. Urea has the advantages of convenient transportation, high energy density and wide sources which widely exists in animal excrement and agricultural wastewater.[12] Using urea as fuel can not only obtain a lot of cheap energy, but also solve the problem of environmental pollution. In order to solve the problem and make rational use of urea to realize waste recycling, urea electrooxidation was proposed as the anode reaction in the hydrogen production unit or direct urea fuel cell (DUFC). Compared with OER, the thermodynamic equilibrium potential of UOR ($CO(NH_2)_2 + H_2O \rightarrow N_2 + 3H_2 + CO_2$) is only 0.37 V, which is much lower than that of oxygen precipitation reaction (1.23 V). Therefore, UOR instead of OER are often used as oxidation reaction of anode, which can not only greatly reduce the energy consumption in the process of hydrogen production, but also play a role in the purification of waste water, showing an important application prospect. However, the electrooxidation of urea is a 6-electron process with slow kinetic reaction rate and complex intermediate products.[43] Therefore, it is very important to understand the mechanism of electrooxidation of urea. Thanks to the development of electrochemical testing technology and *in situ* technology, Botte et al.[44] successfully confirmed the reaction mechanism of urea electrooxidation by electrochemical testing and *in situ* Raman spectroscopy. Usually, urea oxidation goes through a 6-electron transfer process as follows:

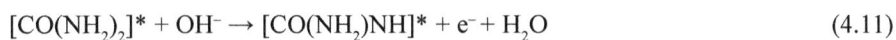

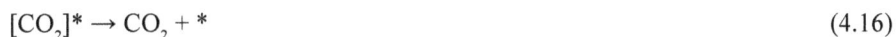

$$CO(NH)_2 + * \rightarrow [CO(NH_2)_2]* \tag{4.10}$$

$$[CO(NH_2)_2]* + OH^- \rightarrow [CO(NH_2)NH]* + e^- + H_2O \tag{4.11}$$

$$[CO(NH_2)NH]* + OH^- \rightarrow [CO(NH_2)N]* + e^- + H_2O \tag{4.12}$$

$$[CO(NH)N]* + OH^- \rightarrow [CON_2]* + e^- + H_2O \tag{4.13}$$

$$[CON_2]* + OH^- \rightarrow [COOH]* + N2 + e^- \tag{4.14}$$

$$[COOH]* + OH^- \rightarrow [CO_2]* + e^- + H_2O \tag{4.15}$$

$$[CO_2]* \rightarrow CO_2 + * \tag{4.16}$$

In this process, the adsorbed urea molecules gradually dehydrogenate to form a variety of intermediate substances and gas molecules such as carbon dioxide and nitrogen. Finally, these gas molecules desorb from the active sites and re-expose the active sites. It is well known that a chemical reaction often consists of many elementary reactions, and the overall chemical reaction rate is determined by the reaction with the slowest rate among many elementary reactions, which is called the rate determining step. In order to improve the catalytic efficiency of urea electro oxidation, it is very important to find the decisive step of urea electro oxidation. The energy barrier of chemical reaction will be available in each step of the elementary reaction. The size of the barrier determines the difficulty of the reaction. Through theoretical calculation, it is found that the desorption of CO_2 is a rate determining step with biggest barrier of chemical reaction.[45] Therefore, in order to design a catalyst with excellent urea oxidation activity, it is necessary to regulate the adsorption energy between the catalyst and carbon dioxide from the point of view of chemical thermodynamics.

4.3.2 UOR catalysts

The first catalyst for urea electrooxidation appeared in 1970s, when platinum was still used as a catalyst for urea electrooxidation in biochemical sensing. Although platinum, rhodium and other precious metals show considerable catalytic performance, their high price limits the large-scale industrial application of hydrogen production from overall urea oxidation. Therefore, it is of great significance to develop a kind of UOR catalyst with low price, excellent performance and good durability. At present, most of the reported catalysts for urea oxidation are nickel-based catalysts. Due to the special electronic structure of nickel atom, nickel-based catalysts have high catalytic activity for urea oxidation. Through some electrochemical means and *in situ* technology, NiOOH is confirmed as the active site of urea oxidation. As a kind of classical catalyst, porous carbon supported metal nanoparticles have been applied in many fields. Nickel nanoparticles with a diameter of about 2–5 nm uniformly distributed on the support were first prepared for UOR to explore the catalytic process of urea electrooxidation. Muralidharan et al.[46] synthesized Ni/CN by calcining Ni based MOF precursor at high temperature for urea electrooxidation under alkaline conditions. Due to the porous structure and excellent conductivity of the catalyst, it shows high activity for urea electrooxidation. The current density can reach 63.5 mA/cm² at 1.5V voltage and remains stable for a long time. Compared with the traditional noble gold-based catalysts, nickel-based catalysts have higher catalytic activity and lower cost, which greatly promotes the application of nickel-based catalysts in urea electrooxidation. Although nickel metal catalysts show considerable catalytic activity, the high initial potential and catalyst poisoning limit the application of metal catalysts. In order to solve these problems, alloy catalysts have been proposed. Through the interaction of alloys, the electronic structure of nickel atoms is optimized to reduce the initial potential and improve the catalyst poisoning. Hyun and their colleagues[47] reported a kind of PdNi alloy catalyst supported on porous carbon (Fig. 5). Through the electronic effect between the two metals, it not only effectively reduced the initial potential of urea oxidation, but also promoted the process of electro hydrogen evolution reaction. Finally, only 1.33 V voltage and 0.071 V overpotential were needed for UOR and HER to drive a current density of 10 mA/cm², respectively.

Based on the recognition that NiOOH is an active site and considering the hydrogen evolution activity of catalysts, some nickel-based compounds including oxides, hydroxides, nitrides, phosphides and sulfides have been reported to have

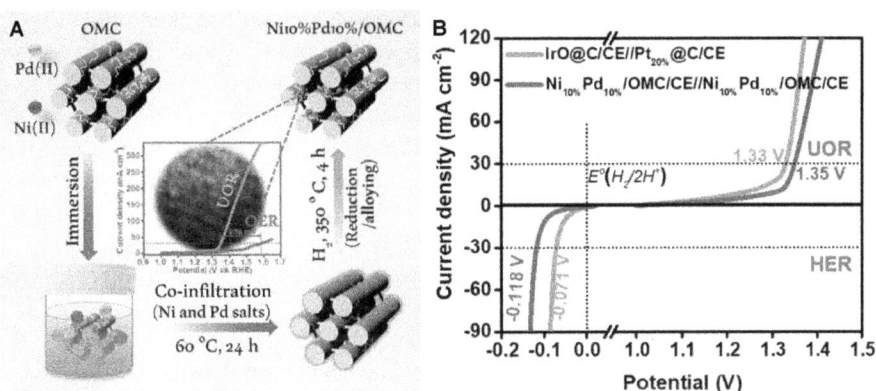

Figure 4.5. (a) Schematic illustration for preparation of the bimetallic $Ni_{(10\%)}Pd_{(10\%)}$/OMC electrocatalyst (b) LSV plots for the overall urea electrolysis system of $Ni_{(10\%)}Pd_{(10\%)}$/OMC/CE//$Ni_{(10\%)}Pd_{(10\%)}$/OMC/CE and $IrO@C/CE//Pt_{(20\%)}@C/CE$ in N_2-saturated 1 M KOH with 0.33 M urea at scan rate = 5 mV·s^{-1}.[47] Reproduced with permission from Ref. 47. Copyright 2019, American Chemical Society.

ultra-high urea oxidation activity. In the past few years, there have been many reports about the application of nickel oxide and nickel hydroxide in the electrochemical oxygen evolution. In the OER, nickel will undergo different valence state changes, among which NiOOH is the active species for urea oxidation, which indicates that nickel oxide and nickel hydroxide will have catalytic activity for the oxidation of urea. Liu et al. reported methanol-directed one-step growth of single-layered $Ni(OH)_2$ nanosheets on a conductive carbon cloth (CC) for highly efficient electrocatalytic urea oxidation.[48] Due to the ultra-thin $Ni(OH)_2$ nanosheets and good conductivity of carbon cloth, the catalyst exhibits excellent electrochemical oxidation performance of urea where only 0.5 V (vs. Ag/AgCl) is required when a current density of 435.6 mA/cm^2 is reached. Qiu et al. reported Ni/NiO@NC nanosheets with Schottky heterointerface for efficient urea oxidation reaction.[49] Because of the self-driven charge redistribution at the heterointerface and the synergistic effect between N-doped C and Ni/NiO, the catalyst only needs 1.35 V voltage to achieve 10 mA/cm^2 current density and can maintain stability for at least 12 hours. Although NiO and $Ni(OH)_2$ have excellent oxidation performance for urea, their application in total electrolysis of urea is limited due to their poor electro hydrogen evolution activity. In order to solve this problem, it is natural for people to use nickel-based compounds with excellent electro hydrogen evolution activity for urea total electrolysis water. Wang and co-workers prepared yolk-shell-structural Ni_2P/C nanosphere hybrids (Ni_2P/C-YS) via a hydrothermal method and subsequent phosphating treatment under different temperature ranging from 250°C to 450°C for UOR applications.[50] Due to the special core-shell structure and the interaction between Ni_2P and conductive support, only 1.366 V is needed to drive a current density of 50 mA/cm^2 and no obvious attenuation was found after 23 h stability test.

Feng et al. demonstrated a kind of excellent catalysis for HER and UOR catalyzed by nickel sulfide with N doping and heterostructure construction.[51] Due to the large specific surface area, the electronic effect of nitrogen doping and the synergistic effect of heterojunction, the N-NiS/NiS$_2$ catalysts prepared by this method show excellent performance in hydrogen evolution and urea oxidation. In alkaline HER and UOR, only 185 mV overpotential and 1.47 V voltage are needed to achieve 10 mA/cm^2 and 100 mA/cm^2, respectively. Considering that the catalyst has excellent HER and UOR performance, the urea electrolyzed water unit with N-NiS/NiS$_2$ as cathode and anode catalysts is assembled to produce hydrogen. Only 1.62 V cell voltage is needed to achieve a current density of 10 mA/cm^2, which is better than most reported non precious metal-based catalysts and shows a good application prospect.

Sun et al. reported a porous Ni$_3$N nanosheet array on carbon cloth (Ni$_3$N NA/CC) as a high-performance and durable electrocatalyst for urea oxidation and HER.[52] Because the porous two-dimensional structure can fully expose the active sites, the final catalyst only needs 1.35 V voltage to achieve a current density of 10 mA/cm^2. Considering the excellent hydrogen evolution performance of nitride, the catalyst will be a good bifunctional catalyst, which can be used in the total electrolysis of urea to produce hydrogen. As expected, when the catalyst is used in urea total electrolysis, only 1.44 V cell voltage is needed to achieve a current density of 10 mA/cm^2, which is superior to most of the catalysts reported so far and has good application prospects. These nickel-based compounds, including sulfide, nitride, phosphide and recently reported NiMoO$_4$, not only have super high urea oxidation activity, but also are excellent HER catalysts. Their bifunctional properties make them have the potential to be used as catalysts for urea total electrolysis.

Combined with modern advanced testing technology and theoretical calculation, the mechanism of urea electro oxidation has been thoroughly studied. On this basis, the types of catalysts are increasing, and the catalytic performance is also improving. Although great progress has been made in the electrooxidation of urea in recent years, the catalytic activity, stability and anti-toxicity ability still need to be improved to meet the needs of industrial production. As for how to achieve these goals, we have to start from the structure and composition of the catalyst. Firstly, on the premise that NiOOH is the active site, we can change the electronic structure of the catalyst by introducing some metal or non-metal atoms into the nickel-based catalyst to improve the catalytic activity or stability of a single catalytic active site. The adsorption state of intermediates on the catalyst surface can be changed by doping, so as to promote the desorption process of products and make the whole catalytic process go smoothly. Next, we need to design the microstructure of the catalyst reasonably, such as three-dimensional frame with open structure, core shell structure, two-dimensional porous structure, so that the active sites of the catalyst can be more exposed on the surface to improve the number of active sites of the catalyst. In addition, considering the poor conductivity of some common nickel-based compounds (oxides, hydroxides, sulfides), nickel-based compounds can be compounded with conductive carbon cloth or foam nickel to reduce the resistance of the reaction process. This will help to accelerate the kinetic reaction rate of catalytic reaction.

In the future, our main attention should be put on practical application, including urea total electrolysis hydrogen production and direct urea fuel cell. In the process of hydrogen production from urea electrolysis, it is still a big problem how to maintain the long-term stability under the practical application of high current. This problem may be solved in the near future by means of reasonable design of catalyst structure and optimization of mass transfer conditions. However, there are still many challenges for alkaline direct methanol fuel cell, including alkaline proton exchange membrane, low power density and cell assembly technology. It is believed that in the near future, these problems will be solved with the joint efforts of the majority of scientific researchers. The maturity of these two technologies will have a positive effect on global environmental governance and energy supply.

4.3.3 Conclusions

In general, the current development of UOR electrocatalysts is mainly based on Ni-based materials, and is mainly limited to applications in alkaline conditions. Through the designation of Ni-based compounds control and construction of hetero-interfaces, activity of UOR catalysts has been significantly developed till now. However, considering the actual application, future research on UOR catalysts should focus on the development and application of highly selective catalysts in the actual urea-containing wastewater systems.

4.4 Perspectives

Both MOR and UOR may have an important role in the future hydrogen energy technology. Therefore, the development of new and inexpensive catalysts for practical application needs has very important practical significance. Based on the study and summary of previous research results, the following perspective are put forward for the future development of MOR and UOR catalysts:

a. Considering the poor stability of Ru metal, a high-stability co-catalyst that replaces Ru should be found to be synergistic with Pt.

b. The MOR activity of Pt-based catalyst can be promoted by precisely controlling the crystal plane exposed.

c. The requirement of activity and stability of the practical alkaline fuel cell should be considered for the developing of non-Pt MOR catalysts.

d. Considering the high cost of pure urea, the design and development of UOR catalysts must be carried out based on urea-containing wastewater.

e. Precious metals can be introduced to improve the selectivity and stability of UOR for urea-containing wastewater.

References

[1]   Song, C. S. 2002. Fuel processing for low-temperature and high-temperature fuel cells-Challenges, and opportunities for sustainable development in the 21st century. *Catal. Today*, 77: 17–49.

[2]   Steele, B. C. H. and A. Heinzel. 2001. Materials for fuel-cell technologies. *Nature*, 414: 345–352.

[3]   Lee, S. J., S. Mukerjee, J. McBreen, Y. W. Rho, Y. T. Kho and T. H. Lee. 1998. Effects of nafion impregnation on performances of PEMFC electrodes. *Electrochim. Acta*, 43: 3693–3701.

[4]   Jiang, R. Z. and D. R. Chu. 2001. Stack design and performance of polymer electrolyte membrane fuel cells. *J. Power Sources*, 93: 25–31.

[5]   Pokojski, M. 2000. The first demonstration of the 250-kW polymer electrolyte fuel cell for stationary application (Berlin). *J. Power Sources*, 86: 140–144.

[6]   Arico, A. S., S. Srinivasan and V. Antonucci. 2001. DMFCs: From fundamental aspects to technology development. *Fuel Cells*, 1: 133–161.

[7]   Ren, X. M., P. Zelenay, S. Thomas, J. Davey and S. Gottesfeld. 2000. Recent advances in direct methanol fuel cells at Los Alamos National Laboratory. *J. Power Sources*, 86: 111–116.

[8]   Taneda, K. and Y. Yamazaki. 2006. Study of direct type ethanol fuel cells—Analysis of anode products and effect of acetaldehyde. *Electrochim. Acta*, 52: 1627–1631.

[9]   Yeom, J., R. S. Jayashree, C. Rastogi, M. A. Shannon and P. J. A. Kenis. 2006. Passive direct formic acid microfabricated fuel cells. *J. Power Sources*, 160: 1058–1064.

[10]   Serov, A. and C. Kwak. 2009. Progress in development of direct dimethyl ether fuel cells. *Applied Catalysis B-Environmental*, 91: 1–10.

[11]   Peled, E., V. Livshits and T. Duvdevani. 2002. High-power direct ethylene glycol fuel cell (DEGFC) based on nanoporous proton-conducting membrane (NP-PCM). *J. Power Sources*, 106: 245–248.

[12]   Wang, L., Y. Zhu, Y. Wen, S. Li, C. Cui, F. Ni, Y. Liu, H. Lin, Y. Li, H. Peng and B. Zhang. 2021. Regulating the local charge distribution of Ni active sites for the urea oxidation reaction. *Angew. Chem. Int. Ed.*, 60: 10577–10582.

[13]   Wasmus, S. and A. Kuver. 1999. Methanol oxidation and direct methanol fuel cells: A selective review. *J. Electroanal. Chem.*, 461: 14–31.

[14]   Ling, Y., X. Yu, Q. Zhang, W. Cai, F. Luo and Z. Yang. 2017. High stability and performance of PtRu electrocatalyst derived from double polymer coatings. *Int. J. Hydrogen Energ.*, 42: 11803–11812.

[15]   Ling, Y., Z. Yang, J. Yang, Y. Zhang, Q. Zhang, X. Yu and W. Cai. 2018. PtRu nanoparticles embedded in nitrogen doped carbon with highly stable CO tolerance and durability. *Nanotechnology*, 29: 055402.

[16]   Ren, X., Q. Lv, L. Liu, B. Liu, Y. Wang, A. Liu and G. Wu. 2020. Current progress of Pt and Pt-based electrocatalysts used for fuel cells. *Sustain. Energy Fules*, 4: 15–30.

[17]   Long, X., P. Yin, T. Lei, K. Wang and Z. Zhan. 2020. Methanol electro-oxidation on Cu@ Pt/C core-shell catalyst derived from Cu-MOF. *Appl. Catal., B*, 260: 118187.

[18]   Sriphathoorat, R., K. Wang, S. Luo, M. Tang, H. Du, X. Du and P. K. Shen. 2016. Well-defined PtNiCo core-shell nanodendrites with enhanced catalytic performance for methanol oxidation. *J. Mater. Chem. A*, 4: 18015–18021.

[19]   Gong, L., Z. Yang, K. Li, W. Xing, C. Liu and J. Ge. 2018. Recent development of methanol electrooxidation catalysts for direct methanol fuel cell. *J. Energy Chem.*, 27: 1618–1628.

[20]   Dinh, H. N., X. M. Ren, F. H. Garzon, P. Zelenay and S. Gottesfeld. 2000. Electrocatalysis in direct methanol fuel cells: in-situ probing of PtRu anode catalyst surfaces. *J. Electroanal. Chem.*, 491: 222–233.

[21]   Liang, H., X. Zhang, Q. Wang, Y. Han, Y. Fang and S. Dong. 2018. Shape-control of Pt-Ru nanocrystals: Tuning surface structure for enhanced electrocatalytic methanol oxidation. *J. Am. Chem. Soc.*, 140: 1142–1147.

[22]   Jin, Y. 2017. WO3 modified graphene supported Pt electrocatalysts with enhanced performance for oxygen reduction reaction. *Int. J. Electrochem. Sci.*, 12: 6535–6544.

[23]   Chen, W., J. Xue, Y. Bao and L. Feng. 2020. Surface engineering of nano-ceria facet dependent coupling effect on Pt nanocrystals for electro-catalysis of methanol oxidation reaction. *Chem. Eng. J.*, 381: 122752.

[24]   Tan, Q., C. Shu, J. Abbott, Q. Zhao, L. Liu, T. Qu, Y. Chen, H. Zhu, Y. Liu and G. Wu. 2019. Highly dispersed Pd-CeO$_2$ nanoparticles supported on N-doped core–shell structured mesoporous carbon for methanol oxidation in alkaline media. *ACS Catal.*, 9: 6362–6371.

[25]   Askari, M. B., P. Salarizadeh, M. Seifi and S. M. Rozati. 2019. Ni/NiO coated on multi-walled carbon nanotubes as a promising electrode for methanol electro-oxidation reaction in direct methanol fuel cell. *Solid State Sci.*, 97: 106012.

[26]   Nazal, M. K., O. S. Olakunle, A. Al-Ahmed, A. S. Sultan and S. J. Zaidi. 2019. Methanol electrooxidation in alkaline medium by Ni based binary and ternary catalysts: Effect of iron (Fe) on the catalyst performance. *Russ. J. Electrochem.*, 55: 61–69.

[27]   Levy, R. B. and M. Boudart. 1973. Platinum-like behavior of tungsten carbide in surface catalysis. *Science*, 181: 547–549.

[28]   Weigert, E. C., M. B. Zellner, A. L. Stottlemyer and J. G. Chen. 2007. A combined surface science and electrochemical study of tungsten carbides as anode electrocatalysts. *Top. Catal.*, 46: 349–357.

[29]   Sheng, T., X. Lin, Z.-Y. Chen, P. Hu, S.-G. Sun, Y.-Q. Chu, C.-A. Ma and W.-F. Lin. 2015. Methanol electro-oxidation on platinum modified tungsten carbides in direct methanol fuel cells: A DFT study. *PCCP*, 17: 25235–25243.

[30]   Nie, M., S. Du, Q. Li, M. Hummel, Z. Gu and S. Lu. 2020. Tungsten carbide as supports for trimetallic AuPdPt electrocatalysts for methanol oxidation. *J. Electrochem. Soc.*, 167: 044510.

[31]   Singh, R. N., A. Singh, D. Mishra, Anindita and P. Chartier. 2008. Oxidation of methanol on perovskite-type La2-xSrxNiO4 (0 ≤ x ≤ 1) film electrodes modified by dispersed nickel in 1M KOH. *J. Power Sources*, 185: 776–783.

[32]   Balasubramanian, A., N. Karthikeyan and V. V. Giridhar. 2008. Synthesis and characterization of LaNiO3-based platinum catalyst for methanol oxidation. *J. Power Sources*, 185: 670–675.

[33]   Yaqoob, L., T. Noor, N. Iqbal, H. Nasir, N. Zaman, L. Rasheed and M. Yousuf. 2020. Development of an efficient non-noble metal based anode electrocatalyst to promote methanol oxidation activity in DMFC. *ChemistrySelect*, 5: 6023–6034.

[34]   Noor, T., N. Zaman, H. Nasir, N. Iqbal and Z. Hussain. 2019. Electro catalytic study of NiO-MOF/rGO composites for methanol oxidation reaction. *Electrochim. Acta*, 307: 1–12.

[35]   Zhu, H., G. Gao, M. Du, J. Zhou, K. Wang, W. Wu, X. Chen, Y. Li, P. Ma, W. Dong, F. Duan, M. Chen, G. Wu, J. Wu, H. Yang and S. Guo. 2018. Atomic-scale core/shell structure engineering induces precise tensile strain to boost hydrogen evolution catalysis. *Adv. Mater.*, 30: 1707301.

[36]   Pu, Z., J. Zhao, I. S. Amiinu, W. Li, M. Wang, D. He and S. Mu. 2019. A universal synthesis strategy for P-rich noble metal diphosphide-based electrocatalysts for the hydrogen evolution reaction. *Energy Environ. Sci.*, 12: 952–957.

[37]   Zheng, Z., L. Yu, M. Gao, X. Chen, W. Zhou, C. Ma, L. Wu, J. Zhu, X. Meng, J. Hu, Y. Tu, S. Wu, J. Mao, Z. Tian and D. Deng. 2020. Boosting hydrogen evolution on MoS2 via co-confining selenium in surface and cobalt in inner layer. *Nat. Commun.*, 11: 3315.

[38]   Zhang, J.-Y., H. Wang, Y. Tian, Y. Yan, Q. Xue, T. He, H. Liu, C. Wang, Y. Chen and B. Y. Xia. 2018. Anodic hydrazine oxidation assists energy-efficient hydrogen evolution over a bifunctional cobalt perselenide nanosheet electrode. *Angew. Chem. Int. Ed.*, 57: 7649–7653.

[39]   Li, Y., Y. Sun, Y. Qin, W. Zhang, L. Wang, M. Luo, H. Yang and S. Guo. 2020. Recent advances on water-splitting electrocatalysis mediated by noble-metal-based nanostructured materials. *Adv. Energy Mater.*, 10: 1903120.

[40]   Chen, W., Z. Lei, T. Zeng, L. Wang, N. Cheng, Y. Tan and S. Mu. 2019. Structurally ordered PtSn intermetallic nanoparticles supported on ATO for efficient methanol oxidation reaction. *Nanoscale*, 11: 19895–19902.

[41]   Lin, C., L. Lu, H. Zhu, Y. Chen, H. Yu, Y. Li and L. Wang. 2017. Improved ethanol electrooxidation performance by shortening Pd–Ni active site distance in Pd–Ni–P nanocatalysts. *Nat. Commun.*, 8: 14136.

[42]   Liu, Z., C. Zhang, H. Liu and L. Feng. 2020. Efficient synergism of NiSe2 nanoparticle/NiO nanosheet for energy-relevant water and urea electrocatalysis. *Appl. Catal., B*, 276: 119165.

[43]   Zhang, Q., F. M. D. Kazim, S. Ma, K. Qu, M. Li, Y. Wang, H. Hu, W. Cai and Z. Yang. 2021. Nitrogen dopants in nickel nanoparticles embedded carbon nanotubes promote overall urea oxidation. *Appl. Catal., B*, 280: 119436.

[44]   Vedharathinam, V. and G. G. Botte. 2014. Experimental investigation of potential oscillations during the electrocatalytic oxidation of urea on Ni catalyst in alkaline medium. *J. Phys. Chem. C*, 118: 21806–21812.

[45]   Daramola, D. A., D. Singh and G. G. Botte. 2010. Dissociation rates of urea in the presence of NiOOH catalyst: A DFT analysis. *J. Phys. Chem. A*, 114: 11513.

[46]   Maruthapandian, V., S. Kumaraguru, S. Mohan, V. Saraswathy and S. Muralidharan. 2018. An insight on the electrocatalytic mechanistic study of pristine Ni MOF (BTC) in alkaline medium for enhanced OER and UOR. *Chem. Electro. Chem.*, 5: 2795–2807.

[47]   Muthuchamy, N., S. Jang, C. P. Ji, S. Park and H. P. Kang. 2019. Bimetallic NiPd nanoparticle-incorporated ordered mesoporous carbon as highly efficient electrocatalysts for hydrogen production via overall urea electrolysis. *ACS Sustain. Chem. Eng.*, 7: 15526–15536.

[48]   Lin, C., Z. Gao, F. Zhang, J. Yang, B. Liu and J. Jin. 2018. *In situ* growth of single-layered α-Ni(OH)2 nanosheets on a carbon cloth for highly efficient electrocatalytic oxidation of urea. *J. Mater. Chem. A*, 6: 13867–13873.

[49]   Ji, X., Y. Zhang, Z. Ma and Y. Qiu. 2020. Oxygen vacancy-rich Ni/NiO@NC nanosheets with schottky heterointerface for efficient urea oxidation reaction. *Chem. Sus. Chem.*, 13: 5004–5014.

[50]   Zhang, Y. and C. Wang. 2021. Yolk-shell nanostructural Ni2P/C composites as the high performance electrocatalysts toward urea oxidation. *Chin. Chem. Lett.*, 32: 2222–2228.

[51]   Liu, H., Z. Liu, F. Wang and L. Feng. 2020. Efficient catalysis of N doped NiS/NiS$_2$ heterogeneous structure. *Chem. Eng. J.*, 397: 125507.

[52]   Liu, Q., L. Xie, F. Qu, Z. Liu, G. Du, A. M. Asiri and X. Sun. 2017. A porous Ni$_3$N nanosheet array as a high-performance non-noble-metal catalyst for urea-assisted electrochemical hydrogen production. Inorg. *Chem. Front.*, 4: 1120–1124.

5

Carbon Materials-based Electrocatalysts for Oxygen Reduction Reaction

Xuedong He, Feng Zhou, Lijie Zhang, Shuang Pan,* Huile Jin,* Yihuang Chen* and Shun Wang*

‖‖

5.1 Introduction

At present, the energy that mankind relies on mainly comes from fossil resources.[1–5] With the consumption of fossil resources and environmental problems becoming increasingly prominent, it is particularly important to develop a new type of energy.[6,7] Among them, the proton exchange membrane fuel cell (PEMFC), as an emerging energy source, possesses the advantages of high energy density, highly-efficient energy conversion and no emission of pollutants, which results in a wide range of applications of PEMFC, such as the new energy vehicles.[8–10] For example, Toyota firstly launched commercial fuel cell vehicle, named as Toyota Mirai in 2016. The maximum endurance of Mirai reaches more than 500 kilometers, which is comparable to many electric vehicles based on Li-ion batteries (LIBs).[11] In addition, the emerging metal-air batteries (MABs) are also considered to be a promising energy source due to its superior theoretical capacity density over that of LIBs as well as its less cost and safer operation.[12–15] Both PEMFC and MAB involve the oxygen reduction reaction (ORR), which however suffers from a sluggish kinetics.[16–18] In other words, an efficient ORR catalyst is a key factor in determining the performance of the relative devices.

Currently, the ORR benchmark catalysts are composed of dispersed Pt NPs on carbon black (i.e., Pt/C); however, the cost of noble metal Pt accounts for ≈ 55% of

College of Chemistry and Materials Engineering, Institute of New Materials and Industrial Technologies, Wenzhou University, Wenzhou 325035, PR China

* Corresponding authors: shuangpan@wzu.edu.cn; huilejin@wzu.edu.cn; yhchen@wzu.edu.cn; shunwang@wzu.edu.cn

fuel cell stacks.[19] Therefore, it is necessary to reduce Pt loading or even completely replace Pt with the abundant and cheap metals. In fact, although Pt is considered state-of-the-art catalysts, its poor stability can only allow a limited lifetime for fuel cells.[19,20] Thus, it's urgent to develop electrocatalysts for ORR with low cost, long durability and high performance. In recent years, carbon-based materials have been widely studied.[21-23] Owing to their high activity and low cost, carbon-based materials are widely used in ORR,[24-27] carbon dioxide reduction (CO_2RR),[28-30] water splitting,[31-34] and nitrogen reduction reaction (NRR).[35-37] After a long period of rapid development, the research on carbon-based ORR electrocatalysts has become more and more in-depth. In the past few years, the design principles and engineering strategies of carbon-based materials have been developed, and the electronic structure has been subtly modified to promote satisfactory ORR activity. In addition, under the joint promotion of theory and experiment, the specific mechanism of carbon-based materials in the ORR process has been further understood.[38-40] These results will guide the design of a new generation of heterogeneous catalysts such as hydrogen evolution reaction (HER), oxygen evolution reaction (OER), CO_2RR and NRR.[41-44]

This chapter briefly summarizes the latest progress of carbon-based materials in ORR. First, the main mechanism of the ORR reaction is introduced. Then, various typical carbon-based materials including platinum group metal (PGM)-based carbon materials, transition metal (TM)-based carbon materials, and metal-free carbon materials, are mainly discussed. Finally, development of carbon-based materials in ORR is pointed.

5.2 ORR reaction mechanism

In the past few decades, impressive progress has been made in ORR reaction.[45] ORR can generate hydrogen peroxide through two-electron (i.e., $2e^-$) reaction, and generate OH^- or H_2O in alkaline or acidic solution through four-electron (i.e., $4e^-$) reaction, respectively[46]. Kulkarni et al.[47] established a model to calculate electrochemical reaction mechanism under an external electric field. The thermodynamic Gibbs free energy of each elementary step of ORR under different electrode potentials was performed on Pt. Under acidic conditions, the pathway of ORR reaction can be presented as follows:

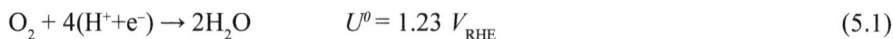

$$O_2 + 4(H^+ + e^-) \rightarrow 2H_2O \qquad U^0 = 1.23\ V_{RHE} \qquad (5.1)$$

$$O_2 + 2(H^+ + e^-) \rightarrow H_2O_2 \qquad U^0 = 0.70\ V_{RHE} \qquad (5.2)$$

where U^0 is the standard equilibrium potential for the reactions, and V_{RHE} stands for the potential relative to the reversible hydrogen electrode (RHE).

In neutral or weakly alkaline solution, the pathway of ORR reaction changes as follows:

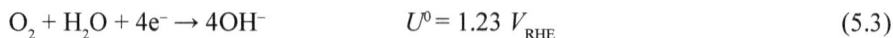

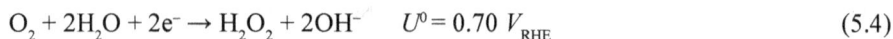

$$O_2 + H_2O + 4e^- \rightarrow 4OH^- \qquad U^0 = 1.23\ V_{RHE} \qquad (5.3)$$

$$O_2 + 2H_2O + 2e^- \rightarrow H_2O_2 + 2OH^- \qquad U^0 = 0.70\ V_{RHE} \qquad (5.4)$$

If the pH > 11.7, the expression for this reaction is

$$O_2 + H_2O + 2e^- \rightarrow HO_2^- + OH^- \qquad U^0 = 0.76 \; V_{RHE} \tag{5.5}$$

The four-electron pathway to H_2O is more favorable than two-electron pathway. There is no doubt that the $4e^-$ pathway is more conducive for fuel cells and metal-air batteries. Not only a higher output voltage can be generated through pathways of $4e^-$, but the hydrogen peroxide obtained by $2e^-$ pathway will inevitably degrade the catalysts.[48,49] Nevertheless, the two-electron pathway of partial reduction of oxygen to H_2O_2 products is inevitable during the ORR. There is about 2% selectivity for $2e^-$ pathway on state-of-the-art PGM catalysts.[50,51] Density functional theory (DFT) calculations can reveal the changed value of Gibbs free energy (U) in each elementary step of ORR at different potentials.[47,52] According to the dissociation of O_2 molecule or not, the pathways of ORR can be divided into associative and dissociative mechanism, which are widely accepted. The associative mechanism contains three intermediates (Eq. 5.6, Eq. 5.7, Eq. 5.8) after forming OOH intermediate. The dissociative mechanism involves two intermediates (Eq. 5.9, Eq. 5.10, Eq. 5.11) after dissociation of O-O bond. For the aforementioned mechanism, the adsorption energy of intermediate O (ΔE_O) can be regarded as an important descriptor for estimating ORR activity.

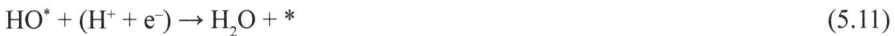

Associative: $O_2 + * \rightarrow O_2^*$ \hfill (5.6)

$$O_2^* + (H^+ + e^-) \rightarrow HO_2^* \tag{5.7}$$

$$HO_2^* + (H^+ + e^-) \rightarrow H_2O + O^* \tag{5.8}$$

Dissociative: $1/2 \; O_2 + * \rightarrow O^*$ \hfill (5.9)

$$O^* + (H^+ + e^-) \rightarrow HO^* \tag{5.10}$$

$$HO^* + (H^+ + e^-) \rightarrow H_2O + * \tag{5.11}$$

where * stands for the active sites of catalysts.

In fact, the reduction of O_2 often occurs through both of dissociation and association simultaneously in a real over-potential environment. Studies have found that the activation energy of oxygen dissociation on Pt is much greater than that of the OOH dissociation activation energy formed after the transfer of protons and electrons, indicating that the formation of OOH is prior to dissociation in ORR process. The first part of electron transfer and proton transfer is considered as the rate-determining step (RDS) of the entire reaction.[53,54] Another important factor affecting the ORR reaction rate is the adsorption energy of intermediate species.[55,56] The next step of the reaction will continue only when the intermediate species adsorb to the surface of the catalyst. Shao et al. studied the oxygen reduction activity of different catalysts through a combination of experiment and theory.[57] They found that the relationship between ORR activity of metal catalysts and the adsorption energy of intermediate species presented a "volcano" shape (Fig. 5.1), which indicates that

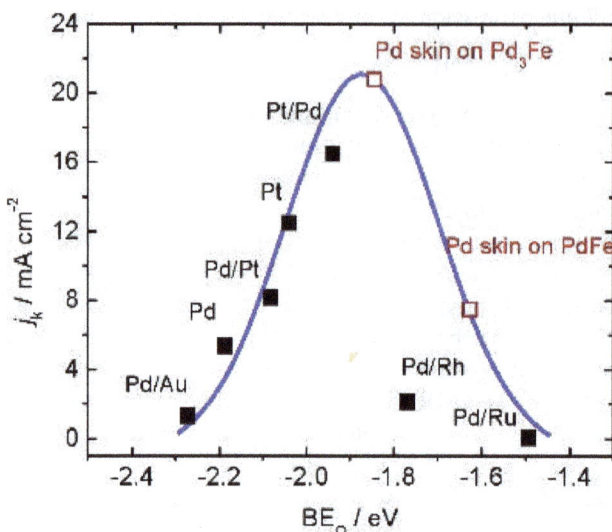

Figure 5.1. The relationship between the catalytic activity of metal catalysts and the adsorption strength of intermediate species. Reproduced with permission from Ref. 57. Copyright 2007, American Chemical Society.

advanced catalysts need to have an appropriate adsorption strength to intermediate species.

Therefore, theoretical calculations based on DFT and microdynamics are also utilized for screening and crafting high-activity carbon-based catalysts, which can design and advance the next generation of ORR electrocatalysts.

5.3 PGM-based carbon materials

5.3.1 PGM nanoparticle-based carbon materials

PGM nanoparticle-based carbon materials possess excellent performance in oxygen reduce reaction, hydrogen evolution reaction and carbon monoxide oxidation due to the concerted effect of high activities of PGM, good electrical conductivity of carbon materials and their synergistic enhancement.[51,58–60] ORR involves four-electron and two-electron transfer. Fortunately, the proportion of H_2O_2 is not high in most of PGM-containing catalysts.[61] Using the DFT to calculate the adsorption energy of O* and HO* on different metals, Nørskov et al.[52] proposed "volcanic curve", and obtained their ORR activity, in which the Pt was nearly located at the top of the curve (Fig. 5.2). Although Pt is considered state-of-the-art catalysts for ORR, there is still much space for Pt-based materials. Adjusting metal tensile/squeeze strain and electron delocalization represent effective strategies for optimizing Pt-based carbon materials.

By introducing other metals to form Pt alloy, the adsorption energy of intermediate in ORR was effectively changed.[62] Cooperating Pt with Ni to form Pt-Ni alloy was proved to be a promising strategy. Tian et al.[63] used electrochemical corrosion method to control the near-surface structure and components of the

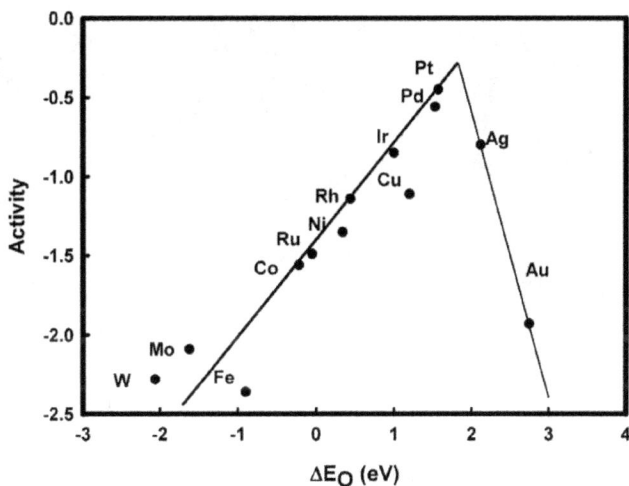

Figure 5.2. Oxygen reduction activity of different metals. Reproduced with permission from Ref. 52. Copyright 2004, American Chemical Society.

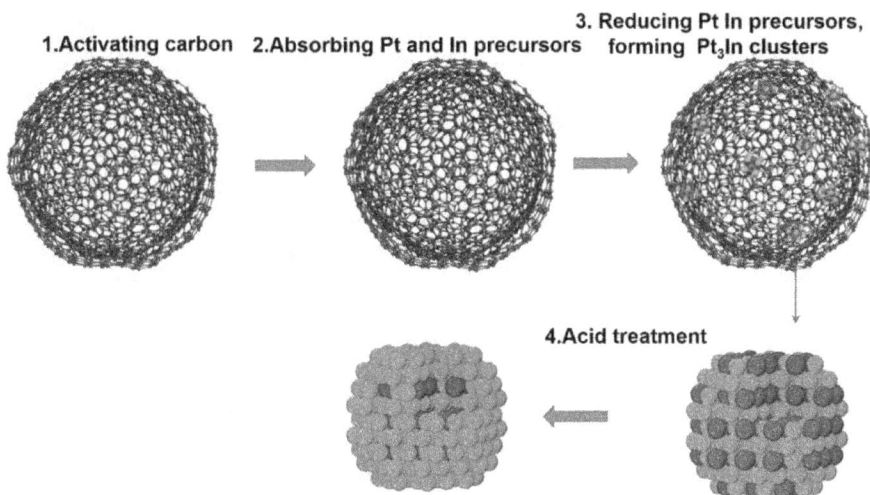

Figure 5.3. Schematic illustration of the synthetic procedure of Pt$_3$In/C catalysts. Reproduced with permission from Ref. 63. Copyright 2020, Wiley-VCH.

Pt-based carbon catalyst, yielding a series of Pt-based nano-cage structure with zero-dimensional structure (Fig. 5.3), which realized the effective combination of the characteristics such as high-stability of zero-dimensional structure and high-activity of hollow structure.

Metal moiety can be modified with defects. The most notable example is that Li et al.[20] prepared carbon-supported jagged Pt nanowires (J-PtNWs) by introducing defects, which demonstrated unprecedented superb performance in ORR with a half wave potential ($E_{1/2}$) of 0.935 V. Briefly, PtNi nanowires were synthesized, followed by etching of metal Ni to obtain defect-rich J-PtNWs (Fig. 5.4) with ultra-high ORR

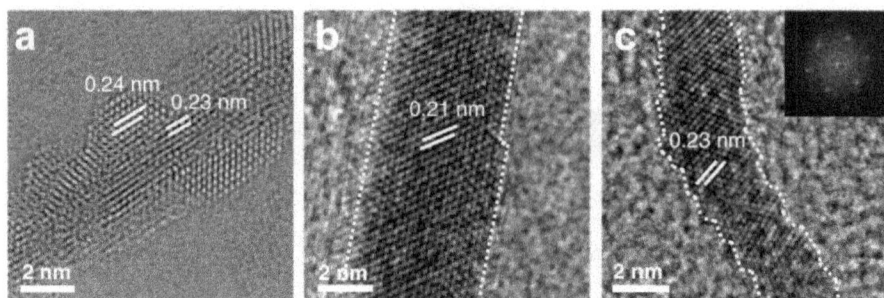

Figure 5.4. The evolution process of ultrafine jagged Pt nanowires. Reproduced with permission from Ref. 20. Copyright 2016, Science.

activity. The corresponding catalytic activity was increased by 33 times compared with the Pt/C catalyst at 0.9 V_{RHE}, and the mass activity reached 13.6 A mg^{-1}. Recently, Goddard et al. developed a bridge nanocluster model for DFT calculations with detailed insights into the high performance of jagged Pt nanowires.[64] They used quantum mechanics to prove that the concave Pt (111) surface greatly reduced the energy barrier of $O^* + H_2O^* \rightarrow 2OH^*$ process.

Despite the excellent performance of PGM, limited resources and high price have significantly restricted their practical application. Therefore, strategies to reduce the loading of precious metal have been developed. Commercial Pt/C materials are often used as the benchmark catalysis for ORR. While the combination of Pt and carbon black effectively reduces the loading of Pt metal, metal Pt still accounts for up to 20%.[65,66] In addition, a worse disadvantage of Pt/C catalyst lies in its poor durability under PEMFC working conditions.[19,67] Four major reasons are well acknowledged for Pt/C attenuation mechanism: (1) migration and agglomeration of Pt nanoparticles through Oswald ripening process; (2) dissolution and deposition of Pt nanoparticles; (3) carbon corrosion; (4) Pt poisoning.[68, 69] To this end, many studies were reported to enhance the activity and stability of PGM carbon-based catalysts. The carbon layer coating Pt can effectively prevent the dissolution and agglomeration of metal particles. By using glucose as both the carbon source and reducing agent, Wen et al.[70] firstly reduced and deposited Pt nanoparticles into the pores of mesoporous silica (SBA-15), followed by carbonization and removal of SBA-15 to obtain a porous carbon-encapsulated Pt@C catalyst. Although carbon coating improves the stability of the PGM catalyst, it greatly reduces electrochemical active surface area (ECSA). Wang et al.[71] prepared low-Pt-loaded carbon-based materials with assistance of peptide. The size of Pt nanoparticles was controlled to a few nanometers and the electronic structure of Pt was adjusted due to N doping into carbon skeleton. Although the Pt loading was lower, the mass activity (MA) and specific activity (SA) of Pt@NiNC were higher compared with commercial Pt/C as shown in Fig. 5.5.

The chemical stability and conductivity of polyaniline (PANI) can be used to inhibit the Ostwald ripening of Pt nanoparticles during the operation of the fuel cell via coating PANI on the Pt/C catalyst, which can also increase the ECSA to increase the three-phase reaction interface.[72] While coating a layer of carbon on the Pt can effectively prevent the aggregation of Pt nanoparticles, it limits the transfer of oxygen on the platinum surface. In order to prevent the aggregation behavior of Pt

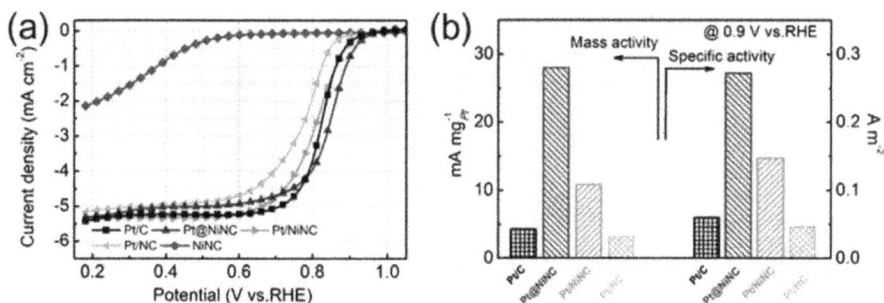

Figure 5.5. (a) LSV curves of Pt/C, Pt@NiNC, Pt/NiNC, Pt/NC and NiNC in O_2-saturation solution. (b) Comparison of MA and SA for the above electrodes. (a and b) Reproduced with permission from Ref. 71. Copyright 2020, the Royal Society of Chemistry.

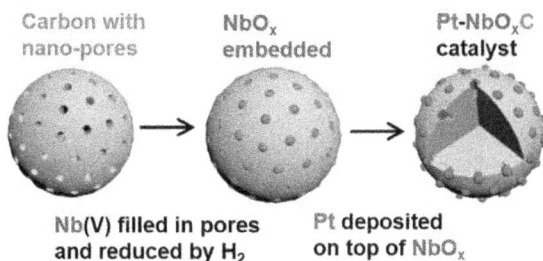

Figure 5.6. Schematic illustration of the synthetic procedure of Pt-NbO_xC catalysts. Reproduced with permission from Ref. 73. Copyright 2020, Elsevier.

during operation, Ma et al.[73] embedded NbO_x nanoparticles in carbon nanopores as nails, followed by deposition of Pt on NbO_x (Fig. 5.6). Due to the uniform dispersion of Pt on the NbO_x, NbO_x can be prevented from leaching out from catalysts. After 5000 cycles of cyclic voltammograms (CVs), $E_{1/2}$ did not change significantly.

Through the above methods, carbon and PGM nanoparticles can be effectively combined to prevent the aggregation and growth, but severe carbon corrosion restricts the long-term operation of PGM nanoparticle-based carbon materials. Carbon will be oxidized to CO_2 in water with voltage of above 0.207 V, which will be slowly oxidized to CO at 0.5 V.[74] Carbon corrosion will aggravate the growth and aggregation of nanoparticles and change the hydrophobicity of carbon-based materials. In addition, CO formed by carbon corrosion can poison the catalyst. Carbon support functionalization is an effective means to improve the overall stability. Strasser et al.[75] studied a series of Pt nanoparticles on N-doped carbons, which exhibited exceptional catalytic performance and stabilities. N-Vulcan obtained by Vulcan XC 72r powders underwent oxidation and ammonolysis (Fig. 5.7a). The fact that there was no significant variation in the size of the particles among the catalysts indicated that the carbon supports were responsible for the significant differences in long-term cycling stability (Fig. 5.7b). From high-temperature differential electrochemical mass spectrometry experiments, they found that the nitrogen-functionalized catalyst revealed the lowest carbon corrosion due to the highest chemical stability of the N-doped carbon lattice.

a

b Pt/N-Vulcan 400°C   Pt/Vulcan **c**

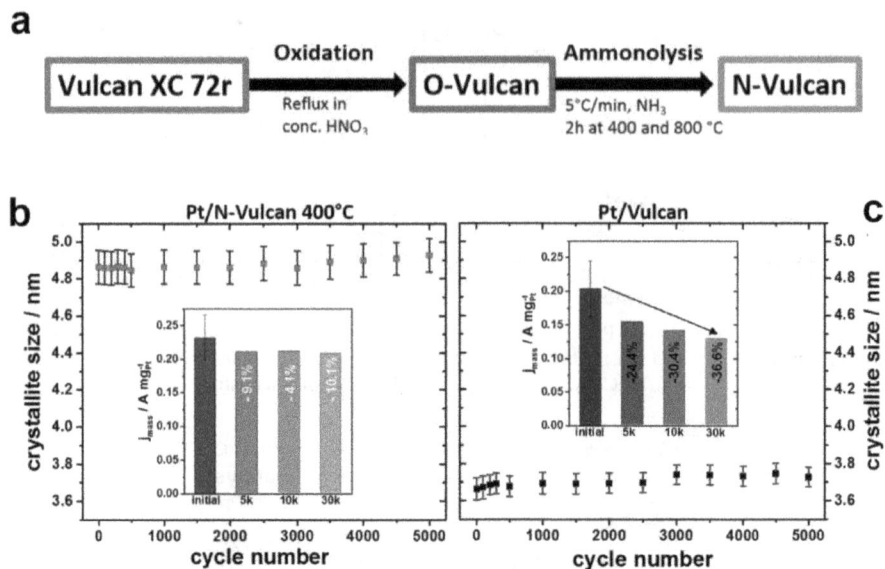

Figure 5.7. (a) Schematic illustration of the carbon modification procedure. Crystallite sizes of obtained Pt/N-Vulcan 400°C (b) and Pt/Vulcan (c). Inlets in both graphs showing the mass activity in 0.1 M HClO$_4$. (a-c) Reproduced with permission from Ref. 75. Copyright 2018, American Chemical Society.

At present, the catalytic activity of ORR in PGM-based nanoparticle catalysts is no longer a challenge, but cost and durability are the biggest obstacles restricting the further commercialization of fuel cells. The principle of novel catalyst design must consider cost-effectiveness and high stability in order to successfully demonstrate its advantages in the marketplace.

5.3.2 Single-atom PGM-based carbon materials

In recent years, single-atom carbon materials have been extensively studied due to their high atom utilization.[76] In general, PGM single atoms show superb advantages as they have high-efficiency catalytic performance and durability compared to transition metal single atoms.[77–79] The activity of Ru is very close to that of Pt during electrocatalytic process, but Ru is easier to agglomerate.[80,81] To this end, confinement of Ru in nanopores through a carbon substrate has shown the feasibility to improve the stability. Xiao et al. developed a single-atom active site of Ru for efficient ORR.[82] As shown in Fig. 5.8, K-edge extended X-ray absorption fine structure (EXAFS) was used to determine the structure of the resultant catalyst. The lack of the characteristic peak for Ru-Ru at 2.5 Å indicated the atomic dispersion of Ru on the nitrogen-doped carbon matrix in prepared material, which was also confirmed by spectra of the wavelet transform K-edge EXAFS. Compared with Ru foil (9 Å), the maximum intensity of Ru single-atom site catalysts (Ru-SSC) is 4 Å. The prepared Ru-SSC exhibited high SA and MA far exceeding those of Pt/C. Meanwhile, the corresponding activity of Ru-SSC was less attenuated after 2000 cycles of CV compared with Pt/C.

Liu et al. reported that single-atom Pt was used for efficient ORR catalysis.[83] The Pt loading was only 0.4 wt%, and the material had very good resistance to CO

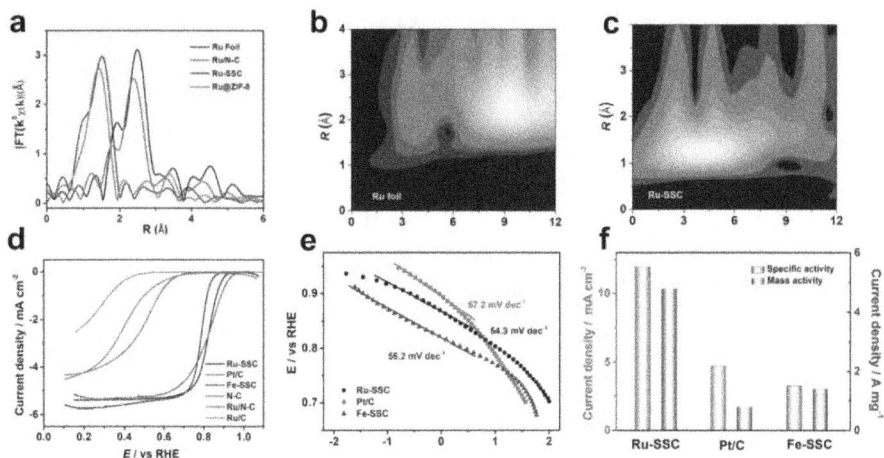

Figure 5.8. (a) Fourier transforms of k³-weighted Ru K-edge EXAFS spectra, (b, c) wavelet transforms for the k³-weighted Ru K-edge EXAFS signals of Ru foil and Ru-SSC, (d) LSV curve of synthesized catalysts and (e) Tafel slopes, (f) SA and MA comparison among Ru-SSC, Pt/C, and Fe-SSC. (a–f) Reproduced with permission from Ref. 82. Copyright 2019, American Chemical Society.

and methanol poisoning (Fig. 5.9a). After 10k CVs, the LSV curve had no obvious negative shift (Fig. 5.9b). In order to further determine the active sites of the as-prepared Pt catalyst, relativistic DFT calculations were utilized. It is believed that pyridinic-N or initial graphene was the most possible point anchoring single-atom Pt as shown in Fig. 5.9c. Though PGM materials have extremely high catalytic activity, the cost and durability are the biggest obstacles restricting their popularization. Nevertheless, while the stability of PGM-based materials is still insufficient in PEMFC as mentioned earlier, the stability of non-PGM-based materials is far inferior to that of PGM-based materials.[8,84] In other words, PGM-based materials are the most promising candidate for PEMFC application currently. Therefore, reducing the Pt loading is a more feasible option. However, for most carbon materials with a high degree of graphitization, sp2 carbon is chemically inert, and there are not enough sites to anchor Pt nanoparticles. The carbon support can be modified by introducing sulfhydryl (-SH),[85] amino (-NH₃)[86] or other N, P and metal compounds[87, 88] on the original carbon by covalent grafting, thereby changing the surface properties of the carbon for improved activity and stability. However, the practical application of the carbon support is currently only carbon black, which therefore is particularly urgent to develop a stable, inexpensive, and anti-oxidant carrier.

5.4 TM-based carbon materials

TM-based carbon material catalysts have also been widely studied for their application in the ORR field because transition metals are cheaper and more abundant than precious metals. Transition metals can be divided into transition metal oxides, phosphides, sulfides, and nitrides. However, TM-based carbon materials still have a long way to go, as its performance in acidic ORR is not satisfactory.

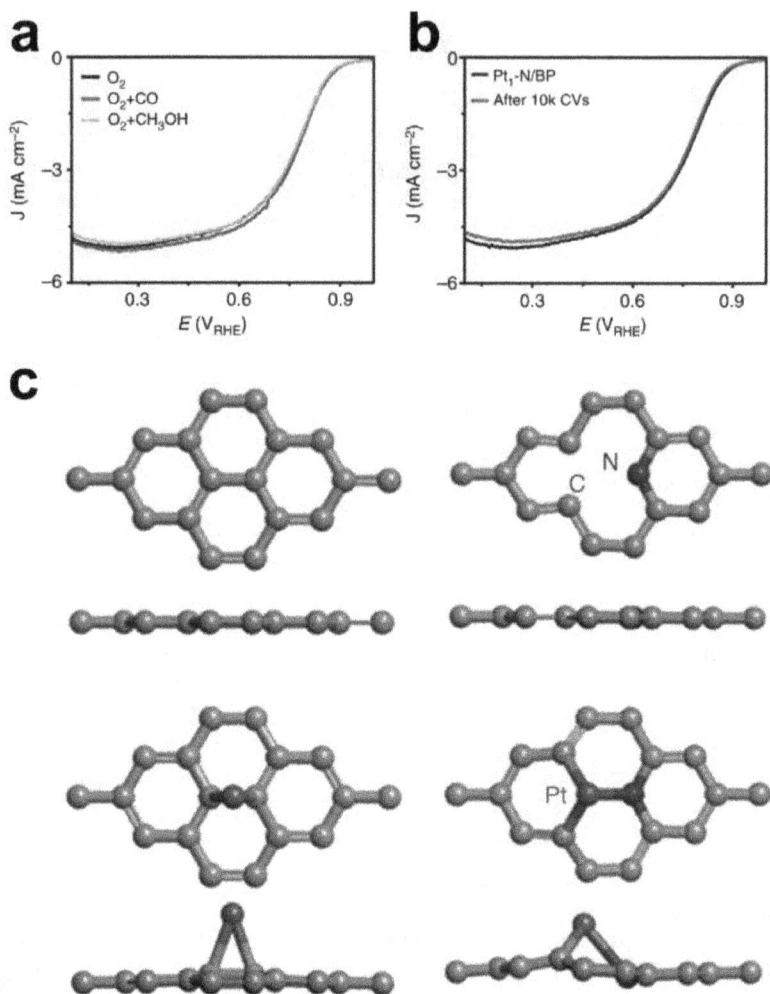

Figure 5.9. (a) The tolerance of catalyst to CO (saturated) and methanol (0.5 M) in O_2-saturated 0.1 M $HClO_4$. (b) Long-term operation stability of catalyst in O_2-saturated 0.1 M $HClO_4$. (c) Configuration of Pt in carbon. (a–c) Reproduced with permission from Ref. 83. Copyright 2017, Nature Publishing Group.

5.4.1 Transition metal oxide-based carbon materials

In fact, many transition metal compounds are excellent catalysts for ORR potentially.[89] For example, spinel metal oxide is one of the promising candidates due to its high activity, long life and low cost. However, due to the lack of sufficient specific surface area for most transition metal compounds, their performance is usually unsatisfying. Moreover, since transition metal oxides have large energy band gaps, they often have limited electrical conductivity, which significantly hinders their wide applications in electrocatalysis. In this regard, carbon materials are usually to improve the conductivity. Xiong et al.[90] prepared an efficient ORR catalyst by adjusting manganese-doped cobalt iron oxide (MCF) spinel nanocrystals, which were mixed with carbon nanotube as electrocatalysts. As shown in Fig. 5.10a, the

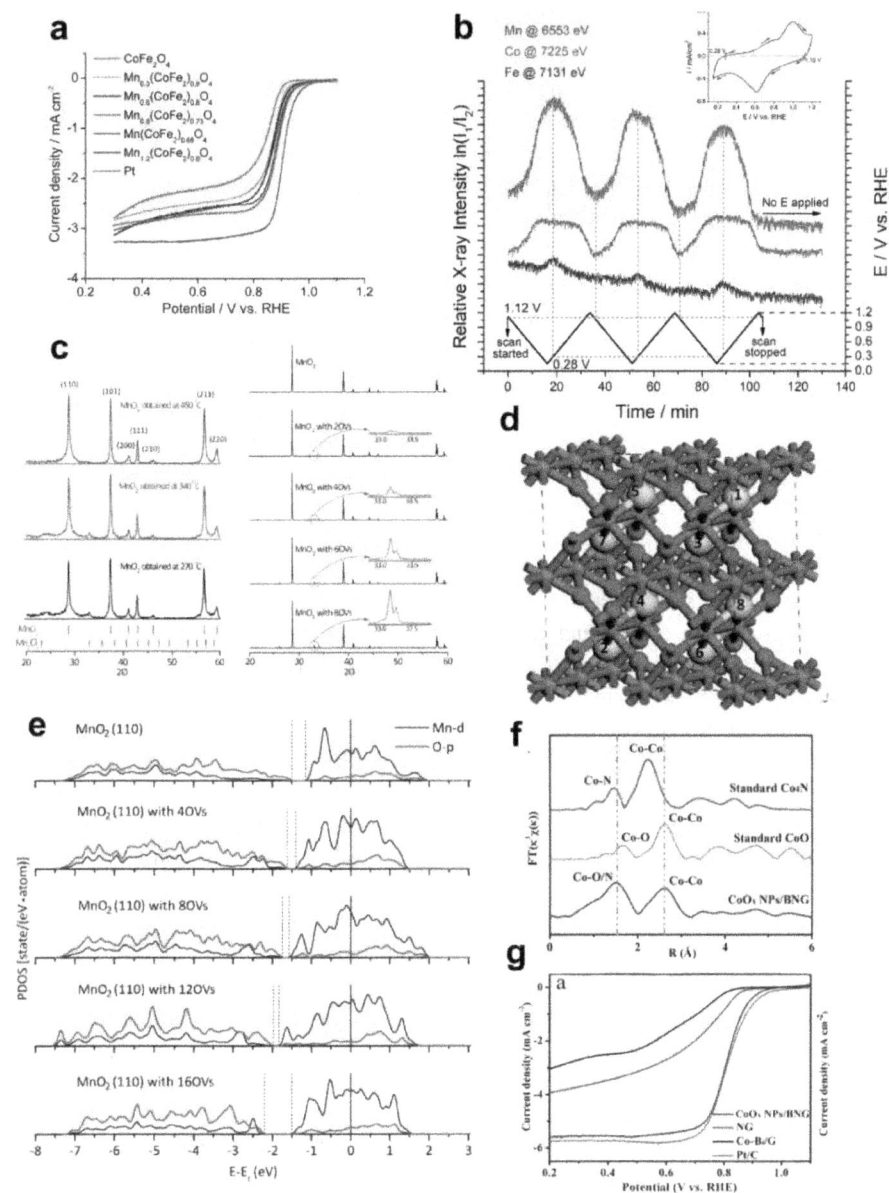

Figure 5.10. (a) ORR polarization curves in O_2-saturated 1 M NaOH. (b) Periodic changes of the relative X-ray intensities, as a function of the cyclic potential from 1.2 to 0.15 V_{RHE}. (a and b) Reproduced with permission from Ref. 90. Copyright 2019, American Chemical Society. (c) XRD spectra of MnO_2 and simulated XRD spectra of β-MnO_2 crystals with different numbers of OVs. (d) β-MnO_2 model. (e) PDOS of β-MnO_2. (c–e) Reproduced with permission from Ref. 91. Copyright 2015, American Chemical Society. (f) EXAFS spectra of CoO_x NPs/BNG. (g) ORR polarization curves in 0.1 M KOH solution. (f and g) Reproduced with permission from Ref. 92. Copyright 2017, Wiley-VCH.

$E_{1/2}$ reached 0.89 V in 1 M NaOH solution, which was only 0.02 V lower than that of commercial Pt/C. Through *in situ* X-ray absorption spectroscopy (XAS), it was found that Mn and Co served as catalytically active center sites, while iron stabilized the spinel structure (Fig. 5.10b).

Some other strategies have shown the feasibility to solve the aforementioned issues. For example, Li et al. [91] improved the catalysis of ORR by adjusting the oxygen vacancies (OVs) in manganese dioxide. In oxygen atmosphere, through different thermal treatment temperatures, different concentrations of OVs were introduced, as confirmed in Fig. 5.10c. The obtained MnO_2 was purely oriented β-MnO_2 at high temperatures (450°C), while at medium and low temperatures (i.e., 340°C and 270°C) a new peak appeared at 2θ of 33.2° (Fig. 5.10c), indicating that a new crystal phase occurred. The characteristics of this emerged phase were consistent with Mn_2O_3 but different from β-MnO_2, suggesting that pyrolysis at insufficient temperature resulted in low oxidation state Mn cations. Based on XRD data, a large unit cell of $Mn_{16}O_{32}$ with 48 atoms was constructed, and the concentrations of OVs were 1/16, 1/8, 3/16, 1/4, respectively (Fig. 5.10d). The introduction of OVs improved the conductivity of MnO_2 and made it a P-type semiconductor. According to the projected density of states (PDOS) diagram (Fig. 5.10e), as the OVs increased, the d orbitals of metal moved to a lower energy, resulting in a decrease in the band gap. When it increased to 16 OVs, the metal d orbital in the conduction band became narrower and the band gap increased. The conductivity of the oxide was successfully controlled by OVs, and then the oxide was loaded on carbon for ORR testing under alkaline conditions. Consistent with the above analysis, certain OVs improved the conductivity and ultimately improved the catalytic activity. The advantage of combining transition metals with carbon materials is to increase the ECSA and stability of transition metals. Loading transition metal oxides on graphene is also considered a feasible strategy for preparing high-performance ORR catalysts. Owing to the couple of oxide and carbon with sp2 hybridization, Tong et al. prepared CoO_x NPs/BNG by compounding CoO_x with B and N double-doped graphene oxide.[92] As shown in Fig. 5.10f, EXAFS was mainly composed of Co-O/N and Co-Co bonds, indicating that Co-O and Co-N bonds existed simultaneously. While the Co-Co bond of CoO_x NPs/BNG was consistent with that of CoO, the Co-O bond of CoO_x NPs/BNG was significantly shorter than that of CoO but larger than Co-N bond of CoN_4. The Co-Co/Co-O ratio in CoO_x NPs/BNG was lower than standard CoO, indicating that Co coordination is absent. As shown in Fig. 5.10g, the high-efficiency ORR performance was attributed to the introduction of many defects by the doping of both B and N, and the sufficient active sites provided by the abundant oxygen vacancies and Co-N-C.

5.4.2 Other transition metal compound-based carbon materials

Except transition metal oxides, transition metal phosphides, sulfides and nitrides are also promising candidates for ORR catalysts. Diao et al.[93] enhanced the ORR activities of Co_2P-based carbon materials via structural optimization and electronic reconfiguration (Fig. 5.11a). The DFT modeling indicated that Cu doping rendered more positive sites on adjacent Co sites with reduced surface binding energy between

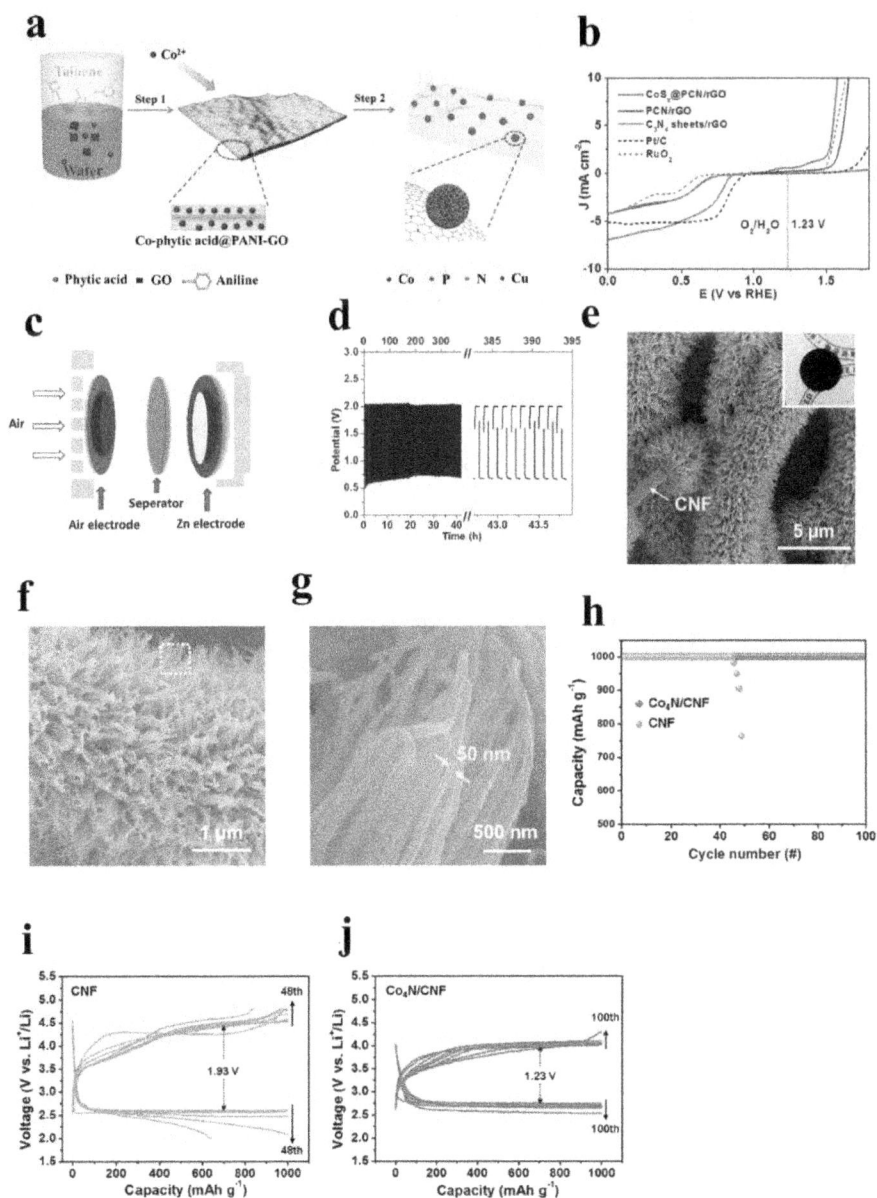

Figure 5.11. (a) Schematic illustration of the fabrication process of Cu-Co$_2$P@2D-NPC. Reproduced with permission from Ref. 93. Copyright 2019, the Royal Society of Chemistry. (b) ORR polarization curves in O$_2$-saturated 0.1 M KOH solution. (c) Diagram of ZABs structure. (d) Galvanostatic discharge/charge cycling curves of CoS$_x$@PCN/rGO. (b–d) Reproduced with permission from Ref. 97. Copyright 2018, Wiley-VCH. (e–g) SEM and photograph image (inset) of Co$_4$N/CNF. (h) Cycling performance of CNF and Co$_4$N/CNF and corresponding discharge/charge profiles of (i) CNF and (j) Co$_4$N/CNF electrodes. (e–j) Reproduced with permission from Ref. 99. Copyright 2018, American Chemical Society.

the active sites and the adsorbed intermediate, leading to boosted ORR performance. Wang et al.[94] crafted a bifunctional catalyst containing Cu_3P coated with a carbon material co-doped with N and P via a direct pyrolysis of a new type of Cu-based metal organic framework (MOF), which possessed both excellent HER and ORR catalytic activities. While transition metal phosphides are relatively stable compared to transition metal oxides, their ORR performance is relatively poor in acidic condition.

In addition to oxides and phosphides, transition metal sulfides are also widely used in ORR catalysis. Traditionally, transition metal sulfides are widely used in HER, but they can be utilized in oxygen reduction catalysis through special preparation methods.[95,96] Niu et al.[97] prepared porous $g\text{-}C_3N_4$ by a top-down method, which was then compounded with reduced graphene oxide (rGO) and CoS_x to form $CoS_x@PCN/rGO$ with high performance in both ORR and OER under alkaline conditions (Fig. 5.11b). Owing to the excellent bifunctional performance of $CoS_x@PCN/rGO$, corresponding ZABs were successfully assembled, which was a sandwich structure as shown in Fig. 5.11c. The ZABs delivered superb stability over 43.8-hour and narrow potential gap (Fig. 5.11d).

In transition metal nitrides, the electronic energy level of N 2p orbitals is higher than that of O 2p orbitals. While metal nitrides can obtain high conductivity by reducing the band gap or overlapping the conductive band with the valence band, they tend to aggregate during pyrolysis process, resulting in the relatively inferior performance compared to their oxide counterpart. To this end, Dong et al.[98] supported cobalt-doped titanium nitride on nitrogen-doped rGO ($TiCoN_x$-N-rGO). The synergy between binary transition metal nitride and N-rGO improved the ORR activity of $TiCoN_x$-N-rGO under alkaline conditions, outperforming that of Pt/C. Similarly, Yoon et al.[99] prepared a brush-shaped Co_4N catalyst anchored on nitrogen-doped carbon nanofibers (Co_4N/CNF; Fig. 5.11e–g). The prepared Co_4N possessed an fcc structure with nitrogen atoms in the center of the unit cell, which exhibited higher conductivity than metal oxides. Swagelok-type cell was used for lithium-air battery test. Compared with CNF, Co_4N/CNF had lower overpotential and stronger stability (Fig. 5.11h–j). According to DFT calculation, LiO_2 were more easily adsorbed on the Co_4N/CNF surface, which effectively reduced the overpotential of the battery. Conversely, Li_2CO_3 and $LiRCO_3$ (R = alkyl) that cannot participate effectively during recharging/discharging cycles were formed in the case of CNF, resulting in higher cycling performance of Co_4N/CNF than that of CNF. Based on the previous reports, metal nitrides usually display excellent ORR performance in alkaline conditions with relatively disappointing performance under acidic conditions. Recently, Qiao et al.[100] reported that CrN loading on glucose-derived carbon materials achieved good ORR performance under acidic condition. This was attributed to the combination of carbon materials and CrN which greatly increased the electron transfer rate.

On the other hand, metal carbides such as Fe_3C,[101] WC,[102,103] Co_3C,[104] etc., are encapsulated in a nitrogen-doped carbon frame, which have attracted widespread attention due to their strong hydrogen adsorption capacity. Schuhmann et al.[105] reported a new assembly strategy to synthesize $Fe_3C@CNTs$ via using MOF as a template. Through electrochemical evaluation, $Fe_3C@CNTs$ delivered a potential of 0.83 (0.63) V at −3 mA cm⁻² in alkaline (acidic) media, which was comparable to

0.86 (0.75) V of 20% Pt/C. Carbides, especially iron carbides, exhibit limited ORR performance compared with currently reported single-atom catalysts under acidic conditions.[24,84]

Though TM-based carbon materials often suffer from insufficient stability and poor performance in acidic electrolytes, their ORR performance alkaline conditions is generally comparable to that of commercial Pt/C materials. Besides, because TMs have relatively good activity on OER, TM-based carbon materials are considered the most promising candidate of precious metal materials for applications in energy conversion devices such as zinc-air batteries or Li-O$_2$ batteries.

5.4.3 Transition metal-nitrogen-carbon-based carbon materials

In recent years, transition metal-nitrogen-carbon (TM-N-C) materials have been the most promising ORR catalysts that can replace Pt-based materials under acidic conditions. It is preferential to use a single atom to describe TM-N-C. Note that the single atom does not refer to an isolated single atom, but a state where the atomic dispersion level is a single atom (e.g., there is no M-M metal bond). Therefore, the biggest difference between TM-N-C and the other TM-based carbon materials in the previous chapters is that the former is dispersed in the entire carbon matrix by means of MN$_x$ at the atomic level, while the latter mostly contains nanoparticles coated with carbon materials.[9] In fact, the ORR performance of carbon materials under alkaline conditions is already comparable to commercial Pt/C. Unfortunately, although the cathode activation overvoltage in the alkaline fuel cell (AFC) is significantly lower than that of the acidic fuel cell at the same temperature, the fuel for AFC must be hydrogen as others may react with the alkaline electrolyte, and the electrolyte must be renewed regularly during operation due to the consumption from residual CO$_2$ in air.[106,107] These disadvantages greatly limit the practical application of AFC. On the other hand, it is well-known that PEMFC can only operate under acidic conditions where hydrogen ions are used as carriers. The invention of Nafion in the last century dispenses with the need for liquid electrolyte in PEMFC during operation, leading to much higher tolerance to gases such as CO$_2$ than that of AFC.[108] However, the excessive dependence of cathode catalysts on Pt-based materials, which are expensive and easily poisoned by CO and CH$_3$OH, limits the development of PEMFC.

Fortunately, recently there have been reports about the performance of TM-N-C catalysts comparable to Pt-based catalysts under acidic conditions. For example, Jiang et al.[109] prepared Fe-N-C based on MOF ligand strategy. Under the condition of 0.1 M HClO$_4$, E$_{1/2}$ reached 0.776 V, which was only 5 mV lower than that of the benchmark Pt/C. TM-N-C had good selectivity as well as high activity due to its single active site. In ORR reaction, traditional TM-N-C catalysis has a high selectivity to pathway of 4e$^-$. However, in recent studies, the selectivity can be controllably adjusted by changing the different coordination structures of TM-N-C.[110–112] It is generally believed that the catalytic activity of TM-N-C is mainly determined by the type of central atom and the chemical environment of the central atom as follows.

(i) The type of central atom. The type of TM atom greatly affects the activity, selectivity and stability of the catalytic reaction. This is because the d orbital of the metal interacts with the p orbital of oxygen, leading to different absorbed energy. For

example, Hossain et al.[113] adjusted the central atom type of TM-N-C (M = Ni, W, Co) to produce a great difference in catalytic activity. Among them, Co had the best ORR activity, followed by W, and Ni has the worst catalytic activity. Compared with Co, Ni, Cu, Mn metals, Fe-N-C has the highest catalytic activity for ORR. Zheng et al.[106] calculated the ORR activities of Mn, Fe, Co, Ni-based catalysts by DFT. Based on the relationship of the binding energy of pyrrole radicals to the different square-planar metal centers in Fig. 5.12a, FeN_4 delivered a strong tendency to form FeN_5 structure, suggesting a stable square-pyramid framework. As shown in Fig. 5.12b, both U_1 and U_2 on the five-coordination active sites were associated with stable pyrrole binding. FeN_4-Pyrrole (FeN_5) exhibited excellent performance for ORR catalysis with high U_1 value of 0.80 V and low U_2 value of 0.28 V, outperforming that of Pt (111).

Recently, Zitoro et al.[115,116] found that in the determination of the mechanism of the two active sites of Co-NC and Fe-NC, FeN_4C_{12} (−1.84 eV) had a stronger oxygen adsorption capacity and better performance than CoN_4C_{10} (−0.97 eV). Peng et al.[114] prepared TM-N-C catalysts via using melamine and polyaniline as carbon source and nitrogen source, respectively, together with different transition metals (i.e., Mn, Fe, Co, Ni, Cu). ORR performance tests were carried out both in 0.1 M $HClO_4$ and 0.1 M KOH solution (Fig. 5.12c–d). Among the various catalysts, Fe-N-C exhibited the best ORR activities regardless of acidic or alkaline conditions. It can be seen from the Fig. 5.12d that the order of ORR onset potential was Fe > Co > Cu > Mn > Ni, which matched well with that of their active N contents. However, catalytic activity is not the only indicator for determination of highly active catalysts as the stability also plays a vital role. Chen et al.[84] prepared different single-atom (Fe, Co) and

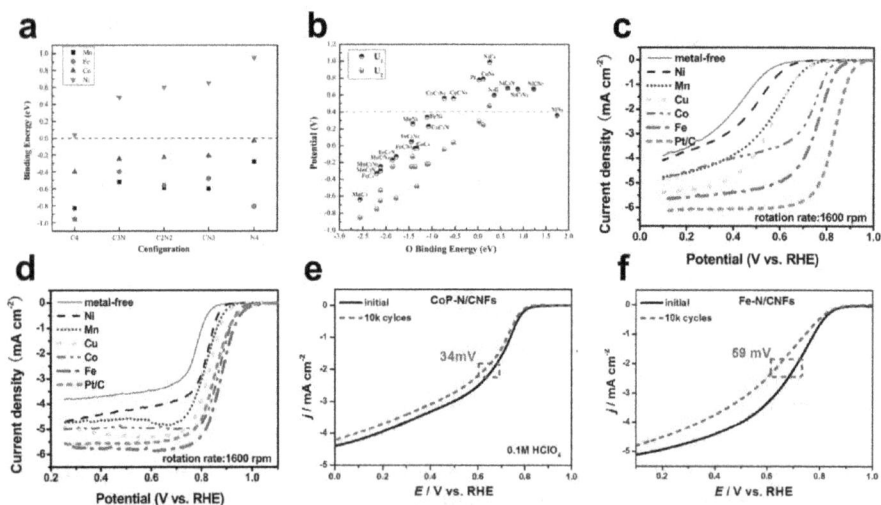

Figure 5.12. (a) The relationship between the binding energy of pyrrole radicals on different square-planar metal centers, (b) U_1 and U_2 values on different square-planar active centers over the oxygen binding energy. ORR polarization curves in O_2-saturated 0.1 M. (a and b) Reproduced with permission from Ref. 106. Copyright 2016, Elsevier. (c) $HClO_4$ and (d) KOH of varied electrodes. (c and d) Reproduced with permission from Ref. 114. Copyright 2014, American Chemical Society. (e, f) LSVs before and after a 10 000-cycle ADT of ORR in 0.1 M $HClO_4$ of varied electrodes. (e and f) Reproduced with permission from Ref. 84. Copyright 2017, American Chemical Society.

N co-doped carbon nanofibers (Fe-N/CNFs, Co-N/CNFs) catalysts. While the half-wave potential of Fe-N/CNFs was more positive than that of Co-N/CNFs, Co-N/CNFs maintained a higher level than Fe-N/CNFs after 10,000 cycles of accelerated durability test (ADT). This revealed that Fe-N/CNFs were less stable (Fig. 5.12e–f) due to the oxidation of the active site by the Fenton effect,[117] which was similar to the recent reports.[118,119] Xie et al. [118] prepared a highly dispersed Co-N-C catalyst from zeolitic imidazolate framework (ZIF-8). Its catalytic oxygen reduction reaction activity was equivalent to that of a similarly synthesized Fe-N-C catalyst, but its durability was enhanced by four times. Degradation mechanism studies suggested Co-N-C was more stable due to the improved resistance to catalyst demetallation and lower activity for Fenton reactions. The TM-N-C catalyst can be situated on the top of the "volcanic curve" by adjusting the coordination method between non-metal and metal elements. For example, Zhang et al.[120] prepared a high-performance CoPt-N-C by coupling Co and Pt, showing 90 mV positive than that of Pt/C in half-wave potential. Bimetal atom catalysts are an attractive strategy, and the synergistic effect of adjacent metal atoms can further improve the activity of single atom catalysts.[121,122] Wang et al.[123] constructed a Fe, Co dual-site catalysts by host-guest strategy. As shown in Fig. 5.13, the presence of a small amount of iron in

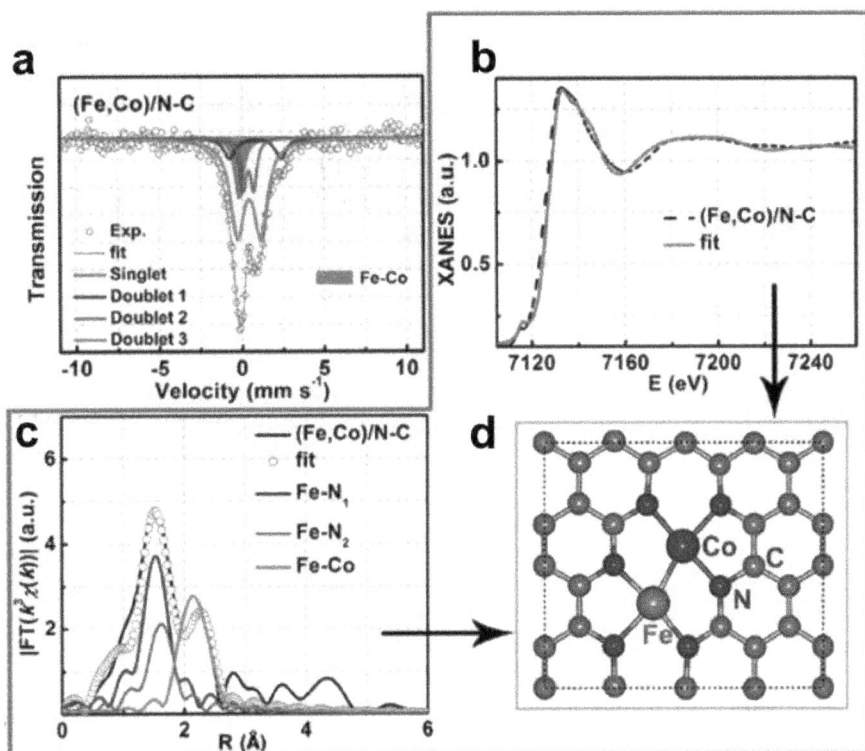

Figure 5.13. (a) Experimental Fe Mössbauer transmission spectra, (b) K-edge X-ray absorption near-edge structure (XANES) spectrum of (Fe,Co)/N-C, (c) Corresponding Fe K-edge EXAFS fittings of (Fe,Co)/N-C, (d) Schematic diagram of architectures of Fe-Co dual-sites. (a–d) Reproduced with permission from Ref. 123. Copyright 2017, American Chemical Society.

(Fe,Co)/N-C proved the existence of Fe-Co bonds. The structure of the Fe, Co bimetallic atom site structure in Fig. 5.13d was obtained by the structural fitting of EXAFS assisted by DFT. In summary, advanced characterization techniques and theoretical calculations have identified MN_x as the active site of ORR electrocatalysis. Related studies have shown that Fe as a central atom has the best performance compared to other transition metals. However, the carbon corrosion caused by the inherent Fenton effect of Fe metal is an urgent problem to be solved. The development of bimetallic atomic sites has shown the feasibility for the further development of high-efficiency PEMFC catalysts.

(ii) Modification of the chemical environment of the central atom. The coordination of metal and the chemical environment of the central atom are the main aspects that affect the performance of the active sites because of the modified adsorption and desorption energies toward ORR intermediates.[124] The chemical environment of N shows significant effects to the TM center. Owing to the development of advanced characterization, the understanding of the TM-N-C structure is further identified, which originated from the porphyrin structure.[125] Most previous work suggested that MN_4 coordination was the active center, but recent studies have found that 4-coordination may not be the optimal structure. Sun et al.[126] embedded CoN_2 and CoN_4 into graphene, and found that the dissociation energy of O_2 and *OOH on the CoN_2 site was lower than that of CoN_4 site using theoretical calculations. Recently, Chen et al.[127] reported a highly efficient ORR W-N_5-C catalyst. In ORR, the central atom of TM-N-C catalyst is generally 3d metal such Fe, Co, Ni, etc., while WNC is generally considered to have poor performance. The reported W-N_5-C catalyst not only had the catalytic ability comparable to that of Pt/C under alkaline conditions, but also exhibited an $E_{1/2}$ of 0.77 V under acidic conditions. Compared with WN_3 and WN_4, WN_5 exhibited superior ORR performance under acidic and alkaline conditions, which was highly consistent with the active volcano diagrams of different tungsten-based ORR electrocatalysts via theoretical simulation. The chemical position of N in TM-N-C is determined by the carbon matrix. Recently, Wan et al.[128] reviewed the structure design of TM-N-C in ORR catalysis, and constructed Fe-N_x-C_y structures according to different coordination numbers and coordination atom environments. In addition to N atom, other light elements (e.g., O) can also coordinate with the central atom. Zitolo et al.[115] judged the catalytic structure of Fe-N-C and found $Fe_{0.5}$ with an FeN_4 moiety possessing one or two oxygen atoms in the axial direction. Inspired by this, Yuan et al.[24] developed a catalyst with iron active sites dual-coordinated by phosphorus and nitrogen (denoted Fe-N/P-C; Fig. 5.14). Fe-N/P-C displayed high activity attributed to the inherent active site of Fe-N_3P. Additionally, the pore structure of carbon based-materials also has a huge impact on catalytic activity. The formation of TM-N-C catalysts such as TM-N_2, TM-N_4C_8, TM-N_4C_{10}, TM-N_4C_{12} is affected by pores. Moreover, pores are particularly important in terms of the number of active site and transport of electron and substance under membrane electrode assembly testing.[129]

It is notable that the formation of TM-N-C structure is still uncertain. Except a few methods, the synthesis of TM-N-C often requires high-temperature pyrolysis. Single atoms may aggregate to form nanoparticles during the high-temperature

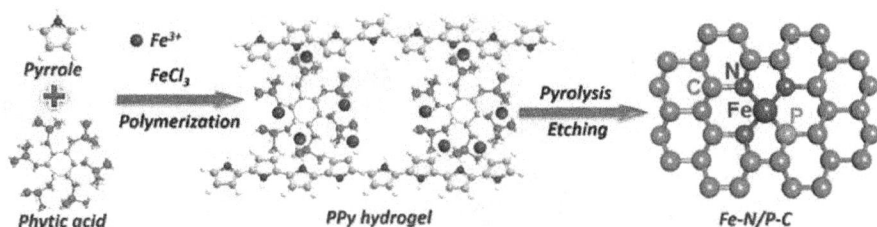

Figure 5.14. Schematic of the synthetic process of the Fe-N/P-C catalyst. Reproduced with permission from Ref. 24. Copyright 2020, American Chemical Society.

pyrolysis process, which reduces the density of active sites, and different metals may affect the pores of carbon materials during the pyrolysis process. At present, it is highly desirable to explore and identify conditions conducive to the formation of single-atom catalysts via *in situ/ex situ* characterization under high-temperature environments, which is difficult though. On the other hand, for MOF-derived carbon materials, relatively high pyrolysis temperatures have been proven to form single-atom structures.[130] This should be attributed to the fact that metal Zn hindered metal agglomeration and the abundant microporous channels limited the migration of metals.[9, 109] TM-N-C catalysts exhibit considerable ORR catalytic activity, although it is still controversial whether metal species directly participate in the formation of active centers. Exploration of the structure-activity relationship of TM-N-C catalyst activity and composition, electronic configuration, surface morphology, utilization of theoretical calculation methods, and development of technologies to increase the activity density are the main directions of future TM-N-C catalyst research.

5.5 Metal-free carbon material catalyst

The catalytic activity of the original carbon for oxygen reduction is not high due to their inert feature to oxygen adsorption and activation. In this regard, intrinsic defect engineering and heteroatom doping can tune the charge distribution and electron spin in sp2 carbon, thereby changing the adsorption and activation of oxygen. Additionally, hierarchical porosity facilitates the mass transfer via exposure of more active sites. These strategies are conducive to excellent catalytic performance, which will be highlighted in this section.

5.5.1 Intrinsic defect engineering

With the development of new catalytic materials and catalyst modification strategies, the research on the defects of catalytic materials has become an important direction in the field of catalysis. The C-C bond forming a planar six-atom ring structure renders the material a variety of defect types including adsorption atom defects, vacancy defects, edge defects, stone-Wales defects and other topological defects.[131] Jin et al.[25] prepared graphene carbon quantum dots (GQDs) supported by graphene nanoribbons (Fig. 5.15) via one-step method of hexabromobenzene and toluene. The resultant carbon materials without heteroatom doping showed superb catalytic activity over Pt/C in 0.1 M KOH solution.

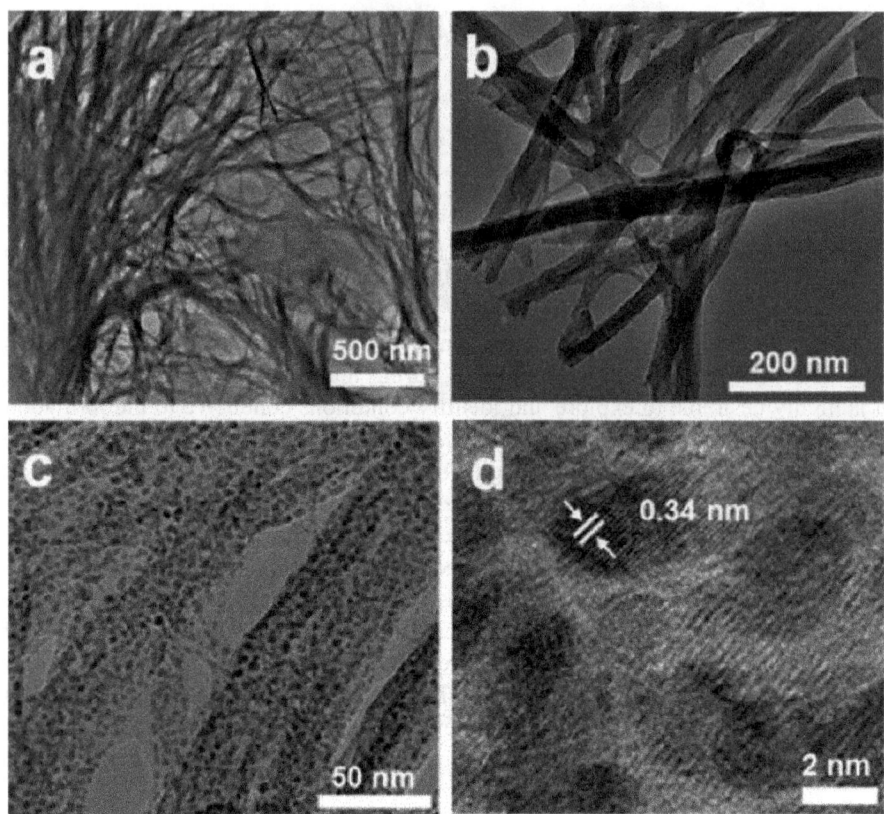

Figure 5.15. SEM image of GQDs supported by graphene nanoribbons. Reproduced with permission from Ref. 25. Copyright 2015, American Chemical Society.

Jiang et al.[132] reported the high ORR activity of the carbon nanocages with a lot of defects including pentagon, hole, zigzag edge and armchair edge as shown in Fig. 5.16a. Specifically, pentagonal defect was modeled by graphene clusters with pentagonal rings in the center and the hole defect was a hexagonal hole of about 0.7 nm in the middle of the graphene lattice while edge defects were simulated by jagged or armchair-shaped graphene nanoribbons. Their results revealed that edge defects and topological defects were crucial for improving the ORR electrocatalytic activity of metal-free nanocarbon materials. In addition, DFT demonstrated showed that the N-doped carbon has a higher ΔE than that of the sp2 graphene. In the past, specific atomic arrangements and configurations were often determined by simulation calculations without intuitive experimental data. With the advancement of technology, we can visually observe the inherent defects of carbon materials under the microscope, especially the rapid development of aberration-corrected high-resolution transmission electron microscope (HR-TEM). In this regard, Jia et al.[133] observed various structural defects (e.g., pentagon, heptagon and octagon) in different combinations at the proximal end of the lattice vacancies (Fig. 5.16b). To further understand the catalytic mechanism of intrinsic defect, a series of DFT calculation

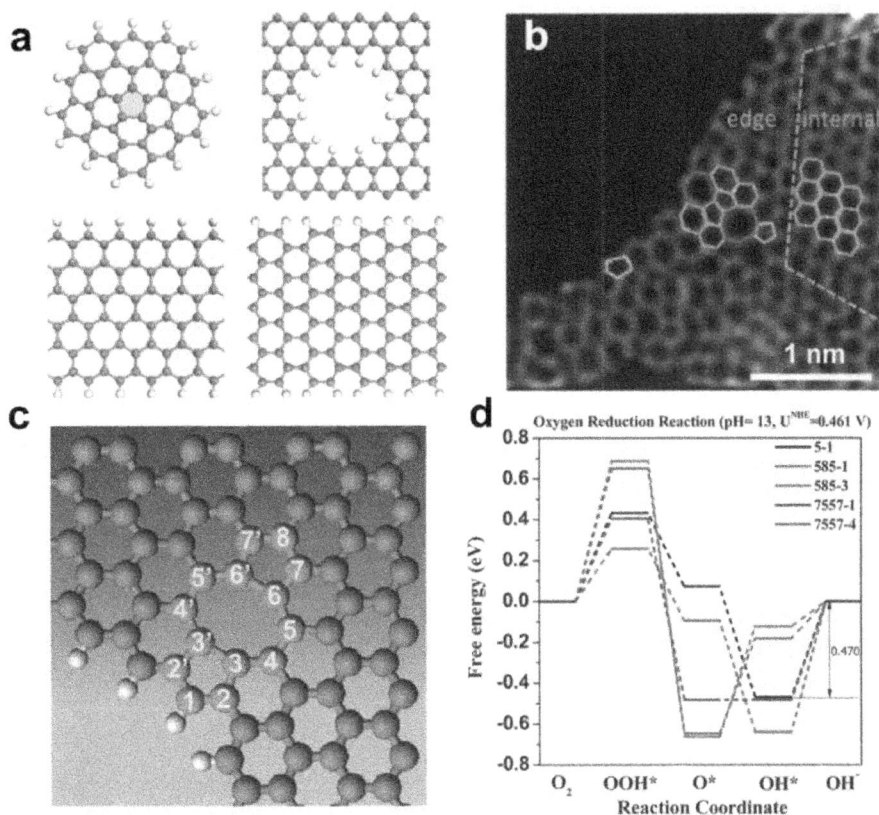

Figure 5.16. (a) Defect models. Reproduced with permission from Ref. 132. Copyright 2015, American Chemical Society. (b) High angle annular dark field image and (c, d) mechanism study for ORR of carbon defects. (b–d) Reproduced with permission from Ref. 133. Copyright 2016, Wiley-VCH.

was conducted to describe the catalytically active sites for the ORR process. The analysis of the frontier molecular orbitals showed that the highest occupied molecular orbital (HOMO) and the lowest unoccupied molecular orbital (LUMO) are mainly distributed on the edge atoms of the graphene holes. The introduction of edge pentagons, 585 defects and 7557 defects (the numbers represent the C atom number within a C ring) contributed to the HOMO/LUMO orbits. Due to the high correlation between the catalytic reaction and the distribution of HOMO/LUMO orbits, edge pentagons, 585 defects and 7557 defects were active for electrocatalytic processes (e.g., 585 defects marked in green in Fig. 5.16c). Finally, the minimum energy path of ORR was calculated, indicating that the most active moieties for ORR were edge 5-1 (Fig. 5.16d).

Similar to the above work, graphene nanoribbons (GNRs) are rich in edge defects. But they are relatively easy to stack, thus burying the active center. Recently, Xue et al.[134] partially unziped multi-walled carbon nanotubes (MWCNTs) to form MWCNTs supported GNRs (GNR@CNT; Fig. 5.17). By opening part of the MWCNT, the rest of the MWCNT acted as the support skeleton. The GNR@CNT with zigzag carbon delivered a peak areal power density of $0.161\,W\,cm^{-2}$ and a peak

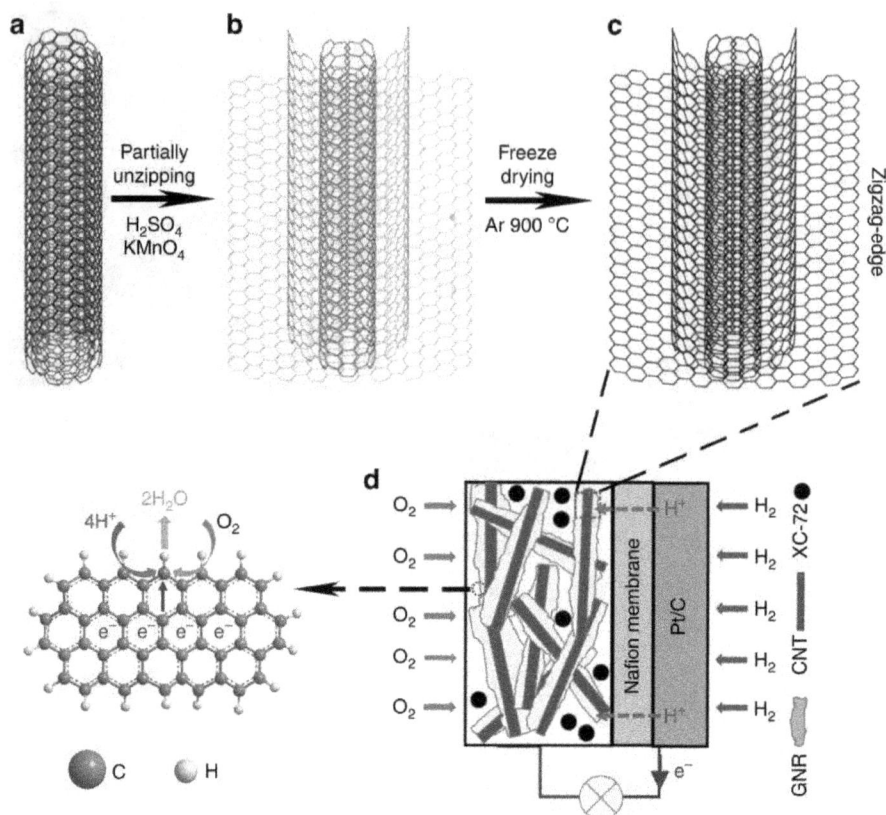

Figure 5.17. Schematic of synthetic strategy and PEMFC application of GNR@CNT. Reproduced with permission from Ref. 134. Copyright 2018, Nature Publishing Group.

mass power density of 520 Wg^{-1} in PEMFC, outperforming most non-precious-metal systems. DFT simulation together with experimental results suggested that the zigzag carbon atom was the most active site for ORR within different types of carbon defects on GNRs in acid conditions.

5.5.2 Heteroatom doping

Another effective way to improve the catalytic activity of ORR is heteroatom doping. In the past, N doping has been widely used in carbon-based catalysts.[135] Obviously, N-doped carbon has two advantages for ORR: (1) N and C have similar atomic radii, which will not cause lattice mismatch; (2) N has one more electron than C, which is conducive to reactions that require electrons.[136] Therefore, N should be an effective dopant to modify sp2 carbon as an ORR electrocatalyst. Indeed, the performance of N-doped carbon is usually better than most other non-metal elements doped carbons prepared by similar routes.[137] In addition, in the case of carbon doped with multiple elements, N is usually the essential one for high activity.[135] Therefore, N doping is the most widely studied regulation mechanism of carbon-based metal-free electrocatalysts while other monodoped ones with nonmetal elements (e.g., S[138]) are relatively less reported. The key factor affecting the catalytic activity of N-doped carbon materials is the N type, including pyridinic-N, pyrrolic-N, graphitic-N and oxidized-N,[139] In addition to the above four types of N, a new type of sp-hybridized N has recently been reported.[140] While N doping represents an effective means to improve the ORR performance under alkaline conditions, the specific positions of N at the carbon plane and their corresponding attribute to the catalytic activity are often controversial.[134] This is because it is impossible to precisely control the content and location of pyridinic N, pyrrolic N, and graphitic N formed by N doping from an experimental point of view as various types of N are often formed after thermal treatment. It is thus difficult to guarantee whether the carbon substrate is changed even though the N content of different types is the same, resulting in debatable comparison in general. Two recent reports can be considered as examples. The graphene catalyst model study proposed by Guo et al.[141] showed that the carbon atom with Lewis basicity adjacent to pyridinic N was the active site of ORR. However, recently Haque et al.[142] controlled the catalyst with or without graphitic N by heating the solution at different temperatures for ORR, which clearly showed that the influence of graphitic N on the catalytic activity was huge while pyridinic and pyrrolic N were ineffective.

On the basis of the improved ORR performance by N doping, heteroatom (e.g., N/P[143], N/S[144, 145], N/B[146] and N/F[147]) co-doping has also been widely reported as a very promising way to improve the adsorption and activation of oxygen. However, most doping activity test conditions are mostly in KOH solution with only a few performed in acid. Recently N/S co-doped carbon has shown excellent ORR performance. For example, Yang et al.[145] found that N/S co-doped carbon not only possessed a higher half-wave potential than Pt/C under alkaline conditions, but also demonstrated unprecedented ORR activities under acidic conditions (Fig. 5.18). Several recent studies also indicated that N/S co-doping is an effective means to improve ORR performance under acidic conditions.[144, 148,149]

Regarding the improvement mechanism of ORR performance in acidic condition via S doping for N/S co-doped carbon materals, it is generally believed that thiophene-S serves as the active site.[150] Li et al.[150] found that through thermal treatment at different temperatures (from 700°C to 1000°C), the oxidized-S moieties were converted to thiophene-S which acted as active sites under acidic conditions, leading to better catalytic activity. First-principal simulations were carried out for S-doped graphene (S-G), S-defect graphene (S-D-G), and N-modified S-defect graphene (N-S-D-G) (Fig. 5.19). Since both S and N were electronegative, the electrons of adjacent C atoms were redistributed. Based on the calculation, the transfer of charge followed the order of S-G ($0.04|e|$), S-D-G ($0.09|e|$) and N-S-D-G (0.18 and $0.28\ |e|$), which matched well with their ORR performance.

Considerable improvements of N and other elements' (e.g., P[143, 151] B,[152] F[147]) co-doped carbon materials have been made in recent years. Zhang et al.[151] prepared an N/P co-doped mesoporous carbon catalyst with a high onset potential of $0.94\ V_{RHE}$ and a half-wave potential of $0.85\ V_{RHE}$ using polyaniline hydrogel as precursors. Han et al.[146] synthesized N/B co-doped hollow carbon materials by simply chlorinating the mixture of $Ti(C_xN_{1-x})$ and TiB_2 in one step and adjusting the content of nitrogen and boron by controlling the composition of $Ti(C_xN_{1-x})$. The ORR activity of N/B co-doped hollow carbon materials was measured using a rotating

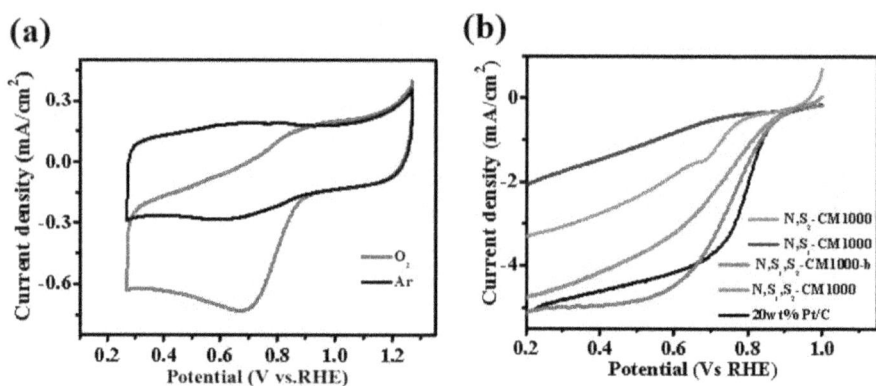

Figure 5.18. CV and LSV of N and S co-doped porous carbon materials in acidic media. Reproduced with permission from Ref. 145. Copyright 2018, Elsevier.

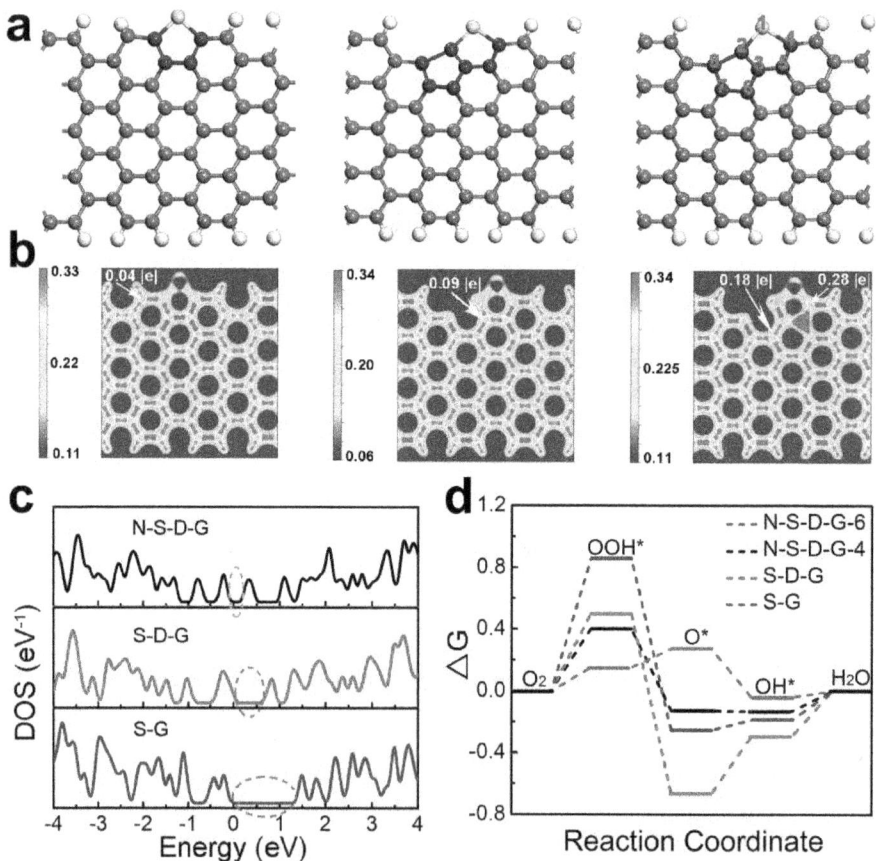

Figure 5.19. Mechanism study of N, S co-doped carbon for ORR. (a) The optimized structures; (b) the simulated electron density; (c) calculated DOS; (d) free-energy diagram. Reproduced with permission from Ref. 150. Copyright 2018, Elsevier.

ring disk electrode (RRDE) in O_2-saturated 0.1 M KOH solution. The corresponding onset potential was 0.89 V_{RHE}, which was slightly lower than that of the commercial Pt/C catalyst. Akula et al.[147] introduced N and F heteroatoms into the graphitic carbon nanofibers, which enlarged the lattice spacing of the graphite flakes with reduced charge-transfer resistance and structural modulation, thus facilitating the faster diffusion of oxygen in the catalytic center for improved ORR catalytic activity.

O doping is often difficult to control during the experiment as it is easy to introduce oxygen-containing group in the high temperature treatment process. It is reported that O doping facilitated the $2e^-$ pathway of ORR process. Though the product (i.e., H_2O_2) of $2e^-$ pathway possesses strong oxidizing properties that can directly or indirectly destroy the catalyst,[24,49] the production of hydrogen peroxide through ORR is also a very innovative method as H_2O_2 is a high-value chemical. Wang et al.[153] disclosed an inexpensive metal-free catalyst with oxygen-containing groups synthesized from commercial carbon black (CB) by plasma method and produces hydrogen peroxide with 100% Faraday efficiency (at 0.60 V_{RHE}). The catalyst had a higher onset potential of 0.80 V_{RHE} and a mass activity of 300 A g^{-1}. The study of Dong et al.[154] further confirmed that oxygen-containing groups in carbon-based materials can change the hydrophilicity and hydrophobicity, and thus adjust the selectivity. As the types and contents of doping elements play a significant role in ORR catalysis, more strategies are needed to precisely control the co-doped structure.

5.5.3 Hierarchical porosity

Hierarchical porous carbon can expose enough active sites and improve the mass transfer in the three-phase reaction, thereby enhancing the catalytic activity. Pores are generally divided into micropores, mesopores, and macropores. It is generally believed that the presence of micropores can increase the specific surface area with fully exposed active sites while mesopores and macropores mainly play an important role in mass transfer.[155] This is because water and oxygen cannot quickly reach the active sites through the micropores to fully access the active sites. ORR only occurs in the three-phase boundary region where electrons, protons, oxygen and water exchange in fuel cells. Proper porosity and micro-nano structure are conducive to both the penetration of water and the transport of ions and oxygen. So far, a lot of efforts have been devoted to the innovation of active sites, and great progress has been made. It has been reported that pores are made by pore-forming agent, hard template, etching, etc. Among them, high temperature pyrolysis with pore-forming agents such as KOH,[156] $ZnCl_2$[157] is the most used and mature method. However, the above reagents are corrosive, and multi-step processing is time-consuming. In addition to the above-mentioned traditional ones, some relatively green, and environmentally friendly pore-forming agents have also been developed, such as K_2FeO_4,[158] $[ZnCO_3]_2 \cdot [Zn(OH)_2]_3$.[159] In most cases, an inexpensive biomass material is active pore-forming agent, sufficient to produce micropores. At present, it is not completely clear how these pore-forming agents produce a number of micropores, but it is certain that the alkali or zinc salt can react with C during the high temperature pyrolysis process. Consequently, the yielding continuous escape of gas facilitates the formation of pores. Adjusting the mesoporous structure through hard template

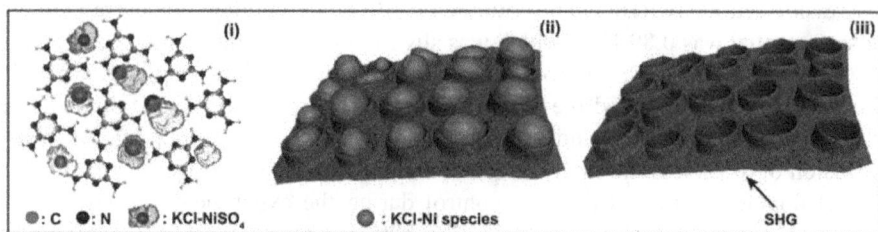

Figure 5.20. Schematic route to stereoscopic holes over the graphitic surface (SHG). Reproduced with permission from Ref. 161. Copyright 2017, Wiley-VCH.

for improved transfer efficiency is also a method to enhance ORR performance.[160] Hu et al.[161] constructed three-dimensional porous graphite sheet and improved the ORR performance through N and S co-doping (Fig. 5.20). These three-dimensional holes provided abundant active sites on the surface for electrochemical reactions.

Template strategy represents another promising route to porous carbon materials. Qian et al.[162] prepared an efficient ORR catalyst containing carbon nanomaterial based on morphology control by NaCl recrystallization solid pyrolysis method. The template effect of molten NaCl connected the ZIF-8 particles to the network of carbon. Through the evaporation and recrystallization of the NaCl/ZIF-8 aqueous solution mixture, ZIF-8 was solidified in the inorganic salt NaCl. The encapsulation effect of the inorganic salt was used to solve problems in the traditional direct carbonization process such as severe disappearance of the active site and structural collapse. The convenient removal of NaCl by water subsequently also avoided the tedious procedure in the conventional template method.

One of the common strategies in nanoengineering to create enough micropores is surface etching. Pyrolysis under NH_3 is an effective method.[163] Carbon defects are generated by NH_3 etching, which can effectively increase the specific surface area. In addition, during the NH_3 activation process, part of the nitrogen is doped into the carbon structure to form highly active pyridinic and graphitic nitrogen, which is conducive to the ORR reaction.[13]

5.6 Conclusion

Development of an alternative ORR catalyst to platinum in fuel cells has become a worldwide goal for decades. In this chapter, we have conducted some discussions on different carbon-based materials. While carbon-based materials have rich prospects as ORR electrocatalysts, there are also challenges. At present, though the TM-N-C catalyst exhibits comparable performance to Pt/C catalyst during half-cell characterization (a three-electrode system), its performance under the membrane electrode assembly test is still far inferior to that of the Pt/C. The design and synthesis of efficient and stable ORR electrocatalysts are still the primary task for the advancement of related energy conversion technologies in fuel cells. For the industrial preparation of carbon material catalysts, reliable quality control still needs to be emphasized to ensure the stability of fuel cell performance. In addition, to apply carbon-based materials in membrane electrode assembly testing, it is necessary to study the deactivation mechanism. Finally, the screening of advanced carbon-based catalysts through DFT must be based on a reasonable model, of which the screening time can be significantly reduced via the use of machine learning and other means.

Acknowledgements

The authors would like to thank the financial support from the National Natural Science Foundation of China (22109120, 62104170, 52072273, and 51872209), Zhejiang Provincial Natural Science Foundation of China (LQ21B030002), Basic Science and Technology Research Project of Wenzhou (G2020007), and Zhejiang Provincial Special Support Program for High-level Talents (2019R52042).

References

[1]   Pan, S., J. Li, Z. Wen, R. Lu, Q. Zhang, H. Jin, L. Zhang, Y. Chen and S. Wang. 2021. Halide Perovskite materials for photo(Electro)chemical applications: Dimensionality, heterojunction, and performance. *Adv. Energy Mater.*, 2004002.

[2]   Zhang, M., M. Ye, W. Wang, C. Ma, S. Wang, Q. Liu, T. Lian, J. Huang and Z. Lin. 2020. Synergistic cascade carrier extraction via dual interfacial positioning of ambipolar black phosphorene for high-efficiency perovskite solar cells. *Adv. Mater.*, 32: e2000999.

[3]   Wang, B., J. Iocozzia, M. Zhang, M. Ye, S. Yan, H. Jin, S. Wang, Z. Zou and Z. Lin. 2019. The charge carrier dynamics, efficiency and stability of two-dimensional material-based perovskite solar cells. *Chem. Soc. Rev.*, 48: 4854–4891.

[4]   Jiang, K., J. Wang, F. Wu, Q. Xue, Q. Yao, J. Zhang, Y. Chen, G. Zhang, Z. Zhu, H. Yan, L. Zhu and H. L. Yip. 2020. Dopant-free organic hole-transporting material for efficient and stable inverted all-inorganic and hybrid perovskite solar cells. *Adv. Mater.*, 32: e1908011.

[5]   Zhang, X., S. Pan, H. Song, W. Guo, S. Zhao, G. Chen, Q. Zhang, H. Jin, L. Zhang, Y. Chen and S. Wang. 2021. Polymer-inorganic thermoelectric nanomaterials: electrical properties, interfacial chemistry engineering, and devices. *Front. Chem.*, 9: 677821.

[6]   Zhao,. M., T. Chen, B. He, X. Hu, J. Huang, P. Yi, Y. Wang, Y. Chen, Z. Li and X. Liu. 2020 Photothermal effect-enhanced photoelectrochemical water splitting of a $BiVO_4$ photoanode modified with dual-functional polyaniline. *J. Mater. Chem. A*, 8: 15976–15983.

[7]   Hu, X., J. Huang, F. Zhao, P. Yi, B. He, Y. Wang, T. Chen, Y. Chen, Z. Li and X. Liu. 2020. Photothermal effect of carbon quantum dots enhanced photoelectrochemical water splitting of hematite photoanodes. *J. Mater. Chem. A*, 8: 14915–14920.

[8]   Huang, D., Y. Luo, S. Li, L. Liao, Y. Li, H. Chen and J. Ye. 2020. Recent advances in tuning the electronic structures of atomically dispersed M–N–C materials for efficient gas-involving electrocatalysis. *Mater. Horiz.*, 7: 970–986.

[9]   He, Y., S. Liu, C. Priest, Q. Shi and G. Wu. 2020. Atomically dispersed metal-nitrogen-carbon catalysts for fuel cells: Advances in catalyst design, electrode performance, and durability improvement. *Chem. Soc. Rev.*, 49: 3484–3524.

[10]   Debe, M. K. 2012. Electrocatalyst approaches and challenges for automotive fuel cells. *Nature*, 486: 43–51.

[11]   Yoshida, T. and K. Kojima. 2015. Toyota MIRAI fuel cell vehicle and progress toward a future hydrogen society. *Interface Magazine*, 24: 45–49.

[12]   Pan, J., Y. Y. Xu, H. Yang, Z. Dong, H. Liu and B. Y. Xia. 2018. Advanced architectures and relatives of air electrodes in Zn-air batteries. *Adv. Sci.*, 5.

[13]   Zhou, T., N. Zhang, C. Wu and Y. Xie. 2020. Surface/interface nanoengineering for rechargeable Zn-air batteries. *Energy Environ. Sci.*, 13: 1132–1153.

[14]   Wang, H.-F. and Q. Xu. 2019. Materials design for rechargeable metal-air batteries. *Matter*, 1: 565–595.

[15]   Wu, J., B. Liu, X. Fan, J. Ding, X. Han, Y. Deng, W. Hu and C. Zhong. 2020. Carbon-based cathode materials for rechargeable zinc-air batteries: From current collectors to bifunctional integrated air electrodes. *Carbon Energy*, 2: 370–386.

[16]   Wang, X., Z. Zhu, L. Chai, J. Ding, L. Zhong, A. Dong, T.-T. Li, Y. Hu, J. Qian and S. Huang. 2019. Generally transform 3-dimensional In-based metal-organic frameworks into 2-dimensional Co,N-doped carbon nanosheets for Zn-air battery. *J. Power Sources*, 440: 227158.

[17]   Zhu, J., M. Xiao, G. Li, S. Li, J. Zhang, G. Liu, L. Ma, T. Wu, J. Lu, A. Yu, D. Su, H. Jin, S. Wang and Z. Chen. 2019. A triphasic bifunctional oxygen electrocatalyst with tunable and synergetic interfacial structure for rechargeable Zn-air batteries. *Adv. Energy Mater.*, 10: 1903003.

[18]   Ren, D., J. Ying, M. Xiao, Y. P. Deng, J. Ou, J. Zhu, G. Liu, Y. Pei, S. Li, A. M. Jauhar, H. Jin, S. Wang, D. Su, A. Yu and Z. Chen. 2019. Hierarchically porous multimetal-based carbon nanorod hybrid as an efficient oxygen catalyst for rechargeable zinc-air batteries. *Adv. Funct. Mater.*, 30: 1908167.

[19]   Bing, Y., H. Liu, L. Zhang, D. Ghosh and J. Zhang. 2010. Nanostructured Pt-alloy electrocatalysts for PEM fuel cell oxygen reduction reaction. *Chem. Soc. Rev.*, 39: 2184–2202.

[20]   Li, M., Z. Zhao, T. Cheng, A. Fortunelli, C. Y. Chen, R. Yu, Q. Zhang, L. Gu, B. V. Merinov, Z. Lin, E. Zhu, T. Yu, Q. Jia, J. Guo, L. Zhang, W. A. Goddard, 3rd, Y. Huang and X. Duan. 2016. Ultrafine jagged platinum nanowires enable ultrahigh mass activity for the oxygen reduction reaction. *Science*, 354: 1414–1419.

[21]   Yuan, Y. and J. Lu. 2019. Demanding energy from carbon. *Carbon Energy*, 1: 8–12.

[22]   Hu, C. and L. Dai. 2019. Doping of carbon materials for metal-free electrocatalysis. *Adv. Mater.*, 31: e1804672.

[23]   Guo, D., H. Wei, X. Chen, M. Liu, F. Ding, Z. Yang, Y. Yang, S. Wang, K. Yang and S. Huang. 2017. 3D hierarchical nitrogen-doped carbon nanoflower derived from chitosan for efficient electrocatalytic oxygen reduction and high performance lithium-sulfur batteries. *J. Mater. Chem. A*, 5: 18193–18206.

[24]   Yuan, K., D. Lutzenkirchen-Hecht, L. Li, L. Shuai, Y. Li, R. Cao, M. Qiu, X. Zhuang, M. K. H. Leung, Y. Chen and U. Scherf. 2020. Boosting oxygen reduction of single iron active sites via geometric and electronic engineering: Nitrogen and phosphorus dual coordination. *J. Am. Chem. Soc.*, 142: 2404–2412.

[25]   Jin, H., H. Huang, Y. He, X. Feng, S. Wang, L. Dai and J. Wang. 2015. Graphene quantum dots supported by graphene nanoribbons with ultrahigh electrocatalytic performance for oxygen reduction. *J. Am. Chem. Soc.*, 137: 7588–7591.

[26]   He, B., Y. Wang, Q. Zhai, P. Qiu, G. Dong, X. Liu, Y. Chen and Z. Li. 2020. From polymeric carbon nitride to carbon materials: Extended application to electrochemical energy conversion and storage. *Nanoscale*, 12: 8636–8646.

[27]   Chai, L., L. Zhang, X. Wang, Z. Hu, Y. Xu, T.-T. Li, Y. Hu, J. Qian and S. Huang. 2020. Cube-shaped metal-nitrogen–carbon derived from metal-ammonia complex-impregnated metal-organic framework for highly efficient oxygen reduction reaction. *Carbon*, 158: 719–727.

[28]   Zhang, H., J. Li, S. Xi, Y. Du, X. Hai, J. Wang, H. Xu, G. Wu, J. Zhang, J. Lu and J. Wang. 2019. A graphene-supported single-atom FeN_5 catalytic site for efficient electrochemical CO_2 reduction. *Angew. Chem.*, 58: 14871–14876.

[29]   Nguyen, D. L. T., Y. Kim, Y. J. Hwang and D. H. Won. 2019 Progress in development of electrocatalyst for CO_2 conversion to selective CO production. *Carbon Energy*, 2: 72–98.

[30]   Yang, C.-H., F. Nosheen and Z.-C. Zhang. 2020. Recent progress in structural modulation of metal nanomaterials for electrocatalytic CO2 reduction. *Rare Met.*, 40: 1412–1430.

[31]   Zhang, M., Q. Dai, H. Zheng, M. Chen and L. Dai. 2018. Novel MOF-Derived Co@N-C bifunctional catalysts for highly efficient Zn-air batteries and water splitting. *Adv. Mater.*, 30: 1705431.

[32]   Zou, H., G. Li, L. Duan, Z. Kou and J. Wang. 2019. *In situ* coupled amorphous cobalt nitride with nitrogen-doped graphene aerogel as a trifunctional electrocatalyst towards Zn-air battery deriven full water splitting. *Appl. Catal., B*, 259: 118100.

[33]   Ding, W.-L., Y.-H. Cao, H. Liu, A.-X. Wang, C.-J. Zhang and X.-R. Zheng. 2020. *In situ* growth of $NiSe@Co_{0.85}Se$ heterointerface structure with electronic modulation on nickel foam for overall water splitting. *Rare Met.*, 40: 1373–1382.

[34]   Pan, Z.-Y., Z. Tang, Y.-Z. Zhan and D. Sun. 2020. Three-dimensional porous $CoNiO_2$@reduced graphene oxide nanosheet arrays/nickel foam as a highly efficient bifunctional electrocatalyst for overall water splitting. *Tungsten*, 2: 390–402.

[35] Zhao, S., X. Lu, L. Wang, J. Gale and R. Amal. 2019. Carbon-based metal-free catalysts for electrocatalytic reduction of nitrogen for synthesis of ammonia at ambient conditions. *Adv. Mater.*, 31: e1805367.

[36] Zhao, C., S. Zhang, M. Han, X. Zhang, Y. Liu, W. Li, C. Chen, G. Wang, H. Zhang and H. Zhao. 2019. Ambient electrosynthesis of ammonia on a biomass-derived nitrogen-doped porous carbon electrocatalyst: Contribution of pyridinic nitrogen. *ACS Energy Lett.*, 4: 377–383.

[37] Wang, J., H. Huang, P. Wang, S. Wang and J. Li. 2021. N, S synergistic effect in hierarchical porous carbon for enhanced NRR performance. *Carbon*, 179: 358–364.

[38] Beall, C. E., E. Fabbri and T. J. Schmidt. 2021. Perovskite oxide based electrodes for the oxygen reduction and evolution reactions: The underlying mechanism. *ACS Catal.*, 11: 3094–3114.

[39] Xiao, Y. and W. Zhang. 2020. High-throughput calculation investigations on the electrocatalytic activity of codoped single metal-nitrogen embedded in graphene for ORR mechanism. *Electrocatalysis*, 11: 393–404.

[40] Singh, S. K., K. Takeyasu and J. Nakamura. 2019. Active sites and mechanism of oxygen reduction reaction electrocatalysis on nitrogen-doped carbon materials. *Adv. Mater.*, 31: e1804297.

[41] Hu, J., A. Liu, H. Jin, D. Ma, D. Yin, P. Ling, S. Wang, Z. Lin and J. Wang. 2015. A versatile strategy for shish-kebab-like multi-heterostructured chalcogenides and enhanced photocatalytic hydrogen evolution. *J. Am. Chem. Soc.*, 137: 11004–11010.

[42] Yu, X.-P., C. Yang, P. Song and J. Peng. 2020. Self-assembly of Au/MoS$_2$ quantum dots core-satellite hybrid as efficient electrocatalyst for hydrogen production. *Tungsten*, 2: 194–202.

[43] Aftab, U., A. Tahira, R. Mazzaro, V. Morandi, M. I. Abro, M. M. Baloch, J. A. Syed, A. Nafady and Z. H. Ibupoto. 2020. Facile NiCo$_2$S$_4$/C nanocomposite: An efficient material for water oxidation. *Tungsten*, 2: 403–410.

[44] Sun, C. B., Y. W. Zhong, W. J. Fu, Z. Q. Zhao, J. Liu, J. Ding, X. P. Han, Y. D. Deng, W. B. Hu and C. Zhong. 2020. Tungsten disulfide-based nanomaterials for energy conversion and storage. *Tungsten*, 2: 109–133.

[45] Bard, A. J. 2017. Standard Potentials in Aqueous Solution, CRC Press.

[46] Prakash, J., D. A. Tryk and E. B. Yeager. 2019. Kinetic investigations of oxygen reduction and evolution reactions on lead ruthenate catalysts. *J. Electrochem. Soc.*, 146: 4145–4151.

[47] Kulkarni, A., S. Siahrostami, A. Patel and J. K. Nørskov. 2018. Understanding catalytic activity trends in the oxygen reduction reaction. *Chem. Rev.*, 118: 2302–2312.

[48] Luo, H., W. J. Jiang, S. Niu, X. Zhang, Y. Zhang, L. P. Yuan, C. He and J. S. Hu. 2020. Self-catalyzed growth of Co-N-C nanobrushes for efficient rechargeable zn-air batteries. *Small*, 16: e2001171.

[49] Zhu, C., Q. Shi, B. Z. Xu, S. Fu, G. Wan, C. Yang, S. Yao, J. Song, H. Zhou, D. Du, S. P. Beckman, D. Su and Y. Lin. 2018. Hierarchically porous M-N-C (M = Co and Fe) single-atom electrocatalysts with robust MN$_x$ active moieties enable enhanced ORR performance. *Adv. Energy Mater.*, 8: 1801956.

[50] Lv, H., D. Li, D. Strmcnik, A. P. Paulikas, N. M. Markovic and V. R. Stamenkovic. 2016. Recent advances in the design of tailored nanomaterials for efficient oxygen reduction reaction. *Nano Energy*, 29: 149–165.

[51] Shao, M., Q. Chang, J. P. Dodelet and R. Chenitz. 2016. Recent advances in electrocatalysts for oxygen reduction reaction. *Chem. Rev.*, 116: 3594–3657.

[52] Nørskov, J. K., J. Rossmeisl, A. Logadottir, L. Lindqvist, J. R. Kitchin, T. Bligaard and H. Jónsson. 2004. Origin of the overpotential for oxygen reduction at a fuel-cell cathode. *J. Phys. Chem. B*, 108: 17886–17892.

[53] Lin, Y., L. Yang, Y. Zhang, H. Jiang, Z. Xiao, C. Wu, G. Zhang, J. Jiang and L. Song. 2018. Defective carbon-CoP nanoparticles hybrids with interfacial charges polarization for efficient bifunctional oxygen electrocatalysis. *Adv. Energy Mater.*, 8: 1703623.

[54] Anderson, A. B., Y. Cai, R. A. Sidik and D. B. Kang. 2005. Advancements in the local reaction center electron transfer theory and the transition state structure in the first step of oxygen reduction over platinum. *J. Electroanal. Chem.*, 580: 17–22.

[55] Perry, S. C. and G. Denuault. 2016. The oxygen reduction reaction (ORR) on reduced metals: Evidence for a unique relationship between the coverage of adsorbed oxygen species and adsorption energy. *Phys. Chem. Chem. Phys.*, 18: 10218–10223.

[56] Lim, D.-H. and J. Wilcox. 2011. DFT-based study on oxygen adsorption on defective graphene-supported Pt nanoparticles. *J. Phys. Chem. C*, 115: 22742–22747.

[57] Shao, M., P. Liu, J. Zhang and R. Adzic. 2007. Origin of enhanced activity in palladium alloy electrocatalysts for oxygen reduction reaction. *J. Phys. Chem. B*, 111: 6772–6775.

[58] Kweon, D. H., M. S. Okyay, S. J. Kim, J. P. Jeon, H. J. Noh, N. Park, J. Mahmood and J. B. Baek. 2020. Ruthenium anchored on carbon nanotube electrocatalyst for hydrogen production with enhanced Faradaic efficiency. *Nat. Commun.*, 11: 1278.

[59] Newton, M. A., D. Ferri, G. Smolentsev, V. Marchionni and M. Nachtegaal. 2015. Room-temperature carbon monoxide oxidation by oxygen over Pt/Al2O3 mediated by reactive platinum carbonates. *Nat. Commun.*, 6: 8675.

[60] Puangsombut, P. and N. Tantavichet. 2018. Effect of plating bath composition on chemical composition and oxygen reduction reaction activity of electrodeposited Pt-Co catalysts. *Rare Met.*, 38: 95–106.

[61] Tang, W. Wu and K. Wang. 2018. Oxygen reduction reaction catalyzed by noble metal clusters. *Catalysts*, 8: 65.

[62] Chong, L., J. Wen, J. Kubal, F. G. Sen, J. Zou, J. Greeley, M. Chan, H. Barkholtz, W. Ding and D. J. Liu. 2018. Ultralow-loading platinum-cobalt fuel cell catalysts derived from imidazolate frameworks. *Science*, 362: 1276–1281.

[63] Wang, Q., Z. L. Zhao, Z. Zhang, T. Feng, R. Zhong, H. Xu, S. T. Pantelides and M. Gu. 2020. Sub-3 nm intermetallic ordered Pt$_3$In clusters for oxygen reduction reaction. *Adv. Sci.*, 7: 1901279.

[64] Chen, Y., T. Cheng and W. A. Goddard Iii. 2020. Atomistic explanation of the dramatically improved oxygen reduction reaction of jagged platinum nanowires, 50 Times Better than Pt. *J. Am. Chem. Soc.*, 142: 8625–8632.

[65] Zhao, D., Z. Zhuang, X. Cao, C. Zhang, Q. Peng, C. Chen and Y. Li. 2020. Atomic site electrocatalysts for water splitting, oxygen reduction and selective oxidation. *Chem. Soc. Rev.*, 49: 2215–2264.

[66] Demarconnay, L., C. Coutanceau and J. M. Léger. 2004. Electroreduction of dioxygen (ORR) in alkaline medium on Ag/C and Pt/C nanostructured catalysts—effect of the presence of methanol. *Electrochim. Acta*, 49: 4513–4521.

[67] Wang, W., Z. Wang, M. Yang, C. J. Zhong and C. J. Liu. 2016. Highly active and stable Pt (111) catalysts synthesized by peptide assisted room temperature electron reduction for oxygen reduction reaction. *Nano Energy*, 25: 26–33.

[68] Jiménez-Morales, I., A. Reyes-Carmona, M. Dupont, S. Cavaliere, M. Rodlert, F. Mornaghini, M. J. Larsen, M. Odgaard, J. Zajac, D. J. Jones and J. Rozière. 2021. Correlation between the surface characteristics of carbon supports and their electrochemical stability and performance in fuel cell cathodes. *Carbon Energy*, DOI: 10.1002/cey2.109.

[69] Meier, J. C., C. Galeano, I. Katsounaros, A. A. Topalov, A. Kostka, F. Schüth and K. J. J. Mayrhofer. 2012. Degradation mechanisms of Pt/C fuel cell catalysts under simulated start–stop conditions. *ACS Catal.*, 2: 832–843.

[70] Wen, Z., J. Liu and J. Li. 2008. Core/Shell Pt/C nanoparticles embedded in mesoporous carbon as a methanol-tolerant cathode catalyst in direct methanol fuel cells. *Adv. Mater.*, 20: 743–747.

[71] Wang, X., S. Yang, Y. Yu, M. Dou, Z. Zhang and F. Wang. 2020. Low-loading Pt nanoparticles embedded on Ni, N-doped carbon as superior electrocatalysts for oxygen reduction. *Catal. Sci. Technol.*, 10: 65–69.

[72] Chen, S., Z. Wei, X. Qi, L. Dong, Y. G. Guo, L. Wan, Z. Shao and L. Li. 2012. Nanostructured polyaniline-decorated Pt/C@PANI core-shell catalyst with enhanced durability and activity. *J. Am. Chem. Soc.*, 134: 13252–13255.

[73] Ma, Z., S. Li, L. Wu, L. Song, G. Jiang, Z. Liang, D. Su, Y. Zhu, R. R. Adzic, J. X. Wang and Z. Chen. 2020. NbOx nano-nail with a Pt head embedded in carbon as a highly active and durable oxygen reduction catalyst. *Nano Energy*, 69: 104455.

[74] Shin, H.-S., O.-J. Kwon and B. S. Oh. 2018. Correlation between performance of polymer electrolyte membrane fuel cell and degradation of the carbon support in the membrane electrode assembly using image processing method. *Int. J. Hydrogen Energy*, 43: 20921–20930.

[75] Schmies, H., E. Hornberger, B. Anke, T. Jurzinsky, H. N. Nong, F. Dionigi, S. Kühl, J. Drnec, M. Lerch, C. Cremers and P. Strasser. 2018. Impact of carbon support functionalization on the electrochemical stability of Pt fuel cell catalysts. *Chem. Mater.*, 30: 7287–7295.

[76] Xiao, M., J. Zhu, G. Li, N. Li, S. Li, Z. P. Cano, L. Ma, P. Cui, P. Xu and G. Jiang. 2019. A single-atom iridium heterogeneous catalyst in oxygen reduction reaction. *Angew. Chem.*, 131: 9742–9747.

[77] Qiao, B., A. Wang, X. Yang, L. F. Allard, Z. Jiang, Y. Cui, J. Liu, J. Li and T. Zhang. 2011. Single-atom catalysis of CO oxidation using Pt_1/FeO_x. *Nat. Chem.*, 3: 634–641.

[78] Lin, J., B. Qiao, N. Li, L. Li, X. Sun, J. Liu, X. Wang and T. Zhang. 2015. Little do more: A highly effective $Pt_{(1)}/FeO_{(x)}$ single-atom catalyst for the reduction of NO by H_2. *Chem. Commun.*, 51: 7911–7914.

[79] Zhang, B., H. Asakura, J. Zhang, J. Zhang, S. De and N. Yan. 2016. Stabilizing a platinum$_1$ single-atom catalyst on supported phosphomolybdic acid without compromising hydrogenation activity. *Angew. Chem.*, 128: 8459–8463.

[80] Mahmood, J., F. Li, S. M. Jung, M. S. Okyay, I. Ahmad, S. J. Kim, N. Park, H. Y. Jeong and J. B. Baek. 2017. An efficient and pH-universal ruthenium-based catalyst for the hydrogen evolution reaction. *Nat. Nanotechnol.*, 12: 441–446.

[81] Wang, Q., M. Ming, S. Niu, Y. Zhang, G. Fan and J.-S. Hu. 2018. Scalable solid-state synthesis of highly dispersed uncapped metal (Rh, Ru, Ir) nanoparticles for efficient hydrogen evolution. *Adv. Energy Mater.*, 8: 1801698.

[82] Xiao, M., L. Gao, Y. Wang, X. Wang, J. Zhu, Z. Jin, C. Liu, H. Chen, G. Li, J. Ge, Q. He, Z. Wu, Z. Chen and W. Xing. 2019. Engineering energy level of metal center: Ru single-atom site for efficient and durable oxygen reduction catalysis. *J. Am. Chem. Soc.*, 141: 19800–19806.

[83] Liu, J., M. Jiao, L. Lu, H. M. Barkholtz, Y. Li, Y. Wang, L. Jiang, Z. Wu, D. J. Liu, L. Zhuang, C. Ma, J. Zeng, B. Zhang, D. Su, P. Song, W. Xing, W. Xu, Y. Wang, Z. Jiang and G. Sun. 2017. High performance platinum single atom electrocatalyst for oxygen reduction reaction. *Nat. Commun.*, 8: 1–10.

[84] Cheng, Q., L. Yang, L. Zou, Z. Zou, C. Chen, Z. Hu and H. Yang. 2017. Single cobalt atom and N codoped carbon nanofibers as highly durable electrocatalyst for oxygen reduction reaction. *ACS Catal.*, 7: 6864–6871.

[85] Chen, S., Z. Wei, L. Guo, W. Ding, L. Dong, P. Shen, X. Qi and L. Li. 2011. Enhanced dispersion and durability of Pt nanoparticles on a thiolated CNT support. *Chem. Commun.*, 47: 10984–10986.

[86] Wang, S., S. P. Jiang and X. Wang. 2008. Polyelectrolyte functionalized carbon nanotubes as a support for noble metal electrocatalysts and their activity for methanol oxidation. *Nanotechnology*, 19: 265601.

[87] Tian, W., Y. Wang, W. Fu, J. Su, H. Zhang and Y. Wang. 2020. PtP_2 nanoparticles on N,P doped carbon through a self-conversion process to core–shell Pt/PtP_2 as an efficient and robust ORR catalyst. *J. Mater. Chem. A*, 8: 20463–20473.

[88] Li, J., Z. Xi, Y. T. Pan, J. S. Spendelow, P. N. Duchesne, D. Su, Q. Li, C. Yu, Z. Yin, B. Shen, Y. S. Kim, P. Zhang and S. Sun. 2018. Fe stabilization by intermetallic $L1_0$-FePt and Pt catalysis enhancement in $L1_0$-FePt/Pt nanoparticles for efficient oxygen reduction reaction in fuel cells. *J. Am. Chem. Soc.*, 140: 2926–2932.

[89] Wang, Z.-Y., S.-D. Jiang, C.-Q. Duan, D. Wang, S.-H. Luo and Y.-G. Liu. 2020. *In situ* synthesis of Co_3O_4 nanoparticles confined in 3D nitrogen-doped porous carbon as an efficient bifunctional oxygen electrocatalyst. *Rare Met.*, 39: 1383–1394.

[90] Xiong, Y., Y. Yang, X. Feng, F. J. DiSalvo and H. D. Abruna. 2019. A strategy for increasing the efficiency of the oxygen reduction reaction in Mn-doped cobalt ferrites. *J. Am. Chem. Soc.*, 141: 4412–4421.

[91] Li, L., X. Feng, Y. Nie, S. Chen, F. Shi, K. Xiong, W. Ding, X. Qi, J. Hu, Z. Wei, L.J. Wan and M. Xia. 2015. Insight into the effect of oxygen vacancy concentration on the catalytic performance of MnO_2. *ACS Catal.*, 5: 4825–4832.

[92] Tong, Y., P. Chen, T. Zhou, K. Xu, W. Chu, C. Wu and Y. Xie. 2017. A bifunctional hybrid electrocatalyst for oxygen reduction and evolution: Cobalt oxide nanoparticles strongly coupled to B,N-Decorated graphene. *Angew. Chem. Int. Ed.*, 56: 7121–7125.

[93]   Diao, L., T. Yang, B. Chen, B. Zhang, N. Zhao, C. Shi, E. Liu, L. Ma and C. He. 2019. Electronic reconfiguration of Co_2P induced by Cu doping enhancing oxygen reduction reaction activity in zinc-air batteries. *J. Mater. Chem. A*, 7: 21232–21243.

[94]   Wang, R., X. Y. Dong, J. Du, J. Y. Zhao and S. Q. Zang. 2018. MOF-derived bifunctional Cu_3P nanoparticles coated by a N,P-codoped carbon shell for hydrogen evolution and oxygen reduction. *Adv. Mater.*, 30: 1703711.

[95]   Guo, Y., T. Park, J. W. Yi, J. Henzie, J. Kim, Z. Wang, B. Jiang, Y. Bando, Y. Sugahara, J. Tang and Y. Yamauchi. 2019. Nanoarchitectonics for transition-metal-sulfide-based electrocatalysts for water splitting. *Adv. Mater.*, 31: e1807134.

[96]   Li, Y., H. Wang, L. Xie, Y. Liang, G. Hong and H. Dai. 2011. MoS_2 nanoparticles grown on graphene: an advanced catalyst for the hydrogen evolution reaction. *J. Am. Chem. Soc.*, 133: 7296–7299.

[97]   Niu, W., Z. Li, K. Marcus, L. Zhou, Y. Li, R. Ye, K. Liang and Y. Yang. 2018. Surface-modified porous carbon nitride composites as highly efficient electrocatalyst for Zn-Air batteries. *Adv. Energy Mater.*, 8.

[98]   Dong, Y., Y. Deng, J. Zeng, H. Song and S. Liao. 2017. A high-performance composite ORR catalyst based on the synergy between binary transition metal nitride and nitrogen-doped reduced graphene oxide. *J. Mater. Chem. A*, 5: 5829–5837.

[99]   Yoon, K. R., K. Shin, J. Park, S. H. Cho, C. Kim, J. W. Jung, J. Y. Cheong, H. R. Byon, H. M. Lee and I. D. Kim. 2018. Brush-like cobalt nitride anchored carbon nanofiber membrane: Current collector-catalyst integrated cathode for long Cycle $Li-O_2$ batteries. *ACS Nano*, 12: 128–139.

[100]   Luo, J., H. Tang, X. Tian, S. Liao, J. Ren, W. Zhao and X. Qiao. 2019. Glucose-derived carbon supported well-dispersed CrN as competitive oxygen reduction catalysts in acidic medium. *Electrochim. Acta*, 314: 202–211.

[101]   Song, L., T. Wang, Y. Wang, H. Xue, X. Fan, H. Guo, W. Xia, H. Gong and J. He. 2017. Porous iron-tungsten carbide electrocatalyst with high activity and stability toward oxygen reduction reaction: From the self-assisted synthetic mechanism to its active-species probing. *ACS. Appl. Mater. Interfaces*, 9: 3713–3722.

[102]   Levy, R. B. and M. Boudart. 1973. Platinum-like behavior of tungsten carbide in surface catalysis. *Science*, 181: 547–549.

[103]   Liu, Y. and W. E. Mustain. 2011. Structural and electrochemical studies of Pt clusters supported on high-surface-area tungsten carbide for oxygen reduction. *ACS Catal.*, 1: 212–220.

[104]   Chen, J. G. 1996. Carbide and nitride overlayers on early transition metal surfaces: Preparation, characterization, and reactivities. *Chem. Rev.*, 96: 1477–1498.

[105]   Aijaz, A., J. Masa, C. Rosler, H. Antoni, R. A. Fischer, W. Schuhmann and M. Muhler. 2017. MOF-templated assembly approach for Fe_3C nanoparticles encapsulated in bamboo-like N-doped CNTs: Highly efficient oxygen reduction under acidic and basic conditions. *Chemistry*, 23: 12125–12130.

[106]   Zheng, Y., D.-S. Yang, J. M. Kweun, C. Li, K. Tan, F. Kong, C. Liang, Y. J. Chabal, Y. Y. Kim, M. Cho, J.-S. Yu and K. Cho. 2016. Rational design of common transition metal-nitrogen-carbon catalysts for oxygen reduction reaction in fuel cells. *Nano Energy*, 30: 443–449.

[107]   Wang, J., Y. Zhao, B. P. Setzler, S. Rojas-Carbonell, C. Ben Yehuda, A. Amel, M. Page, L. Wang, K. Hu, L. Shi, S. Gottesfeld, B. Xu and Y. Yan. 2019. Poly(aryl piperidinium) membranes and ionomers for hydroxide exchange membrane fuel cells. *Nature Energy*, 4: 392–398.

[108]   Mauritz, K. A. and R. B. Moore. 2004. State of understanding of nafion. *Chem. Rev.*, 104: 4535–4585.

[109]   Jiao, L., G. Wan, R. Zhang, H. Zhou, S. H. Yu and H. L. Jiang. 2018. From metal-organic frameworks to single-atom Fe Implanted N-doped porous carbons: Efficient oxygen reduction in both alkaline and acidic media. *Angew. Chem. Int. Ed.*, 57: 8525–8529.

[110]   Li, B. Q., C. X. Zhao, J. N. Liu and Q. Zhang. 2019. Electrosynthesis of hydrogen peroxide synergistically catalyzed by atomic $Co-N_x-C$ sites and oxygen functional groups in noble-metal-free electrocatalysts. *Adv. Mater.*, 31: e1808173.

[111]   Wang, Y., G. I. N. Waterhouse, L. Shang and T. Zhang. 2020. Electrocatalytic oxygen reduction to hydrogen peroxide: From homogenous to heterogenous electrocatalysis. *Adv. Energy Mater.*, 11: 2003323.

[112] Wang, Y., R. Shi, L. Shang, G. I. N. Waterhouse, J. Zhao, Q. Zhang, L. Gu and T. Zhang. 2020. High-efficiency oxygen reduction to hydrogen peroxide catalyzed by nickel single-atom catalysts with tetradentate N_2O_2 coordination in a three-phase flow cell. *Angew. Chem. Int. Ed.*, 59: 13057–13062.

[113] Hossain, M. D., Z. J. Liu, M. H. Zhuang, X. X. Yan, G. L. Xu, C. A. Gadre, A. Tyagi, I. H. Abidi, C. J. Sun, O. L. Wong, A. Guda, Y. F. Hao, X. Q. Pan, K. Amine and Z. T. Luo. 2019. Rational design of graphene-supported single atom catalysts for hydrogen evolution reaction. *Adv. Energy Mater.*, 9: 1803689.

[114] Peng, H., F. Liu, X. Liu, S. Liao, C. You, X. Tian, H. Nan, F. Luo, H. Song, Z. Fu and P. Huang. 2014. Effect of transition metals on the structure and performance of the doped carbon catalysts derived from polyaniline and melamine for ORR Application. *ACS Catal.*, 4: 3797–3805.

[115] Zitolo, A., V. Goellner, V. Armel, M. T. Sougrati, T. Mineva, L. Stievano, E. Fonda and F. Jaouen. 2015. Identification of catalytic sites for oxygen reduction in iron- and nitrogen-doped graphene materials. *Nat. Mater.*, 14: 937–942.

[116] Zitolo, A., N. Ranjbar-Sahraie, T. Mineva, J. Li, Q. Jia, S. Stamatin, G. F. Harrington, S. M. Lyth, P. Krtil, S. Mukerjee, E. Fonda and F. Jaouen. 2017. Identification of catalytic sites in cobalt-nitrogen-carbon materials for the oxygen reduction reaction. *Nat. Commun.*, 8: 1–11.

[117] Martinaiou, I., A. H. A. Monteverde Videla, N. Weidler, M. Kübler, W. D. Z. Wallace, S. Paul, S. Wagner, A. Shahraei, R. W. Stark, S. Specchia and U. I. Kramm. 2020. Activity and degradation study of an Fe-N-C catalyst for ORR in Direct Methanol Fuel Cell (DMFC). *Appl. Catal., B*, 262.

[118] Xie, X., C. He, B. Li, Y. He, D. A. Cullen, E. C. Wegener, A. J. Kropf, U. Martinez, Y. Cheng, M. H. Engelhard, M. E. Bowden, M. Song, T. Lemmon, X. S. Li, Z. Nie, J. Liu, D. J. Myers, P. Zelenay, G. Wang, G. Wu, V. Ramani and Y. Shao. 2020. Performance enhancement and degradation mechanism identification of a single-atom Co–N–C catalyst for proton exchange membrane fuel cells. *Nature Catalysis*, 3: 1044–1054.

[119] Ye, H., L. Li, D. Liu, Q. Fu, F. Zhang, P. Dai, X. Gu and X. Zhao. 2020. Sustained-release method for the directed synthesis of ZIF-Derived Ultrafine Co-N-C ORR catalysts with embedded Co quantum dots. *ACS. Appl. Mater. Interfaces.*, 12: 57847–57858.

[120] Zhang, L., J. Fischer, Y. Jia, X. Yan, W. Xu, X. Wang, J. Chen, D. Yang, H. Liu, L. Zhuang, M. Hankel, D. J. Searles, K. Huang, S. Feng, C. L. Brown and X. Yao. 2018. Coordination of Atomic Co-Pt coupling species at carbon defects as active sites for oxygen reduction reaction. *J. Am. Chem. Soc.*, 140: 10757–10763.

[121] Yang, Y., Y. Qian, H. Li, Z. Zhang, Y. Mu, D. Do, B. Zhou, J. Dong, W. Yan, Y. Qin, L. Fang, R. Feng, J. Zhou, P. Zhang, J. Dong, G. Yu, Y. Liu, X. Zhang and X. Fan. 2020. O-coordinated W-Mo dual-atom catalyst for pH-universal electrocatalytic hydrogen evolution. *Sci. Adv.*, 6: eaba6586.

[122] Fu, J., J. Dong, R. Si, K. Sun, J. Zhang, M. Li, N. Yu, B. Zhang, M. G. Humphrey, Q. Fu and J. Huang. 2021. Synergistic effects for enhanced catalysis in a dual single-atom catalyst. *ACS Catal.*, 11: 1952–1961.

[123] Wang, J., Z. Huang, W. Liu, C. Chang, H. Tang, Z. Li, W. Chen, C. Jia, T. Yao, S. Wei, Y. Wu and Y. Li. 2017. Design of N-coordinated dual-metal sites: A stable and active Pt-free catalyst for acidic oxygen reduction reaction. *J. Am. Chem. Soc.*, 139: 17281–17284.

[124] Zhang, H., H. T. Chung, D. A. Cullen, S. Wagner, U. I. Kramm, K. L. More, P. Zelenay and G. Wu. 2019. High-performance fuel cell cathodes exclusively containing atomically dispersed iron active sites. *Energy Environ. Sci.*, 12: 2548–2558.

[125] Jasinski, R. 1964. A new fuel cell cathode catalyst. *Nature*, 201: 1212–1213.

[126] Sun, X., K. Li, C. Yin, Y. Wang, M. Jiao, F. He, X. Bai, H. Tang and Z. Wu. 2016. Dual-site oxygen reduction reaction mechanism on CoN_4 and CoN_2 embedded graphene: Theoretical insights. *Carbon*, 108: 541–550.

[127] Chen, Z., W. Gong, Z. Liu, S. Cong, Z. Zheng, Z. Wang, W. Zhang, J. Ma, H. Yu, G. Li, W. Lu, W. Ren and Z. Zhao. 2019. Coordination-controlled single-atom tungsten as a non-3d-metal oxygen reduction reaction electrocatalyst with ultrahigh mass activity. *Nano Energy*, 60: 394–403.

[128] Wan, C., X. Duan and Y. Huang. 2020. Molecular design of single-atom catalysts for oxygen reduction reaction. *Adv. Energy Mater.*, 10: 1903815.

[129] Lei, J., H. Liu, D. Yin, L. Zhou, J. A. Liu, Q. Chen, X. Cui, R. He, T. Duan and W. Zhu. 2020. Boosting the loading of metal single atoms via a bioconcentration strategy. *Small*, 16: e1905920.

[130] Guo, L., J. Sun, J. Wei, Y. Liu, L. Hou and C. Yuan. 2020. Conductive metal-organic frameworks: Recent advances in electrochemical energy-related applications and perspectives. *Carbon Energy*, 2: 203–222.

[131] Xie, C., D. Yan, H. Li, S. Du, W. Chen, Y. Wang, Y. Zou, R. Chen and S. Wang. 2020. Defect chemistry in heterogeneous catalysis: Recognition, understanding, and utilization. *ACS Catal.*, 10: 11082–11098.

[132] Jiang, Y., L. Yang, T. Sun, J. Zhao, Z. Lyu, O. Zhuo, X. Wang, Q. Wu, J. Ma and Z. Hu. 2015. Significant contribution of intrinsic carbon defects to oxygen reduction activity. *ACS Catal.*, 5: 6707–6712.

[133] Jia, Y., L. Zhang, A. Du, G. Gao, J. Chen, X. Yan, C. L. Brown and X. Yao. 2016. Defect Graphene as a trifunctional catalyst for electrochemical reactions. *Adv. Mater.*, 28: 9532–9538.

[134] Xue, L., Y. Li, X. Liu, Q. Liu, J. Shang, H. Duan, L. Dai and J. Shui. 2018. Zigzag carbon as efficient and stable oxygen reduction electrocatalyst for proton exchange membrane fuel cells. *Nat. Commun.*, 9: 3819.

[135] Gong, K., F. Du, Z. Xia, M. Durstock and L. Dai. 2009. Nitrogen-doped carbon nanotube arrays with high electrocatalytic activity for oxygen reduction. *Science*, 323: 760–764.

[136] Li, J.-C., P.-X. Hou, M. Cheng, C. Liu, H.-M. Cheng and M. Shao. 2018. Carbon nanotube encapsulated in nitrogen and phosphorus co-doped carbon as a bifunctional electrocatalyst for oxygen reduction and evolution reactions. *Carbon*, 139: 156–163.

[137] Bian, Y., H. Wang, J. Hu, B. Liu, D. Liu and L. Dai. 2020. Nitrogen-rich holey graphene for efficient oxygen reduction reaction. *Carbon*, 162: 66–73.

[138] Yang, Z., Z. Yao, G. Li, G. Fang, H. Nie, H. Liu, Z. Liu, X. Zhou, X. Chen and S. HUang. 2012. Sulfur-doped graphene as an efficient metal-free cathode catalyst for oxygen reduction. *ACS Nano.*, 6: 205–211.

[139] Yang, L., J. Shui, L. Du, Y. Shao, J. Liu, L. Dai and Z. Hu. 2019. Carbon-based metal-free ORR electrocatalysts for fuel cells: Past, present, and future. *Adv. Mater.*, 31: e1804799.

[140] Zhao, Y., J. Wan, H. Yao, L. Zhang, K. Lin, L. Wang, N. Yang, D. Liu, L. Song, J. Zhu, L. Gu, L. Liu, H. Zhao, Y. Li and D. Wang. 2018. Few-layer graphdiyne doped with sp-hybridized nitrogen atoms at acetylenic sites for oxygen reduction electrocatalysis. *Nat. Chem.*, 10: 924–931.

[141] Guo, D., R. Shibuya, C. Akiba, S. Saji, T. Kondo and J. Nakamura. 2016. Active sites of nitrogen-doped carbon materials for oxygen reduction reaction clarified using model catalysts. *Science*, 351: 361–365.

[142] Haque, E., A. Zavabeti, N. Uddin, Y. Wang, M. A. Rahim, N. Syed, K. Xu, A. Jannat, F. Haque, B. Y. Zhang, M. A. Shoaib, S. Shamsuddin, M. Nurunnabi, A. I. Minett, J. Z. Ou and A. T. Harris. 2020. Deciphering the role of quaternary N in O2 reduction over controlled N-doped carbon catalysts. *Chem. Mater.*, 32: 1384–1392.

[143] Xue, X., H. Yang, T. Yang, P. Yuan, Q. Li, S. Mu, X. Zheng, L. Chi, J. Zhu, Y. Li, J. Zhang and Q. Xu. 2019. N,P-coordinated fullerene-like carbon nanostructures with dual active centers toward highly-efficient multi-functional electrocatalysis for CO$_2$RR, ORR and Zn-air battery. *J. Mater. Chem. A*, 7: 15271–15277.

[144] Nong, J., M. Zhu, K. He, A. Zhu, P. Xie, M. Rong and M. Zhang. 2019. N/S co-doped 3D carbon framework prepared by a facile morphology-controlled solid-state pyrolysis method for oxygen reduction reaction in both acidic and alkaline media. *J. Energy Chem.*, 34: 220–226.

[145] Yang, C., H. Jin, C. Cui, J. Li, J. Wang, K. Amine, J. Lu and S. Wang. 2018. Nitrogen and sulfur co-doped porous carbon sheets for energy storage and pH-universal oxygen reduction reaction. *Nano Energy*, 54: 192–199.

[146] Han, J. S., D. Y. Chung, D. G. Ha, J. H. Kim, K. Choi, Y. E. Sung and S. H. Kang. 2016. Nitrogen and boron co-doped hollow carbon catalyst for the oxygen reduction reaction. *Carbon*, 105: 1–7.

[147] Akula, S. and A. K. Sahu. 2020. Structurally modulated graphitic carbon nanofiber and heteroatom (N,F) engineering toward metal-free ORR electrocatalysts for polymer electrolyte membrane fuel cells. *ACS. Appl. Mater. Interfaces.*, 12: 11438–11449.

[148] Li, J.-C., X. Qin, P.-X. Hou, M. Cheng, C. Shi, C. Liu, H.-M. Cheng and M. Shao. 2019. Identification of active sites in nitrogen and sulfur co-doped carbon-based oxygen reduction catalysts. *Carbon*, 147: 303–311.

[149] Zehtab Yazdi, A., E. P. L. Roberts and U. Sundararaj. 2016. Nitrogen/sulfur co-doped helical graphene nanoribbons for efficient oxygen reduction in alkaline and acidic electrolytes. *Carbon*, 100: 99–108.

[150] Li, D., Y. Jia, G. Chang, J. Chen, H. Liu, J. Wang, Y. Hu, Y. Xia, D. Yang and X. Yao. 2018. A defect-driven metal-free electrocatalyst for oxygen reduction in acidic electrolyte *Chem*, 4: 2345–2356.

[151] Zhang, J., Z. Zhao, Z. Xia and L. Dai. 2015. A metal-free bifunctional electrocatalyst for oxygen reduction and oxygen evolution reactions. *Nat. Nanotechnol.*, 10: 444–452.

[152] Lu, Z., J. Wang, S. Huang, Y. Hou, Y. Li, Y. Zhao, S. Mu, J. Zhang and Y. Zhao. 2017. N,B-codoped defect-rich graphitic carbon nanocages as high performance multifunctional electrocatalysts. *Nano Energy*, 42: 334–340.

[153] Wang, Z., Q.-K. Li, C. Zhang, Z. Cheng, W. Chen, E. A. McHugh, R. A. Carter, B. I. Yakobson and J. M. Tour. 2021. Hydrogen peroxide generation with 100% faradaic efficiency on metal-free carbon black. *ACS Catal.*, 11: 2454–2459.

[154] Dong, K., J. Liang, Y. Wang, Z. Xu, Q. Liu, Y. Luo, T. Li, L. Li, X. Shi, A. M. Asiri, Q. Li, D. Ma and X. Sun. 2021. Honeycomb carbon nanofibers: A superhydrophilic O_2-entrapping electrocatalyst enables ultrahigh mass activity for the two-electron oxygen reduction reaction. *Angew. Chem. Int. Ed.*, 60: 10583–10587.

[155] Tang, C., H. F. Wang and Q. Zhang. 2018. Multiscale principles to boost reactivity in gas-involving energy electrocatalysis. *Acc. Chem. Res.*, 51: 881–889.

[156] Muthuswamy, N., M. E. M. Buan, J. C. Walmsley and M. Rønning. 2018. Evaluation of ORR active sites in nitrogen-doped carbon nanofibers by KOH post treatment. *Catal. Today*, 301: 11–16.

[157] Yin, X., H.-T. cHUNG, S. K. Babu, U. Martinez, G. M. Purdy and P. Zelenay. 2017. Effects of MEA fabrication and ionomer composition on fuel cell performance of PGM-Free ORR catalyst. *ECS Trans.*, 77: 1273–1281.

[158] Gong, Y., D. Li, C. Luo, Q. Fu and C. Pan. 2017. Highly porous graphitic biomass carbon as advanced electrode materials for supercapacitors. *Green Chem.*, 19: 4132–4140.

[159] Li, X., Y. Zhao, Y. Yang and S. Gao. 2019. A universal strategy for carbon-based ORR-active electrocatalyst: One porogen, two pore-creating mechanisms, three pore types. *Nano Energy*, 62: 628–637.

[160] Hu, B.-C., Z.-Y. Wu, S.-Q. Chu, H.-W. Zhu, H.-W. Liang, J. Zhang and S.-H. Yu. 2018. SiO_2-protected shell mediated templating synthesis of Fe–N-doped carbon nanofibers and their enhanced oxygen reduction reaction performance. *Energy Environ. Sci.*, 11: 2208–2215.

[161] Hu, C. and L. Dai. 2017. Multifunctional carbon-based metal-free electrocatalysts for simultaneous oxygen reduction, oxygen evolution, and hydrogen evolution. *Adv. Mater.*, 29: 1604942.

[162] Qian, Y., T. An, K. E. Birgersson, Z. Liu and D. Zhao. 2018. Web-like interconnected carbon networks from NaCl-assisted pyrolysis of ZIF-8 for highly efficient oxygen reduction catalysis. *Small*, 14: e1704169.

[163] Zhong, W., J. Chen, P. Zhang, L. Deng, L. Yao, X. Ren, Y. Li, H. Mi and L. Sun. 2017. Air plasma etching towards rich active sites in Fe/N-porous carbon for the oxygen reduction reaction with superior catalytic performance. *J. Mater. Chem. A*, 5: 16605–16610.

6

Metal-Based Heterogeneous Electrocatalysts for Carbon Dioxide Reduction

Jianjun Su,[1] Yun Song,[1] Weihua Guo,[1] Libei Huang,[1] Xiaohu Cao,[1] Yubing Dou,[1] Le Cheng[1] and Ruquan Ye[1,2,]*

6.1 Introduction

In the past two decades, the carbon dioxide (CO_2) emissions have soared to about 2.6% per year. This resulted in the atmospheric CO_2 concentration exceeding 410 parts per million (ppm), which was about 50% higher than the pre-industrial level (280 ppm).[102–104] The climate change due to the global warming (the average temperature is 0.8°C higher than the pre-industrial level) has caused environmental issues.[105] In order to maintain the sustainable development of society, the Intergovernmental Panel on Climate Change (IPCC) recently recommended limiting the increase of temperature to 1.5°C instead of 2.0°C, which requires global actions to control the carbon emissions.[106] According to the Paris Agreement, many countries are striving to gradually reduce greenhouse gas emissions by 2030.[107] Carbon capture and storage (CCS) is a viable strategy to effectively reduce carbon dioxide emissions. However, long-term storage issues, such as CO_2 leakage and the lack of economic incentives, limit its development.[108] The CO_2 reduction reaction (CO_2RR) that converts CO_2 into other value-added carbon products can provide a good choice for the use of the captured CO_2, which not only helps to reduce CO_2 emissions, but also provides useful products that can be used as fuels such as CO, HCOOH, CH_3OH and CH_4.[109] In addition, the driving force of CO_2RR may come from renewable energy sources,

[1] Department of Chemistry, State Key Laboratory of Marine Pollution, City University of Hong Kong, Hong Kong 999077, China.

[2] City University of Hong Kong Shenzhen Research Institute, Shenzhen, Guangdong 518057, China.

* Corresponding author: ruquanye@cityu.edu.hk

such as solar, wind and hydropower, which aligns with the green concept. Therefore, CO_2RR can complete the carbon cycle and may simultaneously solve the problems of global warming and energy crisis.[110]

6.1.1 Fundamentals of metal-based heterogeneous catalysts for CO₂ reduction

The basic electrochemical CO_2 reduction system consists of an anode, a cathode, an electrolyte containing dissolved CO_2, and a proton exchange membrane, as shown in Fig. 6.1a. The cathode provides catalytically active sites for electrochemical CO_2RR, and the anode promotes oxidation reactions, such as oxygen evolution reaction (OER). The membrane separates the oxidation product from the reduction product, while allowing proton (H^+) exchange to maintain the charge balance.[111] The electrolyte is used both as a medium for transferring charged substances (e^-/H^+) and as a medium for dissolving CO_2.[112] For the entire electrochemical CO_2 reduction system, the design of CO_2RR electrocatalyst is the most critical process.[113] It is generally believed that the entire electrochemical CO_2RR process includes CO_2 solvation in the gas phase, adsorption and activation of CO_2 molecules on the surface of the electrocatalyst, and then a multi-step proton-electron coupling transfer reaction to form hydrocarbons and desorption.[114]

Thermodynamically, the half-cell reduction potentials of CO_2 to the main product are within ± 0.2 V of the hydrogen evolution reaction (HER; Fig. 6.1b). However, the low solubility of CO_2, high CO_2 activation barrier and the multi-electron transfer steps lead to slow reaction kinetics of CO_2RR, showing a higher overpotential and low turnover frequency.[115,116] In the past few decades, many transition metal-based catalysts have been reported to achieve effective CO_2RR, ranging from molecular catalysts[117–123] to metal nanoparticles[124,125] and bulk heterogeneous materials.[126–128] Nevertheless, it is still necessary to further improve the performance of the catalyst in terms of activity, selectivity and stability to meet the requirements of practical applications.

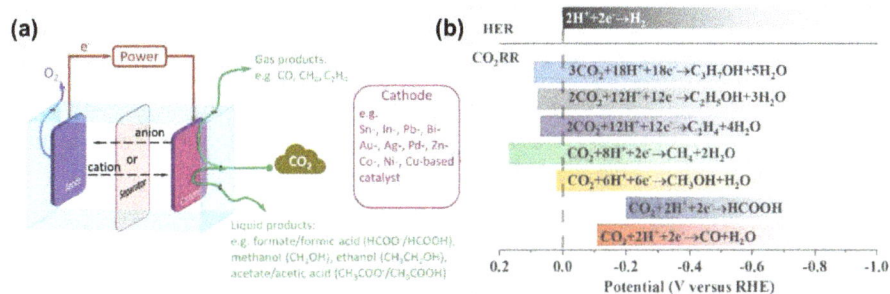

Figure 6.1. (a) A schematic of the reactor for CO_2 electrochemical conversion, typical heterogeneous metal-based catalysts for cathode and anode. Reproduced with permission from Ref. 28. Copyright 2020, Wiley-VCH. (b) The thermodynamic potentials for the major CO_2RR half reactions and the competing HER in aqueous electrolyte under standard conditions (1.0 atm, 25°C, and pH = 7.0). Reproduced with permission from Ref. 130. Copyright 2020, Wiley-VCH.

6.1.2 Evaluation parameters of CO₂RR catalysts

The following parameters, onset potential, current density, faraday efficiency, Tafel slope, turnover number (TON), turnover frequency (TOF) and stability, have been widely used to compare and evaluate the performance of heterogeneous molecular catalysts for CO_2RR. These parameters not only depend on the intrinsic activity of the heterogeneous catalysts, but also the extrinsic factors such as electrolyte and reaction temperature.

6.1.2.1 Onset potential

Since the reduction of CO_2 must overcome the thermodynamic energy barrier, the starting potential is always lower than the standard CO_2 reduction potential. The difference between the onset potential and the standard reduction potential is defined as the overpotential.[131] The onset potential shows the minimum required overpotential to reduce CO_2 to the product.[132] This parameter is highly dependent on the activity of the solid catalyst.

6.1.2.2 Current density

The current density reflects the number of charges related to the reaction in the electrochemical catalysis process, and it indicates the rate of the reaction.[133] Different standards, such as the geometric area of the electrode, the electrochemical surface area (ECSA) and the mass loading of the catalyst, can normalize the current density. The total current density includes the contribution of all reduction reactions. In CO_2RR, combined with FE, the partial current density corresponding to a specific reduction product can be measured.

6.1.2.3 Faradaic efficiency (FE)

Faraday efficiency can be used to obtain selectivity information and is always defined as the proportion of electrons consumed to produce a given product. It can be calculated as the number of moles of electrons consumed divided by the total electrons transferred from the anode to the cathode during the electrolysis process. The formula to determine the Faraday efficiency is shown as follows:

$$\text{Faraday efficiency} = anF/Q \tag{6.1}$$

where α is the number of electrons transferred (for H_2, CO and formic acid production, $\alpha = 2$), n is the number of moles of a given product, F is the Faraday constant 96 485 C mol⁻¹, and Q is the total charge through the cell.

6.1.2.4 Tafel slope

The Tafel equation can be used to fit the partial current density over a certain potential window:

$$\eta = b \log j + a \tag{6.2}$$

where b is the Tafel slope.[134] The Tafel slope reflects the increase in overpotential required for a ten-fold increase in current density. A smaller value corresponds to a sharp increase in current density and is very desirable. In addition, the Tafel slope is also a useful indicator of possible reaction mechanisms and rate-determining steps (RDS).[135] For example, *COOH is usually considered as the intermediate to form CO during CO_2RR. The formation of *COOH intermediates can be carried out through coupled (Eq. 6.5) electron-proton transfer (CEPT) pathways.

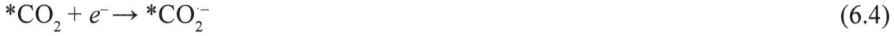

$$* + CO_2 \rightarrow *CO_2 \tag{6.3}$$

$$*CO_2 + e^- \rightarrow *CO_2^- \tag{6.4}$$

For CO formation:

$$*CO_2^- + H^+ \rightarrow *COOH \tag{6.5}$$

$$*COOH + e^- + H^+ \rightarrow *CO + H_2O \tag{6.6}$$

$$*CO_2 \rightarrow * + CO_2 \tag{6.7}$$

For formate formation:

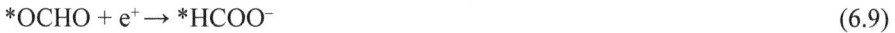

$$*CO_2^- + H^+ \rightarrow *OCHO \tag{6.8}$$

$$*OCHO + e^+ \rightarrow *HCOO^- \tag{6.9}$$

After CO_2 molecules are adsorbed on the catalyst surface, it can proceed through different reaction paths. A Tafel slope close to the theoretical value of 59 mV dec^{-1} indicates the rapid one-electron transfer step (Eq. (6.4)) and the rate-determination step (RDS) chemical reaction step (Eq. (6.5) or Eq. (6.8)). When the Tafel slope is close to 118 mV dec^{-1}, it is likely that the RDS involves an electron transfers to form the adsorbed CO_2 (Eq. (6.6)), and a larger Tafel slope usually indicates that the mass transfer is sluggish.[136]

6.1.2.5 The turnover number (TON) and turnover frequency (TOF)

Turnover (TON) and turnover frequency (TOF) can respectively evaluate the reaction rate related to the actual catalyst and the number of active sites under a specific applied potential.[136] TOF is defined as the number of reduction product generated per electrocatalytic active site per unit time, caculated as follows:

$$TOF = \frac{j_{tot} \times FE_{co}}{2F \times n_{tot}} \tag{6.10}$$

where n_{tot} is the total mole loading amount of catalysts.

TON is defined as the mole of reduction product generated per electrocatalytic active site over a given period of time, calculated as follows:

$$TON = \frac{Q \times FE_{co(average)}}{2F \times n_{tot}} \tag{6.11}$$

where Q is the total reduction charge amount during the electrocatalysis test, and $FE_{co(average)}$ is the average CO faradaic efficiency during the test perieod.

The units of TOF and TON are both s^{-1}. However, when calculating the number of active sites of an electrode, it is usually a fuzzy estimate. Researchers tend to estimate the number of active sites by assuming that all surface atoms of the active material contribute to the target reaction, which can lead to inaccurate results. Although it is difficult to evaluate the actual number of active sites, TOF can still well reflect the actual kinetics of the reaction, thus providing insight into the activity.

6.1.2.6 Stability

Long-term stability is another key factor in evaluating catalysts for commercially relevant electrolysis.[137,138] Stability can measure the retention of reactivity and selectivity over a long period of time, which is usually evaluated by potentiostatic polarization at a potential corresponding to the peak Faraday efficiency. Good catalyst stability will reduce the operating and maintenance costs for industrial applications.[139,140] Stability evaluation in the laboratory is usually in the scale of tens of hours. Thousands of hours, however, are required before the catalyst is considered for practical use.[141]

6.2 Heterogeneous molecular catalysts for CO_2RR

Molecular catalysts are attractive to fundamental research due to their well-defined active sites and precise adjustment capabilities, which help to develop a clear structure-activity relationship for rational catalyst design. Heterogeneous catalysts are usually preferred in practical applications for several reasons: (i) The integration of molecular species into the substrate can enhance the catalytic activity by promoting electron transfer and increasing the utilization of catalytic sites; (ii) through immobilization, a smaller amount of molecular catalyst can be easily used while maintaining high activity; (iii) the aqueous solubility of the molecular catalyst is negligible by functionalization, so that the operating conditions can be extended from the organic phase to the water phase; (iv) the use of heterogeneous molecular catalysts can simplify the separation and recycling of the catalyst after each use; (v) if the catalyst can be used well dispersed on the surface of the substrate, their molecules can be spatially confined to the substrate and kept separated from nearby molecules, thereby preventing any deactivation caused by aggregation or dimerization. Here we will discuss some of the latest developments in this field by considering the intrinsic effects of catalyst structure, the extrinsic effects of catalyst immobilization and the supporting substrates. Key issues such as catalyst loading, electron transfer kinetics, mass transfer effect, activity, selectivity and stability are also discussed (Fig. 6.2).

6.2.1 Catalyst structure

6.2.1.1 Metal atom effect

The structure of the catalyst determines the intrinsic activity of the catalyst and the selectivity of the product. In particular, the central metal atom is considered the most important in homogeneous catalysis. Taking porphyrin-based and phthalocyanine-based catalysts for example, CO is the main product when the metals are Fe, Co,

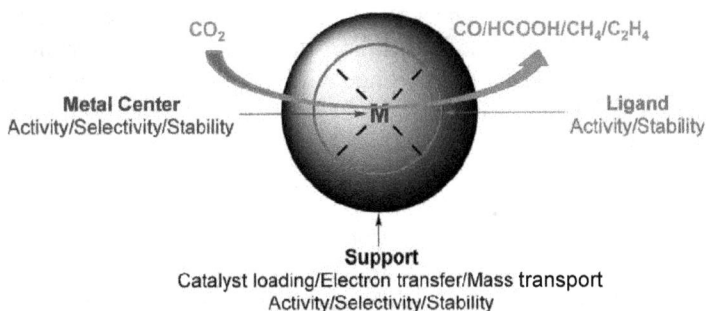

Figure 6.2. Key parameters determining the catalytic performance of a heterogenized molecular catalyst for the electrochemical CO_2RR. Reproduced with permission from Ref. 142. Copyright 2019, Elsevier.

Ni and Zn.[143–146] In the CO_2RR process, the central metal atom is usually the active redox center (i.e., Fe, Co and Ni), but the zinc porphyrin complex is an exception, in which the porphyrin ligand plays this role.[146] On the contrary, if the central metal is In, Sn or Rh, formate will become the main product.[147,148] Similar selectivity of formate is also reported on In and Sn metal electrodes.[149] Density functional theory (DFT) calculations show that electron injection into metalloporphyrin catalysts leads to reduction of metal centers or metal hydrides/ligand hydrides, which depends on the properties of central metal atoms.[150] In the former case, the reduction of CO_2 produces CO, which might be further reduced to methane as detected on cobalt porphyrin,[151] while in the latter case, formate is the more favorable product.

Copper centered molecular catalysts are a special case because they can catalyze the reduction of CO_2 to hydrocarbons such as methane and ethylene, which is similar to those observed on bulk and nanostructured copper metal electrodes.[149,152] Surprisingly, the copper atoms in the center were observed to leached out and aggregate to form small Cu nanoparticles, which are the actual active centers for hydrocarbon production.[153,154] This observation emphasizes the importance of thorough mechanism study to determine the reaction sites of heterogeneous molecular catalysts. Along these lines, it seems to be worth studying whether the metal leached out from the Sn- and In-centered molecular catalyst constitutes the active center for the reduction of CO_2. When Fe, Co, and Ni are used as the center metal atom, deactivation of the catalyst was reported in the long-term experiments.[155] If Fe, Co and Ni nanoparticles are formed, they will be highly active in the HER side reaction. Therefore, the Faraday yield of CO_2 conversion process is reduced.

After immobilization, the activity of metalloporphyrin/phthalocyanine catalyst is still strongly affected by the metal identity. It was reported that cobalt meso tetraphenylporphyrin (CoTPP) and cobalt phthalocyanine (CoPc) complexes exhibit higher CO_2 reduction activity than Fe and Ni.[156,157] This is contrary to the order of reactivity found homogeneous catalysis. In homogeneous case, Fe-centered complex is the best reductant catalyst.[158] According to DFT calculation, the best activity observed on heterogeneous CoPc can be attributed to the medium binding energy of *CO, which benefits the desorption of CO and the formation of *COOH.[156] In contrast, the Fe site is hindered by the strong CO binding, while the COOH formation step at Ni site has high energy barrier. The opposite activity order was found for the heterogeneous and homogeneous catalysts of the same series, which indicates that the interaction between the immobilized catalyst and its support may overturn the inherent activity order, or even change the mechanism. For example, the active component of CoTPP in CO_2RR changes from Co^0 to Co^I going from homogeneous to heterogeneous.[44]

6.2.1.2 Ligand effect

One of the main attractive features of molecular catalysts is the ability to modify the catalysts precisely, which in principle allows the fine control of the binding energy of intermediates and optimization of reaction rate and selectivity.[59] Zhu et al. studied the effects of different peripheral aromatic substituents on the properties of immobilized cobalt porphyrin (CoP) by experimental and theoretical (DFT) methods[160] (Fig. 6.3). CoP is functionalized by eight different groups with different electronegativity, which are catalogued into neutral and positively charged groups. They show that there is a linear relationship between TOF_{CO} as a kinetic descriptor and Hammett substituent constant (σ), which in turn is linearly related to the electronegativity of substituents. They conclude that more electron donating substituents will increase the catalytic activity regardless of the charge of substituents. This is due to the higher electron density around the cobalt atom, which is favorable for the adsorption of CO_2 on the complex. In addition, it was found that the positively charged substituents could improve the catalytic performance by stabilizing the transition state and reducing the energy of electron transfer at RDS.

Others have also observed this trend in CoPc.[161,162] For example, Zhang et al. modified CoPc with eight cyano groups (CoPc-CN) and compared it with unmodified CoPc. CoPc-CN shows better performance than CoPc with FE_{CO} of 98% and current densities of 15 mA cm^{-2} at -0.63 V vs. RHE.[163] Wang et al. proved that the introduction of four electron donating amino groups ($-NH_2$) at the β-position of the phthalocyanine ligand (CoPc-NH_2/CNT) significantly improved its stability for

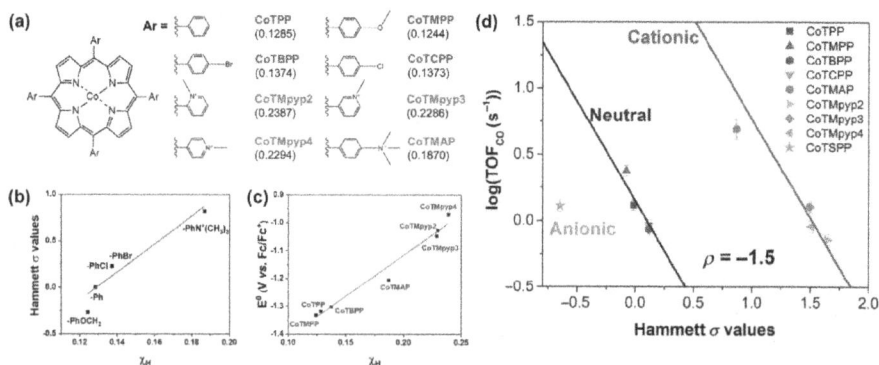

Figure 6.3. (a) The chemical structures of functionalized cobalt porphyrins and the corresponding Mulliken charge population on a probe hydrogen atom (χ_H) for the aromatic substituents (Ar) shown in parentheses. (b) Correlation between calculated χ_H and the corresponding para position Hammett substituent constant. (c) Correlation between calculated χ_H and experimentally measured Co(I/II) redox potentials. (d) TOF_{CO} of cobalt-porphyrin derivatives with various functionalities versus calculated Hammett σ values (-1.02 V vs. SHE, 8×10^{-10} mol cm^{-2}). Reproduced with permission from Ref. 160. Copyright 2019, Royal Society of Chemistry.

CO_2RR to MeOH. This modification makes the reduction potential of CoPc shift negatively and enhances the electrochemical stability of Pc ligand. The synthesized CoPc-NH_2/CNT was electrolyzed at -1.0 V for 12 h with a stable total current density between 30 and 33 mA cm^{-2} and 28% FE_{MeOH} in 0.1 M $KHCO_3$.[164]

Besides the electrons push-pull inductive effect, ligand modification with charged functionalities has also been explored. Robert et al. modified the periphery of the phthalocyanine ring in CoPc with a positively charged trimethylammonium group and three tert-butyl groups.[165] This complex demonstrated 30% larger CO partial current density than the unmodified CoPc in 0.5 M $NaHCO_3$. A high CO selectivity of 95% and large partial current density of 165 mA cm^{-2} were also achieved at -0.92 V vs. RHE in a flow cell configuration. This excellent property is due to the space charge interaction between the positive charge of the trimethylammonium substituent and the partial negative charge of the oxygen atom, which may contribute to the coordination of CO_2 with the metal center and the fracture of the C-O bond during the reduction reaction.

The introduction of a large number of substituents may also hinder the aggregation of macrocycles and increase the exposure of surface active centers through $\pi - \pi$ stacking.[166] Officials et al. modified CoPc with sterically hindered octaalkoxy groups and immobilized the molecules on chemically transformed graphene (CCG/CoPc-a). The addition of alkoxy group inhibited the aggregation of the catalyst. Therefore, compared with the effective TOF (eTOF) of ~ 2 s^{-1} recorded on unmodified CoPc, the eTOF of ~ 5 s^{-1} at -0.59 V vs. RHE indicates a significant increased activity of CO_2RR for the modified CoPc.

6.2.1.3 Electrode support effects

The type of support can affect the binding energy of intermediates by affecting the electron transfer rate or changing the electronic structure of the catalyst. In general, high porosity carriers will improve mass transfer, while high conductivity will allow fast electron transfer. Therefore, the use of more porous and/or more conductive support materials enhanced the performance of the immobilized catalyst for CO_2RR.[167] In addition, in the strong interfacial electric field near the electrode surface, the structure and exact location of the immobilized catalyst will also affect the catalytic process, thus affecting the overall performance of the catalyst.

It is found that the properties of carbon nanotube supported CoPc are better than those of reduced graphene oxide (rGO) or carbon black supported CoPc.[163] The j_{CO} of carbon nanotube supported CoPc is at least three times of that of the rGO or carbon black supported CoPc, and the FE_{CO} is also 10% higher. This improvement is due to the better $\pi - \pi$ interaction between molecular and CNTs, which have a higher degree of sp^2 carbons than rGO or CB.

6.2.2 Immobilization strategies for heterogeneous molecular catalysts

The performance of heterogeneous CO_2RR involving molecular electrocatalysts is highly dependent on how the catalyst is immobilized on the electrode surface. The key parameters of immobilization are catalyst loading and accessibility, electron transfer

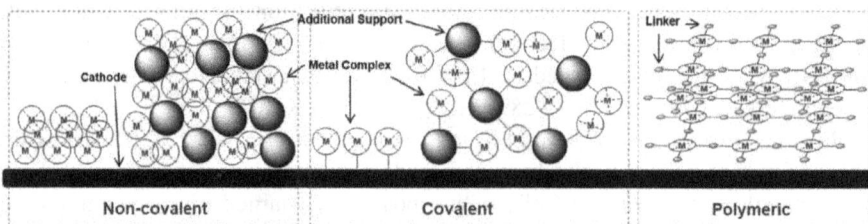

Figure 6.4. Heterogeneous molecular catalysts using three different immobilization strategies for the CO_2RR. Reproduced with permission from Ref. 142. Copyright 2019, Elsevier.

kinetics and mass transport effects. Three main immobilization strategies have been widely applied, namely non-covalent immobilization, covalent immobilization and polymeric immobilization (Fig. 6.4).

6.2.2.1 Non-covalent immobilization

By using $\pi - \pi$ and/or electrostatic interaction, molecular catalysts can be physically adsorbed on electrodes with or without additional support such as carbon black, carbon nanotubes, graphene and metal oxides (Table 6.1). The main advantage of this method is that it does not require the special functionalization of catalyst and electrode. On the other hand, the non-covalently immobilized molecular catalyst easily suffers from leaching problem due to the weak interaction force.[168] In general, the catalyst load can be adjusted in the range of several to hundreds of nmol cm^{-2}. However, it is difficult to control the dispersion of the catalyst because the $\pi - \pi$ and/or electrostatic interaction is much less specific than the chemical bonding. Therefore, the aggregation of catalyst often happens, which hinders the mass transport of CO_2 to the underlying catalyst and the electron transfer.[169] In addition, it is difficult to extract the related kinetic parameters for more complex reaction mechanism, especially under high catalyst loading.

The aggregation also affects the performance of the immobilized catalyst. For example, the high loading leads to high current density but low TOF, while low loading leads to high TOF but low current density.[168] For example, Zhang et al. tested the CO_2RR of cobalt phthalocyanine/carbon nanotube hybrid (CoPc/CNT) in 0.1 M $KHCO_3$ aqueous solution Fig. 6.5. They compared the electrocatalytic activity when the same phthalocyanine was added to reducing graphene oxide (RGO) and carbon black (CB), and the results showed that once adsorbed on CNT, the selectivity towards CO production and the current density increases significantly, reaching 92% and ~ 6 mA cm^{-2} at –0.59 V vs. RHE.[62] Similar trends were also reported by Hu et al. who tested cobalt porphyrin immobilized on CNTs (CoTPP/CNT).[144] The non-covalent immobilization of cobalt meso-tetraphenylporphyrin (CoTPP) on carbon nanotubes (CNTs) was achieved by simply sonicating their dispersion in N,N-dimethylformamide (DMF). The prepared material (CoTPP-CNT) can convert CO_2 into CO with a Faraday efficiency of 83% and a current density of 0.59 mA cm^{-2} at –1.15 V (η = 350 mV) vs. saturated calomel electrode (SCE) in 0.5 M $KHCO_3$. Zhu and colleagues demonstrated the non-covalent immobilization of pyrrolidone-based nickel phthalocyanine (PyNiPc) molecules on CNTs.[170] It is believed that the

Table 6.1. Summary of non-covalently immobilized heterogeneous molecular catalyst for CO_2RR.

Molecular	j (mA cm^{-2})	V vs. RHE	Electrolyte	FE$_{CO}$ (%)	TOF (s^{-1})	References
	−0.33	−0.8	0.1 M HClO$_4$	37.4	0.8	151
	2	-0.73	0.1 M NaH$_2$PO$_4$	89	4.8	171
	N.A.	−1.1	5.0 mM Na$_2$SO$_4$	89	390	172
	~ 6	−0.9	0.5 M KHCO$_3$	88	2.05	161
	49	-0.976	0.5 M KHCO$_3$	CH$_4$(27) C$_2$H$_4$ (17) CO (10)	TOF$_{methane}$ = 4.3, TOF$_{ethylene}$ = 1.8	154
	2.1	−1.7	0.1 M TBAPF$_6$/ DMF/H$_2$O	95	14.4	146
	~ 2.4	−0.7	0.5 M KHCO$_3$	~ 73	2.75	144

Table 6.1 contd. ...

...Table 6.1 contd.

Molecular	j (mA cm^{-2})	V vs. RHE	Electrolyte	FE$_{CO}$ (%)	TOF (s^{-1})	References
	~ 10.0	−0.63	0.1 M KHCO$_3$	92	2.7	173
	~ 15.0	−0.63	0.1 M KHCO$_3$	98	4.1	173
	0.24	−0.59	0.5 M KHCO$_3$	93	0.04	145
	1.65	−0.49	0.1 M K$_2$B$_4$O$_7$/0.2 M K$_2$SO$_4$	87.4	N.A.	174
	3.3	−1.4	0.5 M LiClO$_4$/0.1 M NaHCO$_3$, 1% v/v MeCN	Formate(93)	7.4	175
	~ 1.68	−0.54	0.1 M KCl	98.7	2.9	176
	~ 2.11	−0.54	0.1 M KHCO$_3$	95.0	2.5	177
	10.0	−0.45	0.5 M KHCO$_3$	100	5.9	155
	~ 3	−0.7	0.1 M KHCO$_3$	90.9	5	166

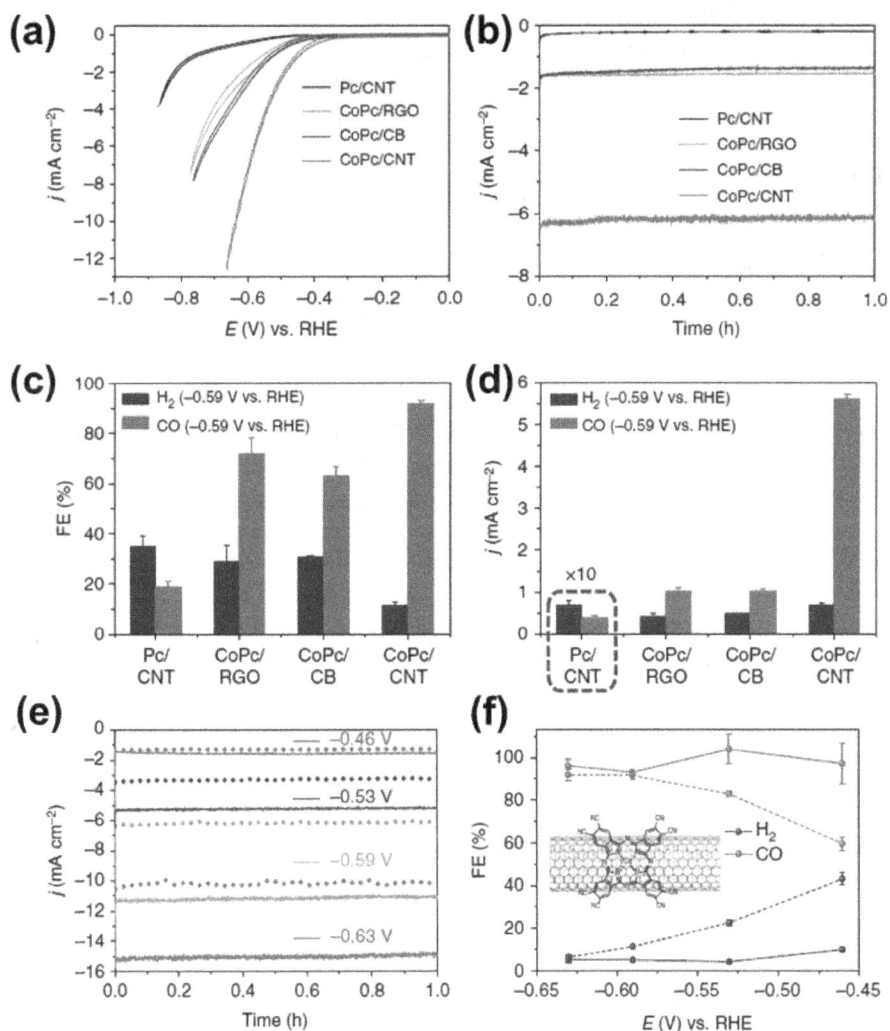

Figure 6.5. (a) Cyclic voltammograms at 5mV s^{-1}, (b) chronoamperograms at -0.59V versus RHE, (c) Faradaic efficiencies of CO_2 reduction products, and (d) partial current densities of CO_2 reduction products for Pc/CNT, CoPc/RGO and CoPc/CB in comparison with CoPc/CNT in 0.1 M KHCO$_3$ solution. (e) Chronoamperograms and (f) Faradaic efficiencies of reduction products at different potentials for CoPc-CN/CNT (solid line) in comparison with CoPc/CNT (dotted line). Reproduced with permission from Ref. 173. Copyright 2017, Nature Publishing Group.

presence of peripheral pyrrolidone groups in PyNiPc can promote monomolecular dispersion, which enhances catalytic site exposure and improves electron transfer. Therefore, the hybrid electrocatalyst can achieve nearly 100% CO selectivity at 0.68 ~ 0.93 V vs. RHE, and has a high activity with a TOF of 13000 h^{-1} at -0.93 V.

In addition to $\pi - \pi$ interactions, non-covalent immobilization can also be assisted by electrostatic interactions. Officials and colleagues prepared the positively charged 5,10,15,20-tetrakis(4-trimethylammoniumphenyl) porphyrinato iron (III) pentachloride (FeTMAP) and negatively charged reduced graphene oxide frameworks.[177] The immobilization of porphyrin molecules on a high surface area carrier not only promotes rapid electron transfer, but also delocalizes the electron density, which causes the Fe^I/Fe^0 redox wave to move forward ($\sim$ 200 mV), thereby generating catalytic activity species (Fe^0). When loaded on a carbon fiber paper electrode, FePGF exhibits a current density of 1.68 mA cm^{-2} (corresponding to a TOF of 2.9 s^{-1}) and CO Faraday efficiency of 98.7% at η = 430 mV for 10 h in 0.1 M KCl.

6.2.2.2 Covalent immobilization

Covalent immobilization of molecular catalysts on substrates by forming chemical bonds has been a hot topic in recent years. Molecular catalysts can be covalently bonded to the electrode surface by various chemical reactions (Table 6.2). For this purpose, catalysts and/or electrodes should be modified with specific functional groups such as diazo,[178,179] amine,[178] alkynyl,[180,181] phosphonic acid[182,183] or thiol.[167] An important advantage of the covalent immobilization is that it establishes a strong connection between the catalyst and the electrode. Because non-covalent immobilization mainly uses $\pi - \pi$ interaction or surface confinement method, the relatively weak van der Waals interaction between catalyst and substrate may lead to leaching of catalyst and deactivation in long-term electrolysis. Non-covalent immobilization may not be effective in electron transfer between molecular catalyst and substrate due to aggregation, which often slows down the electrocatalytic process. Therefore, covalent immobilization technology shows advantages in overcoming the shortcomings of non-covalent immobilization catalyst.

Ye group developed an *in situ* functionalization strategy by first covalently grafting cobalt (II) tetraamino phthalocyanine (CoTAPc) onto carbon nanotubes via a diazo-reaction, followed by a complete methylation reaction[187] (Fig. 6.6a). This is conducive to a 700% increase in CO partial current density compared to that of the physically mixed sample at –0.72 V vs. RHE with highly stable currents (Fig. 6.6b). In a flow cell, this covalently immobilized structure delivers an industrially relevant current density of 239 mA cm^{-2}, CO selectivity of 95.6% at 590 mV overpotential and very low molecular loading of 0.069 mg cm^{-2} (Fig. 6.6d and e). Maurin and Robert[184] successfully immobilized the iron porphyrin derivative (TPP$_{CO2H}$) on the CNT by functionalizing the support with the diazonium salt of p-benzylamine, and then coupling the amine end group of the functionalized support to the carboxylic acid group of the porphyrin. The catalyst obtained a TOF of 178 h^{-1} and 90% CO Faraday efficiency at η = 510 mV in 0.5 M NaHCO$_3$. Using the same diazo reaction, Jiang and his colleagues proposed the concept of conductive molecular wires, which connect CoTPP molecules to the surface of carbon cloth through phenylene linkers.[87] Interestingly, this aromatic linker enhances the transfer of electrons to the Co center, and improves the TOF to 8.3 s^{-1} at η = 500 mV in 0.5 M KHCO$_3$, which is $\sim$ 2 times higher than that of the non-covalent counterpart.

The axial ligand also has a great influence on the electronic state of the metal center and can be used to adjust the electron density on the metal center, thereby reducing the overpotential and tuning the selectivity of the reaction. Zhu and

Table 6.2. Summary of covalently immobilized heterogeneous molecular catalyst for CO_2RR.

Molecular	j (mA cm^{-2})	V vs. RHE	Electrolyte	FE$_{CO}$ (%)	TOF (s^{-1})	References
	~ 0.16	−0.62	0.5 M NaHCO$_3$	95	0.05	184
	0.25	−0.62	DMF/5% H$_2$O	93	245	182
	25.1	−0.60	0.5 M NaHCO$_3$	98.3	1.9	185
	5.5	−0.63	0.2 M KHCO$_3$	98.4	4.9	186
	0.38	−0.63	0.2 M KHCO$_3$	90.0	30.7	186
	11.9	−0.62	0.5 M KHCO$_3$	98.0	8.6	187
	0.21	-0.62	0.5 M KHCO$_3$	94.6	102.9	187

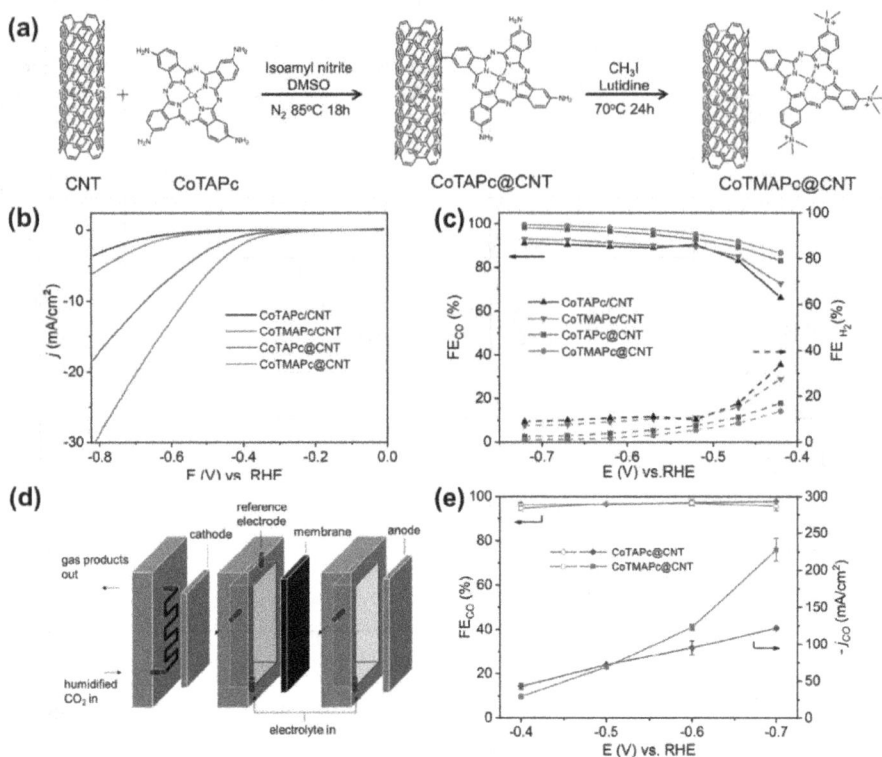

Figure 6.6. (a) Scheme for the synthesis of CoTMAPc@CNT. (b) LSV curves acquired at a scan rate of 10 mV s^{-1}. The catalyst loading is 7×10^{-9} mol cm^{-2}, (c) FE of CO and H$_2$. (d) Graphic illustration of the CO$_2$ flow cell. (e) Current densities and selectivities of CO production at various potentials of CoTAPc@CNT and CoTMAPc@CNT. Reproduced with permission from Ref. 187. Copyright 2021, Royal Society of Chemistry.

colleagues used a diazo reaction to make pyridine-functionalized CNTs bind to CoPc through covalent bond between the nitrogen atom in pyridine and the cobalt center of CoPc (CoPc-py-CNT).[186] The author believes that coordination improves the dispersibility of CoPc on the carbon support and regulates the electronic structure of its active site (Fig. 6.7a). As a result, compared with non-covalently fixed CoPc (CoPc-CNT) and unsupported CoPc, the CO$_2$RR performance of CoPc-py-CNT is improved with CO Faradaic efficiency of > 98% and TOF of 34.5 s^{-1} at –0.63 V vs. RHE in 0.2 M NaHCO$_3$. The same research group also reported a method to directly attach protoporphyrin IX cobalt chloride (CoPPCl) to CNTs through the reaction between chlorine atoms on CoPPCl and OH groups on CNTs[185] (Fig. 6.7b). In this way, the metal active center is close to the support, which not only enhances the interaction between the catalyst and the support, but also promotes the transfer of electrons to the active sites. Compared with the physical mixing control, this covalently grafted method increases TOF by a factor of three at η = 490 mV in 0.5M NaHCO$_3$, and attains a high current density of 25.1 mA cm^{-2}, a CO Faraday efficiency of 98.3% and excellent long-term stability.

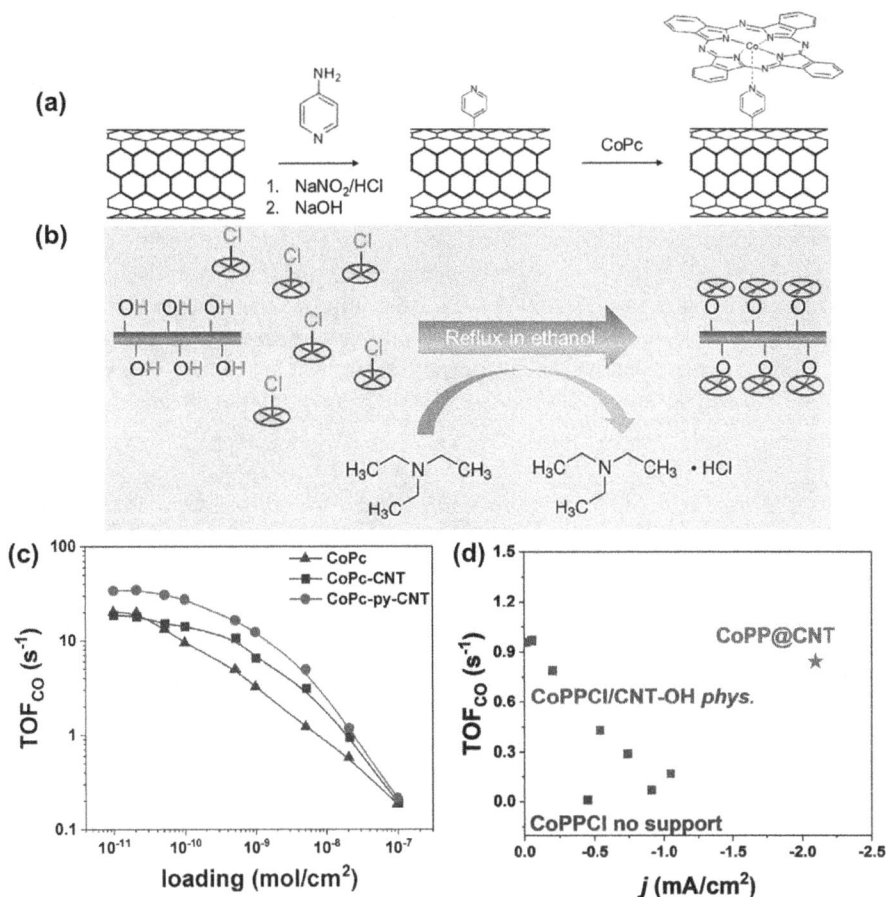

Figure 6.7. Axial ligand tuning strategy of heterogenized molecular catalyst: Synthesis of (a) CoPc-py-CNT and (b) CoPP@CNT composites. (c) TOF_{CO} comparison of CoPc, CoPc-CNT, and CoPc-py-CNT at various CoPc loadings. Reproduced with permission from Ref. 85. Copyright 2019, Elsevier. (d) TOF_{CO} as a function of the current density at -0.55 V vs. RHE for unsupported CoPPCl, physically mixed CoPPCl/CNT-OH, and CoPP@CNT. Reproduced with permission from Ref. 185. Copyright 2019, Wiley-VCH.

6.2.2.3 Polymeric immobilization

In recent years, the method of polymeric assembly into porous frameworks has been developed, such as covalent organic or metal organic frameworks (COFs or MOFs, respectively). This method can also turn molecular complexes into heterogeneous catalysts.[189–195] This strategy provides the following advantages when the framework is constructed: (1) no additional support materials are required; (2) high exposure to active sites due to high porosity; (3) direct adjustment of the electronic properties of active sites and/or organic linkers through structural modification of catalysts.

Lin et al. prepared the model framework (COF-366-Co) by an imine condensation between [5,10,15,20-tetrakis (4-aminophenyl) porphinato] cobalt (CoTAP) and 1,4-benzenedicarboxaldehyde (BDA).[189] An extended COF (COF-367-Co) was also prepared by replacing BDA with biphenyl-4,4′-dicarboxaldehyde (BPDA) (Fig. 6.8a). Scanning electron microscopy (SEM) images show that the as-prepared COFs hold a three-dimensional structure with a rectangular rod-like morphology (~ 50 nm long) (Fig. 6.8b and c). Thus, their Brunner-Emmett-Teller (BET) areas are very high (1360 and 1470 $m^2 g^{-1}$ for COF-366-Co and COF-367-Co, respectively). In the range of applied potentials from -0.57 to -0.97 V vs. RHE, CO is the main reduction product and COF is at least 10% more selective for CO than the molecular cobalt porphyrins. Under electrolysis for 25 h, COF-367-Co showed better catalytic activity than COF-366-Co due to the increased channel width and interlayer distance, which exposed more Co sites in the catalyst. To further improve the catalytic performance of the catalysts, the authors applied a multivariate strategy in the preparation of COF, i.e., diluting the electroactive cobalt porphyrin active sites in the extended COF (COF-367-Co (10%) and COF-367-Co (1%)) using catalytically inactive copper

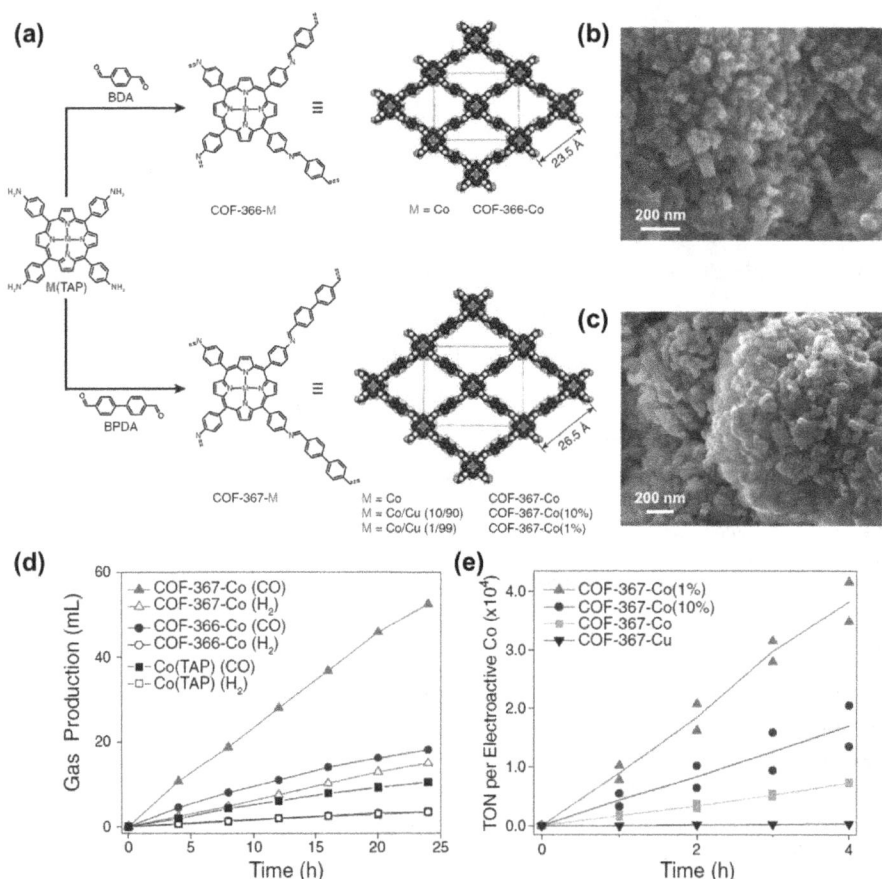

Figure 6.8. (a) The space-filling structural models of COF-366-M and COF-367-M (obtained by using Materials Studio 7.0 and refined with experimental XRD data). SEM images of (b) COF-366-Co and (c) COF-367-Co. (d) Results of long-term bulk electrolysis at −0.67 V versus RHE for the original COF and the expanded COF. (e) TOF_{CO} plots of bulk electrolysis of bimetallic COFs at −0.67 V versus RHE. Reproduced with permission from Ref. 189. Copyright 2015, American Association for the Advancement of Science.

porphyrins to increase the proportion of active sites, thus increasing the TOF. These multivariate Co/Cu COF-367 catalysts showed a large increase in TOF_{CO} with each 10-fold dilution of cobalt loading. The increase in activity is due to the multivariate nature of the bimetallic COFs.

6.2.3 Reduction products beyond two electrons

Transition metal molecular catalysts usually reduce CO_2 to CO as the main product. Sometimes a reduction product with more than two electrons, such as methane and methanol (MeOH), is obtained under large overpotential with low faraday efficiency. It is worth noting that the theoretical calculations of Rossmeisl and colleagues predict that the Co porphyrin that can convert CO_2 to methanol at 0.23 V overpotential, which is surprisingly easier than Cu (211) for methanol production ($\eta = 0.77$ V).[196] It is also predicted that the generation of methane or methanol on Ni porphyrin or Fe porphyrin is extremely challenging.

In 2019, Robert and colleagues[197] demonstrated that in 0.5 M $KHCO_3$, CO_2RR on CoPc can generate methanol, although the selectivity relative to RHE at –0.88 V is only 0.3%, and the methanol partial density is only 0.03 mA cm^{-2}. It has been confirmed that carbon monoxide and formaldehyde are the key reaction intermediates. When using CO as the substrate, the methanol Faraday efficiency increases to 14.3% at –0.64 V in 0.1 M KOH. When using formaldehyde as the substrate, the methanol Faraday efficiency increases to 18.2% at –0.54 V in 0.1 M KOH. Wu et al. reported that CoPc fixed on commercial CNT (CoPc/CNT) can drive CO_2RR to produce methanol[164] (Fig. 6.9a). Consistent with previous findings, CoPc/CNT mainly reduces CO_2 to CO at low and medium potentials. However, under high overpotential, it converts CO_2 to CH_3OH with the maximum Faradaic efficiency of 44% obtained at –0.82 V vs. RHE in 0.1 M $KHCO_3$. Further stability enhancement was achieved via the introduction of amino groups to the Pc ligand (CoPc-NH_2/CNT). This shows that the process of converting CO_2RR to MeOH on CoPc follows the domino catalytic process, in which it first reduces CO_2 to CO through two-electron reduction, and then forms MeOH through a four-electron four-proton process (Fig. 6.9d and e). Theoretical calculations prove that the moderate CoPc-CO binding strength makes the domino catalytic process possible. HAADF-STEM confirmed that another key advantage of the catalyst is the molecular level dispersion of CoPc on CNTs. This indicates that CoPc molecules effectively dispersed on a highly conductive substrate can improve the catalytic reactivity of its CO_2RR to methanol.

6.3 Metal-based single-atom catalysts for CO_2RR

Single atom catalysts (SACs) with separated metal atoms dispersed on a support have become a prosperous field in catalysis science.[198–200] Several experimental methods for preparing SACs have recently been reported, such as pyrolysis, wet chemical synthesis, physical and chemical vapor deposition, electrochemical deposition, and ball milling.[201–204] SACs have low coordination state and a theoretic 100% metal utilization. SACs have an electronic structure completely different from their nanostructured metals. Moreover, there will be a strong interaction between SACs and support, which greatly changes the electronic structure of SACs. To date, various

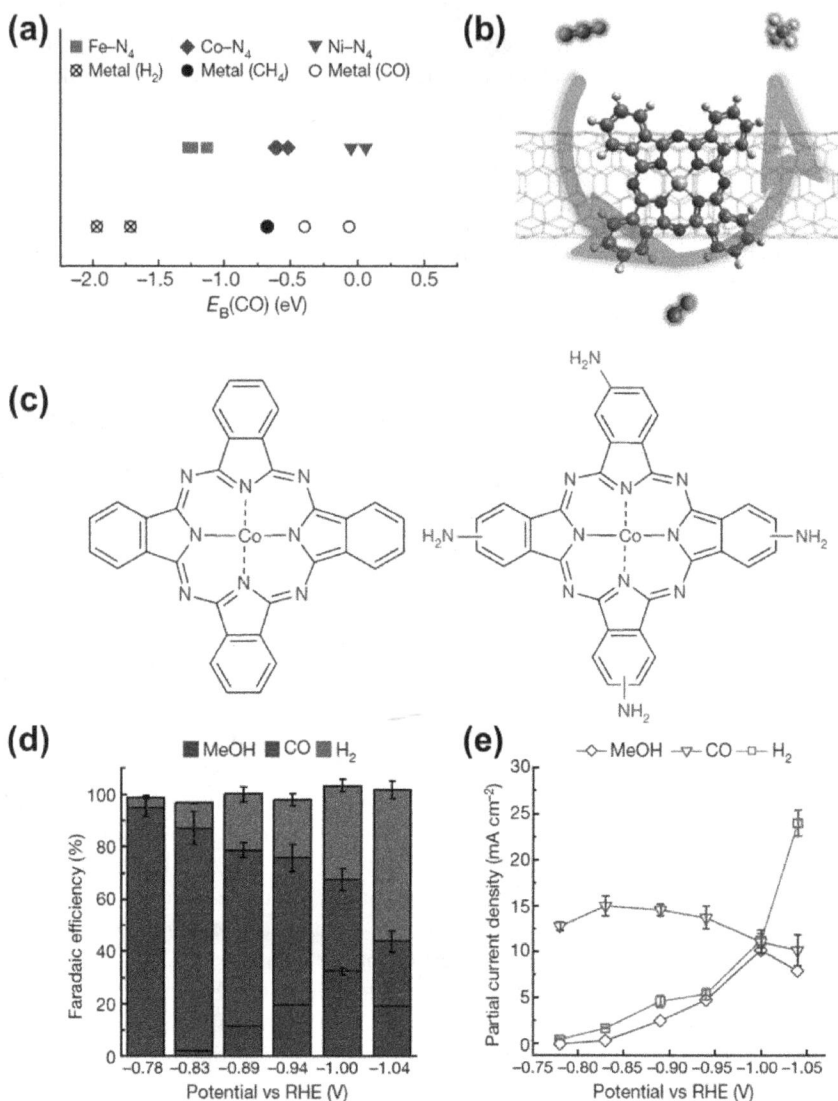

Figure 6.9. (a) Computed CO binding energies on metal surfaces and on M–N$_4$ moieties. (b) Domino process of CO$_2$-to-MeOH conversion via CO, catalysed by CoPc supported on carbon nanotubes (CNT). Color code: hydrogen and cobalt, light grey; carbon, nitrogen, and oxygen, dark grey. See Ref. 63 for the original color. (c) Structural comparison between CoPc and CoPc–NH$_2$. Potential-dependent product selectivity FE (d) and partial current density (e) for CO$_2$ electroreduction catalysed by CoPc–NH$_2$/CNT. Reproduced with permission from Ref. 164. Copyright 2019, Springer Nature.

isolated transition metal atoms (e.g., Fe,[205–214] Co,[215–218] Ni,[112,157,219–230] Cu,[231–234] Zn[235–237] Sn[238]) have been proven to be effective for electrocatalytic CO$_2$ transformation. Table 6.3 shows examples of SACs in CO$_2$RR.

Recently, Chen and co-workers reported that dispersed single-atom Fe sites (Fe^{3+}-N-C) are ultra-active for the electroreduction of CO$_2$ to CO (Fig. 6.10). The

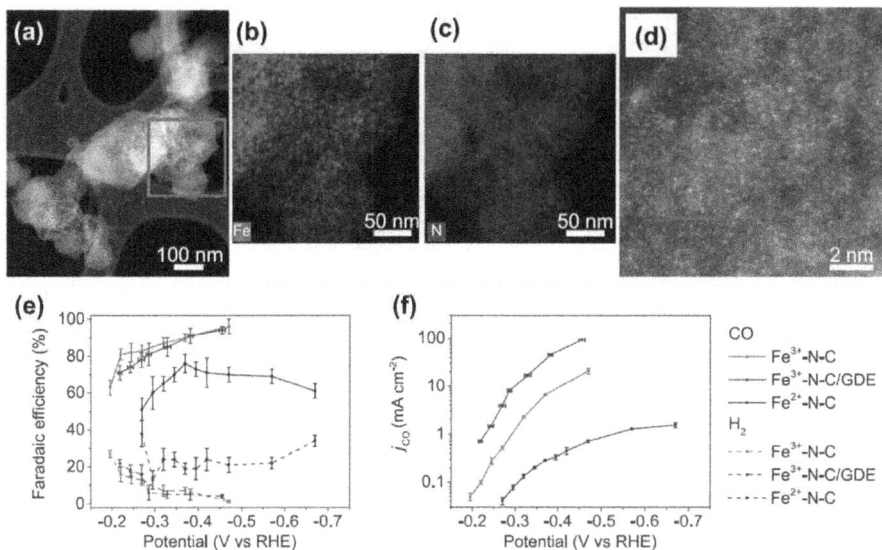

Figure 6.10. Characterizations of Fe^{3+}–N–C. (a) HAADF-STEM image and the corresponding EDS mappings of (b) Fe and (c) N of the region enclosed by the red square. (d) Aberration-corrected HAADF-STEM image. CO_2 electroreduction performance. (e) Faradaic efficiency of CO (solid lines) and H_2 (dashed lines) production and (f) j_{CO} of Fe^{3+}–N–C in an H-cell (light grey) and on a GDE (dark grey), and of Fe^{2+}–N–C in an H-cell (black). Data from the H-cell were obtained by means of chronoamperometry, whereas data from the GDE were obtained by means of chronopotentiometry. Each error bar was the standard deviation determined based on tests of three individual electrodes. Loading was 0.6 mg cm^{-2} for Fe^{3+}–N–C and Fe^{2+}–N–C; 2.5 mg cm^{-2} for Fe^{3+}–N–C/DGE. Reproduced with permission from Ref. 214. Copyright 2019, American Association for the Advancement of Science.

extended X-ray absorption fine structure (EXAFS) showed that the discrete Fe^{3+} ions are coordinated with the pyrrole nitrogen (N) atoms of the N-doped carbon substrates, probably electrically coupled to the conducting carbon, maintaining their +3 oxidation state during the electrocatalytic process. The activity of the SACs is better due to the faster adsorption of CO_2 and weaker adsorption of CO by the Fe^{3+} sites, with a partial current density of 94 mA cm^{-2} at 340 mV overpotential.

For Co SACs, the metal-ligand environments play a key role in improving catalyst activity and selectivity. Wu and co-workers prepared a series of Co-N_x SACs and compared their CO_2RR performance[215] (Fig. 6.11). The results showed that the CO_2 reduction activity of Co-N_2 was superior compared to that of Co-N_3, Co-N_4 and Co NPs in 0.5 M KHCO$_3$ solution, with a FE_{CO} of 94% at 520 mV overpotential and a current density of 18.1 mA cm^{-2}. Theoretical calculations and EXAFS results suggest that this excellent performance is attributed to the effective adsorption and activation of CO_2 molecules on the two-coordination number of Co sites. Their group also designed atomically dispersed Co-N_5 active sites immobilized on polymer-derived hollow N-doped porous carbon spheres (HNPCSs) to convert CO_2 to CO.[218] Co-N_5 SACs exhibited superb performance at –0.79 V (–0.73 V), with FE_{CO} up to 99.4% (99.2%) and FE_{CO} above 90% over the entire potential range from –0.57 to –0.88 V vs. RHE. Meanwhile, the current density and FE_{CO} remained constant after 10 hours of testing. The *in situ* X-ray absorption fine structure (XAFS)

Table 6.3. Summary of the performances of SACs for CO_2RR.

Material	j (mA cm^{-2})	V vs. RHE	Electrolyte	FE_{CO} (%)	References
Fe^{3+}–N–C	94	−0.45	0.5 M $KHCO_3$	∼ 90	214
Fe–SAs/N–C	N.A.	−0.45	0.5 M $NaHCO_3$	99.6	239
Co-N_2	18.1	−0.68	0.5 M $KHCO_3$	95	215
Co-N_5/HNPCSs	6.2	−0.79	0.2 M $NaHCO_3$	99.4	218
Ni SAs/N–C	10.48	−0.9	0.5 M $KHCO_3$	71.9	240
Ni–N–C	12.75	−0.9	0.5 M $KHCO_3$	91.2	241
Ni–N_4–C	28.6	−0.81	0.5 M $KHCO_3$	99	219
A-Ni-NSG	22	−0.5	0.5 M $KHCO_3$	97	112
ZnN_x/C	4.8	−0.43	0.5 M $KHCO_3$	95	236
Cu–N–C	16.2	−1.2	0.1 M $CsHCO_3$	55 (CH_3CH_2OH)	242
$Sn^{\delta+}SA$	11.7	−1.6	0.25 M $KHCO_3$	74.3	131

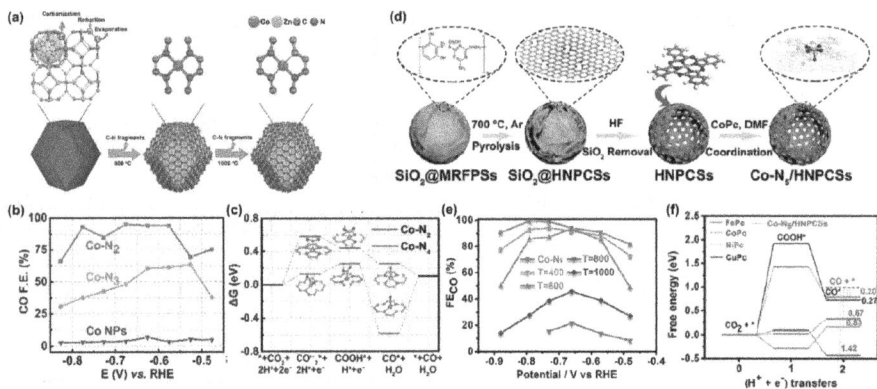

Figure 6.11. (a) The formation process of Co-N_4 and Co-N_2. (b) CO Faradaic efficiencies at different applied potentials of Co-N_2, Co-N_3 and Co NPs. (c) Calculated Gibbs free energy diagrams for CO_2 electroreduction to CO on Co-N_2 and Co-N_4. Reproduced with permission from Ref. 215. Copyright 2018, Wiley-VCH. (d) Schematic illustration of Co-N_5/HNPCSs synthesis. (e) FE_{CO} of different Co−N_x coordination number. (f) Calculated free energy of CO_2RR (dark grey short dash line, the desorption free energy level of CO; light grey dash dot lines, the desorption free energy of CO). Reproduced with permission from Ref. 218. Copyright 2018, American Chemical Society.

results and DFT calculations explain the intrinsic reactivity and properties of the Co-N_5 SACs. The results indicate that the active center of CO-N_5 contributes to the activation of CO_2 and the formation of intermediates *COOH and CO.

6.4 Metal nanostructure engineering for CO_2RR

Transition metal electrodes and main group metal electrodes, such as Cu, Au, Sn, Bi, etc., have been widely explored as electrocatalysts for CO_2RR. Usually, the generation of *CO intermediates are considered as the RDS. In order to achieve high efficiency in this process, a necessary condition is the stabilization of this key intermediate by electrocatalyst. Metal electrodes can be divided into three groups based on their tendency to bind to various intermediates and products (Fig. 6.12). Group 1 includes Sn, Hg, Pb and In,[243,244] which hardly bind to *CO intermediates and therefore form formic acid or formic acid as the main product through an exospheric mechanism. Group 2 includes metals such as Au, Ag, Zn, and Pd, which are tightly bound to *COOH, but have weak affinity for *CO intermediates. Thus, CO can be easily desorbed and appear as the main reaction products.[245] Group 3 only contains Cu, which can bind to *CO intermediates and convert them to various hydrocarbons and alcohols via *COH or *CHO intermediates.[246,247]

Compared with monometallic catalysts, alloys can improve catalytic performance because combining with a second or even multiple metals introduces changes to the electronic structure as well as geometric or ensemble effects on the metal active site. The effect of alloy homogeneity on CO_2RR reactivity was also investigated by Kenis et al.[261] A series of bimetallic CuPd alloy NPs with ordered, disordered, and phase-separated atomic arrangements, as well as two additional

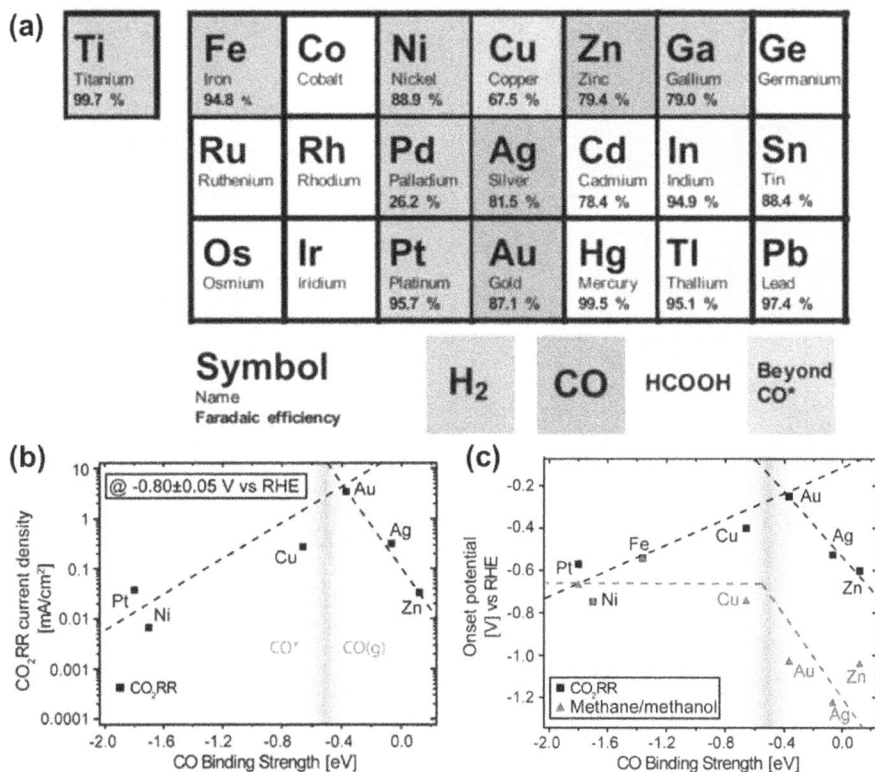

Figure 6.12. (a) Major product classification of metal catalysts for electroreduction of CO_2 and the respective FEs. Four groups are identified: H2, formic acid, CO, and beyond CO* (see Ref. 148 for original color) that is hydrocarbons and multicarbon oxygenates. Reproduced with permission from Ref. 248. Copyright 2017, Wiley-VCH. (b) Volcano plot of partial current density for CO_2RR at –0.8 V vs CO binding strength. (c) Three distinct onset potentials plotted vs CO binding energy: the HER, the overall CO_2RR, and methane or methanol. Reproduced with permission from Ref. 249. Copyright 2014, American Chemical Society.

disordered arrangements (Cu_3Pd and $CuPd_3$), were studied to determine the key factors in the selective generation of C_1 or C_2 products in CO_2RR (Fig. 6.13). They found that ordered CuPd catalysts had the highest selectivity for C_1 products (> 80%) compared to disordered and phase-separated CuPd catalysts, while phase-separated CuPd and Cu_3Pd catalysts had higher selectivity for C_2 products (> 60%) than $CuPd_3$ or ordered CuPd. Based on these findings, the authors proposed that the dimerization probability of C_1 intermediates with adjacent Cu atoms surfaces are higher, implying that geometric effects rather than electronic effects may be critical in determining the selectivity of Cu-Pd alloy catalysts. A variety of metal electrocatalysts for CO_2RR are summarized in Table 6.4.

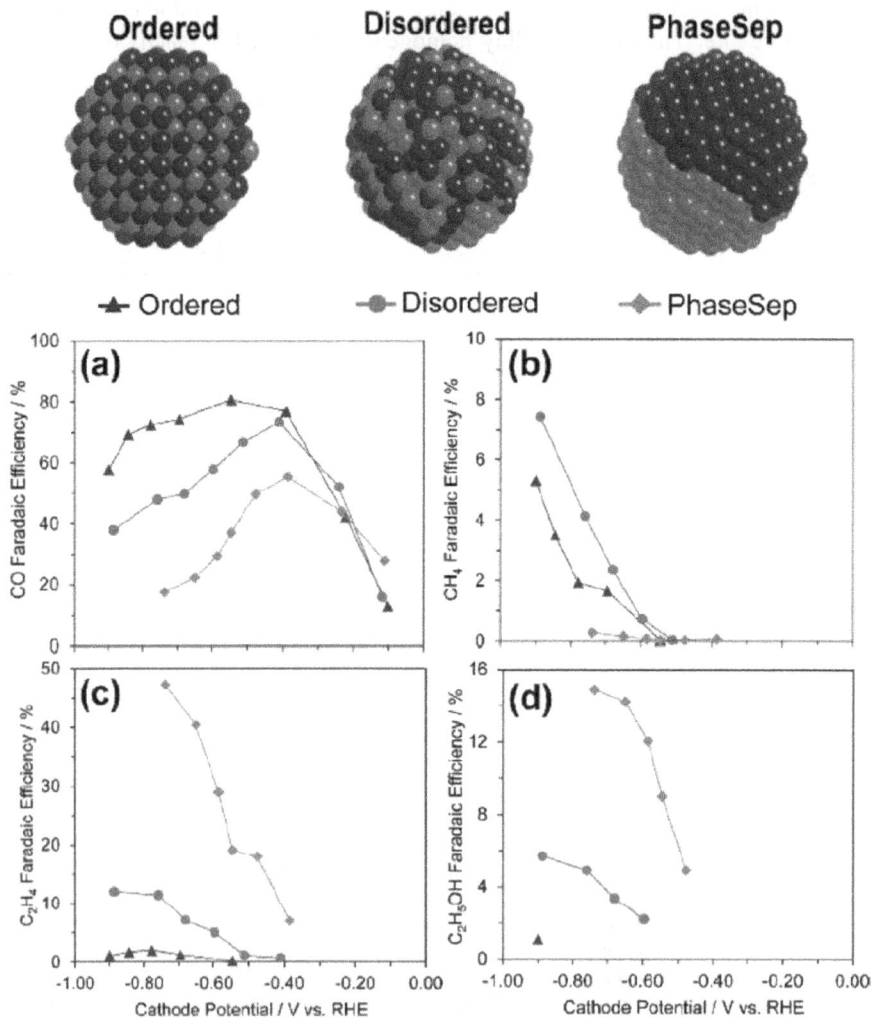

Figure 6.13. Faradaic efficiencies for (a) CO; (b) CH_4; (c) C_2H_4; (d) C_2H_5OH for bimetallic Cu-Pd catalysts with different mixing patterns: ordered, disordered, and phase-separated. Reproduced with permission from Ref. 261. Copyright 2017, American Chemical Society.

Table 6.4. Summary of metal electrocatalysts for CO_2RR.

Material	j (mA cm^{-2})	V vs. RHE	Electrolyte	Main product	References
Monometallic Catalysts					
Au nanowire	N.A.	−0.35	0.5 M KHCO$_3$	94.0 (CO)	250
Pd NP	22.0	−0.4	0.5 M KHCO$_3$	97.0 (HCOO$^-$)	251
Cu-12	232	−0.83	1 M KHCO$_3$	72 (C$_2$H$_4$)	252
N-ND/Cu	N.A.	−0.5	0.5 M KHCO$_3$	34.7 (CH$_3$COOH)	253
Cu-SA/NPC	N.A.	−0.36	0.1 M KHCO$_3$	36.7 (CH$_3$COCH$_3$)	254
Bi nanotube	40.6	−1.1	0.5 M KHCO$_3$	97 (HCOO$^-$)	255
Bi-ene	30.4	−0.9	0.5 M KHCO$_3$	98.6 (HCOO$^-$)	256
Alloys catalysts					
CuSn$_3$	34.7	−0.5	0.1 M KHCO$_3$	95 (HCOO$^-$)	257
BiSn	52.1	−1.1	0.5 M KHCO$_3$	96 (HCOO$^-$)	258
PdAg	3.2	−0.27	0.1 M NaHCO$_3$	94 (HCOO$^-$)	259
Cu-Al	400	−1.5	1 M KOH	80 (C$_2$H$_4$)	260

6.5 Conclusions

In this chapter, we summarize the advances for the rational design of transition metal-based heterogeneous catalysts for the electrochemical conversion of CO_2 to high-value carbon products. With a particular focus on structural effects, we provide a critical comparison of the design principles guiding the development of different types of catalysts, including molecular systems, single-atom and nanostructured catalysts. The performance of molecular catalyst is evaluated in two aspects—one is the intrinsic effect from the catalyst structure and the other is the extrinsic effect from the immobilization strategy (including non-covalent, covalent and periodic) and support material. The main strategies employed for the design of single-atom and nanostructured catalysts are also discussed. Compared to the bulk metal surfaces, molecular catalysts exhibit intrinsically higher selectivity for CO_2RR due to the specificity of their active sites. For both molecular and SACs catalysts, earth-rich first row transition metals (e.g., Mn, Fe, Ni, Co) are commonly used for efficient and selective CO_2RR in organic and aqueous media.

Although considerable efforts have been made in catalyst development and in understanding the catalytic mechanism theoretically and experimentally, there are still some challenges in this field that need to be addressed. First, in many cases, the knowledge gained from homogeneous catalysis through appropriate ligand modification can facilitate the development of more active and selective heterogeneous molecular catalysts. However, the relationship between these two cases is not trivial considering that the global catalytic performance exhibited by heterogeneous catalysts is determined by a combination of intrinsic and extrinsic effects from the catalyst, the support and their interactions. Therefore, more fundamental studies, including DFT calculations and operando spectroscopic characterization, are needed to reveal the role of each factor in heterogeneous CO_2RR. Second, along with more detailed mechanistic studies and investigation of new strategies for catalyst optimization based on structural effects, efforts should be made to address the technical challenges associated with the application of heterogeneous catalysts in commercial devices. In particular, the use of standard electrolyzer or flow cell setups to test the long-term performance of transition metal catalysts should be strongly encouraged to explore their practical applications.

References

[102]   Lewis, N. S. 2016. Aspects of science and technology in support of legal and policy frameworks associated with a global carbon emissions-control regime. *Energy Environ. Sci.*, 9: 2172–2176.

[103]   Nguyen, D. L. T., Y. Kim, Y. J. Hwang and D. H. Won. 2020. Progress in development of electrocatalyst for CO_2 conversion to selective CO production. *Carbon Energy*, 2: 72–98.

[104]   Shen, H., Z. Gu and G. Zheng. 2019. Pushing the activity of CO_2 electroreduction by system engineering. *Sci. Bull.*, 64: 1805–1816.

[105]   Panzone, C., R. Philippe, A. Chappaz, P. Fongarland and A. Bengaouer. 2020. Power-to-liquid catalytic CO_2 valorization into fuels and chemicals: focus on the Fischer-Tropsch route. *J. CO_2 Util.*, 38: 314–347.

[106]   Zhang, Z., S. Y. Pan, H. Li, J. Cai, A. G. Olabi, E. J. Anthony and V. Manovic. 2020. Recent advances in carbon dioxide utilization. Renew. Sustain. *Energy Rev.*, 125: 109799.

[107]   McNutt, M. 2015. Climate warning, 50 years later. *Science*, 350: 721–721.

[108]   Chen, A. and B. L. Lin. 2018. A simple framework for quantifying electrochemical CO_2 fixation. *Joule* 2: 594–606.

[109]   Wei, J., Q. Ge, R. Yao, Z. Wen, C. Fang, L. Guo, H. Xu and J. Sun. 2017. *In situ* click chemistry generation of cyclooxygenase-2 inhibitors. *Nat. Commun.*, 8: 1.

[110]   Bin Song, W. Zhu, J. Fu, Y. Chen, L. Liu, J. R. Zhang, Y. Lin and J. J. Zhu. 2020. Electrode materials engineering in electrocatalytic CO_2 reduction: Energy input and conversion efficiency. *Adv. Mater.*, 32: 1903796.

[111]   Delacourt, C., P. L. Ridgway, J. B. Kerr and J. Newman. 2008. Design of an electrochemical cell making syngas ($CO + H_2$) from CO_2 and H_2O reduction at room temperature. *J. Electrochem. Soc.*, 155, B42.

[112]   Bin Yang, H., S. F. Hung, S. Liu, K. Yuan, S. Miao, L. Zhang, X. Huang, H. Y. Wang, W. Cai, R. Chen, J. Gao, X. Yang, W. Chen, Y. Huang, H. M. Chen, C. M. Li, T. Zhang and B. Liu. 2018. Atomically dispersed Ni(I) as the active site for electrochemical CO_2 reduction. *Nat. Energy*, 3: 140–147.

[113]   Endrödi, B., E. Kecsenovity, A. Samu, F. Darvas, R. V. Jones, V. Török, A. Danyi and C. Janáky. 2019. Multilayer electrolyzer stack converts carbon dioxide to gas products at high pressure with high efficiency. *ACS Energy Lett.*, 4: 1770–1777.

[114]   Sa, Y. J., C. W. Lee, S. Y. Lee, J. Na, U. Lee and Y. J. Hwang. 2020. Catalyst-electrolyte interface chemistry for electrochemical CO_2 reduction. *Chem. Soc. Rev.*, 49: 6632–6665.

[115]   Ross, M. B., P. De Luna, Y. Li, C. T. Dinh, D. Kim, P. Yang and E. H. Sargent. 2019. Designing materials for electrochemical carbon dioxide recycling. *Nat. Catal.*, 2: 648–658.

[116]   Vasileff, A., C. Xu, Y. Jiao, Y. Zheng and S. Z. Qiao. 2018. Surface and interface engineering in copper-based bimetallic materials for selective CO_2 electroreduction. *Chem.*, 4: 1809–1831.

[117]   Francke, R., B. Schille and M. Roemelt. 2018. Homogeneously catalyzed electroreduction of carbon dioxide-methods, mechanisms, and catalysts. *Chem. Rev.*, 118: 4631–4701.

[118]   Fukuzumi, S., Y. M. Lee, H. S. Ahn and W. Nam. 2018. Mechanisms of catalytic reduction of CO_2 with heme and nonheme metal complexes. *Chem. Sci.*, 9: 6017–6034.

[119]   Franco, F., S. Fernández and J. Lloret-Fillol. 2019. Advances in the electrochemical catalytic reduction of CO_2 with metal complexes. *Curr. Opin. Electrochem.*, 15: 109–117.

[120]   Zhang, B. and L. Sun. 2019 Artificial photosynthesis: Opportunities and challenges of molecular catalysts. *Chem. Soc. Rev.*, 48: 2216–2264.

[121]   Jiang, C., A. W. Nichols and C. W. Machan. 2019. A look at periodic trends in d-block molecular electrocatalysts for CO_2 reduction. *Dalt. Trans.*, 48: 9454–9468.

[122] Elouarzaki, K., V. Kannan, V. Jose, H. S. Sabharwal and J. M. Lee. 2019. Recent trends, benchmarking, and challenges of electrochemical reduction of CO_2 by molecular catalysts. *Adv. Energy Mater.*, 9: 1900090.

[123] Takeda, H., C. Cometto, O. Ishitani and M. Robert. 2017. Electrons, photons, protons and earth-abundant metal complexes for molecular catalysis of CO_2 reduction. *ACS Catal.*, 7: 70–88.

[124] Zhang, L. Z. J. Zhao and J. Gong. 2017. Nanostructured materials for heterogeneous Electrocatalytic CO_2 reduction and their related reaction mechanisms. *Angew. Chem. Int. Ed.*, 56: 11326–11353.

[125] Yin, Z., G. T. R. Palmore and S. Sun. 2019. Electrochemical reduction of CO_2 catalyzed by metal nanocatalysts. *Trends Chem.*, 1: 739–750.

[126] Zhang, W., Y. Hu, L. Ma, G. Zhu, Y. Wang, X. Xue, R. Chen, S. Yang and Z. Jin. 2018. Progress and perspective of electrocatalytic CO_2 reduction for renewable carbonaceous fuels and chemicals. *Adv. Sci.*, 5: 1700275.

[127] Zhao, G., X. Huang, X. Wang and X. Wang. 2017. Progress in catalyst exploration for heterogeneous CO_2 reduction and utilization: a critical review. *J. Mater. Chem. A*, 5: 21625–21649.

[128] Zhou, J. H. and Y. W. Zhang. 2018. Metal-based heterogeneous electrocatalysts for reduction of carbon dioxide and nitrogen: Mechanisms, recent advances and perspective. *React. Chem. Eng.*, 3: 591–625.

[129] Li, M., S. Garg, X. Chang, L. Ge, L. Li, M. Konarova, T. E. Rufford, V. Rudolph and G. Wang. 2020. Toward excellence of transition metal-based catalysts for CO_2 electrochemical reduction: an overview of strategies and rationales. *Small Methods*, 4: 2000033.

[130] Li, M., H. Wang, W. Luo, P. C. Sherrell, J. Chen and J. Yang. 2020. Heterogeneous single-atom catalysts for electrochemical CO_2 reduction reaction. *Adv. Mater.*, 32: 2001848.

[131] Zu, X., X. Li, W. Liu, Y. Sun, J. Xu, T. Yao, W. Yan, S. Gao, C. Wang, S. Wei and Y. Xie. 2019. Efficient and robust carbon dioxide electroreduction enabled by atomically dispersed $Sn^{\delta+}$ sites. *Adv. Mater.*, 31: 1808135.

[132] Liu, S., X. F. Lu, J. Xiao, X. Wang and X. W. Lou. 2019. Bi_2O_3 nanosheets grown on multi-channel carbon matrix to catalyze efficient CO_2 electroreduction to HCOOH. *Angew. Chem. Int. Ed.*, 58: 13828–13833.

[133] Kibria, M. G., J. P. Edwards, C. M. Gabardo, C. T. Dinh, A. Seifitokaldani, D. Sinton and E. H. Sargent. 2019. Electrochemical CO_2 Reduction into chemical feedstocks: From mechanistic electrocatalysis models to system design. *Adv. Mater.*, 31: 1–24.

[134] Fletcher, S. 2009. Tafel slopes from first principles. *J. Solid State Electrochem.*, 13: 537–549.

[135] Wuttig, A., Y. Yoon, J. Ryu and Y. Surendranath. 2017. bicarbonate is not a general acid in Au-catalyzed CO_2 electroreduction. *J. Am. Chem. Soc.*, 139: 17109–17113.

[136] Lee, C. W., N. H. Cho, S. W. Im, M. S. Jee, Y. J. Hwang, B. K. Min and K. T. Nam. 2018. New challenges of electrokinetic studies in investigating the reaction mechanism of electrochemical CO_2 reduction. *J. Mater. Chem. A*, 6: 14043–14057.

[137] Xia, C., P. Zhu, Q. Jiang, Y. Pan, W. Liang, E. Stavitsk, H. N. Alshareef and H. Wang. 2019. Continuous production of pure liquid fuel solutions via electrocatalytic CO_2 reduction using solid-electrolyte devices. *Nat. Energy*, 4: 776–785.

[138] Fan, L., C. Xia, P. Zhu, Y. Lu and H. Wang. 2020. U1 snRNP regulates cancer cell migration and invasion *in vitro*. *Nat. Commun.*, 11: 1.

[139] Leow, W. R., Y. Lum, A. Ozden, Y. Wang, D. H. Nam, B. Chen, J. Wicks, T. T. Zhuang, F. Li, D. Sinton and E. H. Sargent, Chloride-mediated selective electrosynthesis of ethylene and propylene oxides at high current density. Science, 2020, **368**, 1228-1233.

[140] H. Cheng, S. Liu, J. Zhang, T. Zhou, N. Zhang, X. S. Zheng, W. Chu, Z. Hu, C. Wu and Y. Xie. 2020. Surface nitrogen-injection engineering for high formation rate of CO_2 reduction to formate. *Nano Lett.*, 20: 6097–6103.

[141] García de Arquer, F. P., C. T. Dinh, A. Ozden, J. Wicks, C. McCallum, A. R. Kirmani, D. H. Nam, C. Gabardo, A. Seifitokaldani, X. Wang, Y. C. Li, F. Li, J. Edwards, L. J. Richter, S. J. Thorpe, D. Sinton and E. H. Sargent. 2020. CO_2 electrolysis to multicarbon products at activities greater than 1 A cm^{-2}. *Science*, 367: 661–666.

[142] Hu, X. M., S. U. Pedersen and K. Daasbjerg. 2019. Supported molecular catalysts for the heterogeneous CO_2 electroreduction. *Curr. Opin. Electrochem.*, 15: 148–154.

[143] Sonkar, P. K., V. Ganesan, R. Gupta, D. K. Yadav and M. Yadav. 2018. Nickel phthalocyanine integrated graphene architecture as bifunctional electrocatalyst for CO_2 and O_2 reductions. *J. Electroanal. Chem.*, 826: 1–9.

[144] Hu, X. M., M. H. Rønne, S. U. Pedersen, T. Skrydstrup and K. Daasbjerg. 2017. Enhanced catalytic activity of cobalt porphyrin in CO_2 electroreduction upon immobilization on carbon materials. *Angew. Chem. Int. Ed.*, 56: 6468–6472.

[145] Maurin, A. and M. Robert. 2016. Noncovalent immobilization of a molecular iron-based electrocatalyst on carbon electrodes for selective, efficient CO_2-to-CO conversion in water. *J. Am. Chem. Soc.*, 138: 2492–2495.

[146] Wu, Y., J. Jiang, Z. Weng, M. Wang, D. L. J. Broere, Y. Zhong, G. W. Brudvig, Z. Feng and H. Wang. 2017. Electroreduction of CO_2 catalyzed by a heterogenized Zn–porphyrin complex with a redox-innocent metal center. *ACS Cent. Sci.*, 3: 847–852.

[147] Birdja, Y. Y., R. E. Vos, T. A. Wezendonk, L. Jiang, F. Kapteijn and M. T. M. Koper. 2018. Effects of substrate and polymer encapsulation on CO_2 electroreduction by immobilized indium(III) protoporphyrin. *ACS Catal.*, 8: 4420–4428.

[148] Birdja, Y. Y., J. Shen and M. T. M. Koper. 2017. Influence of the metal center of metalloprotoporphyrins on the electrocatalytic CO_2 reduction to formic acid. *Catal. Today*, 288: 37–47.

[149] Hori, Y., H. H. I. Wakebe, T. Tsukamoto and O. Koga. 1994. Electrocatalytic process of CO selectivity in electrochemical reduction of CO_2 at metal electrodes in aqueous media. *Electrochim. Acta*, 39: 1833–1839.

[150] Göttle, A. J. and M. T. M. Koper. 2018. Determinant role of electrogenerated reactive nucleophilic species on selectivity during reduction of CO_2 catalyzed by metalloporphyrins. *J. Am. Chem. Soc.*, 140: 4826–4834.

[151] Shen, J., R. Kortlever, R. Kas, Y. Y. Birdja, O. Diaz-Morales, Y. Kwon, I. Ledezma-Yanez, K. J. P. Schouten, G. Mul and M. T. M. Koper. 2015. Electrocatalytic reduction of carbon dioxide to carbon monoxide and methane at an immobilized cobalt protoporphyrin. *Nat. Commun.*, 6: 8177.

[152] Zhao, J., S. Xue, J. Barber, Y. Zhou, J. Meng and X. Ke. 2020. An overview of Cu-based heterogeneous electrocatalysts for CO_2 reduction. *J. Mater. Chem. A*, 8: 4700–4734.

[153] Weng, Z., Y. Wu, M. Wang, J. Jiang, K. Yang, S. Huo, X. F. Wang, Q. Ma, G. W. Brudvig, V. S. Batista, Y. Liang, Z. Feng and H. Wang. 2018. Structural absorption by barbule microstructures of super black bird of paradise feathers. *Nat. Commun.*, 9: 1.

[154] Weng, Z., J. Jiang, Y. Wu, Z. Wu, X. Guo, K. L. Materna, W. Liu, V. S. Batista, G. W. Brudvig and H. Wang. 2016. Electrochemical CO_2 reduction to hydrocarbons on a heterogeneous molecular Cu catalyst in aqueous solution. *J. Am. Chem. Soc.*, 138: 8076–8079.

[155] Wang, M., L. Chen, T. C. Lau and M. Robert. 2018. A hybrid Co quaterpyridine complex/carbon nanotube catalytic material for CO_2 reduction in water. *Angew. Chem. Int. Ed.*, 57: 7769–7773.

[156] Zhang, Z., J. Xiao, X. J. Chen, S. Yu, L. Yu, R. Si, Y. Wang, S. Wang, X. Meng, Y. Wang, Z. Q. Tian and D. Deng. 2018. Reaction Mechanisms of well-defined metal–N_4 sites in electrocatalytic CO_2 reduction. *Angew. Chem. Int. Ed.*, 57: 16339–16342.

[157] Hu, X. M., H. H. Hval, E. T. Bjerglund, K. J. Dalgaard, M. R. Madsen, M. M. Pohl, E. Welter, P. Lamagni, K. B. Buhl, M. Bremholm, M. Beller, S. U. Pedersen, T. Skrydstrup and K. Daasbjerg. 2018. Selective CO_2 reduction to CO in water using earth-abundant metal and nitrogen-doped carbon electrocatalysts. *ACS Catal.*, 8: 6255–6264.

[158] Bhugun, I., D. Lexa and J. M. Savéant. 1996. Catalysis of the electrochemical reduction of carbon dioxide by iron(0) porphyrins: Synergystic effect of weak brönsted acids. *J. Am. Chem. Soc.*, 118: 1769–1776.

[159] Azcarate, I., C. Costentin, M. Robert and J. M. Savéant. 2016. Dissection of electronic substituent effects in multielectron-multistep molecular catalysis. Electrochemical CO_2-to-CO conversion catalyzed by iron porphyrins. *J. Phys. Chem. C*, 120: 28951–28960.

[160] Zhu, M., D. T. Yang, R. Ye, J. Zeng, N. Corbin and K. Manthiram. 2019. Inductive and electrostatic effects on cobalt porphyrins for heterogeneous electrocatalytic carbon dioxide reduction. *Catal. Sci. Technol.*, 9: 974–980.

[161] Morlanés, N., K. Takanabe and V. Rodionov. 2016. Simultaneous reduction of CO_2 and splitting of H_2O by a single immobilized cobalt phthalocyanine electrocatalyst. *ACS Catal.*, 6: 3092–3095.

[162] Abe, T., H. Imaya, T. Yoshida, S. Tokita, D. Schlettwein, D. Wöhrle and M. Kaneko. 1997. Electrochemical CO_2 reduction catalysed by cobalt octacyanophthalocyanine and its mechanism. *J. Porphyr. Phthalocyanines*, 01: 315–321.

[163] Zhang, X., Z. Wu, X. Zhang, L. Li, Y. Li, H. Xu, X. Li, X. Yu, Z. Zhang, Y. Liang and H. Wang. 2017. *In situ* click chemistry generation of cyclooxygenase-2 inhibitors. *Nat. Commun.*, 8: 1.

[164] Wu, Y., Z. Jiang, X. Lu, Y. Liang and H. Wang. 2019. Domino electroreduction of CO_2 to methanol on a molecular catalyst. *Nature*, 575: 639–642.

[165] Wang, M., K. Torbensen, D. Salvatore, S. Ren, D. Joulié, F. Dumoulin, D. Mendoza, B. Lassalle-Kaiser, U. Işci, C. P. Berlinguette and M. Robert. 2019. CO_2 electrochemical catalytic reduction with a highly active cobalt phthalocyanine. *Nat. Commun.*, **10**: 3602 .

[166] Choi, J., P. Wagner, S. Gambhir, R. Jalili, D. R. Macfarlane, G. G. Wallace and D. L. Officer. 2019. Steric modification of a cobalt phthalocyanine/graphene catalyst to give enhanced and stable electrochemical CO_2 reduction to CO. *ACS Energy Lett.*, 4: 666–672.

[167] Clark, M. L., A. Ge, P. E. Videla, B. Rudshteyn, C. J. Miller, J. Song, V. S. Batista, T. Lian and C. P. Kubiak. 2018. CO_2 reduction catalysts on gold electrode surfaces influenced by large electric fields. *J. Am. Chem. Soc.*, 140: 17643–17655.

[168] Zhu, M., R. Ye, K. Jin, N. Lazouski and K. Manthiram. 2018. Elucidating the reactivity and mechanism of CO_2 electroreduction at highly dispersed cobalt phthalocyanine. *ACS Energy Lett.*, 3: 1381–1386.

[169] Ren, S., D. Joulié, D. Salvatore, K. Torbensen, M. Wang, M. Robert and C. P. Berlinguette. 2019. Molecular electrocatalysts can mediate fast, selective CO_2 reduction in a flow cell. *Science*, 365: 367–369.

[170] Ma, D. D., S. G. Han, C. Cao, X. Li, X. T. Wu and Q. L. Zhu. 2020. Remarkable electrocatalytic CO_2 reduction with ultrahigh CO/H_2 ratio over single-molecularly immobilized pyrrolidinonyl nickel phthalocyanine. *Appl. Catal. B Environ.*, 264: 118530.

[171] Kramer, W. W. and C. C. L. McCrory. 2016. Polymer coordination promotes selective CO_2 reduction by cobalt phthalocyanine. *Chem. Sci.*, 7: 2506–2515.

[172] Aoi, S., K. Mase, K. Ohkubo and S. Fukuzumi. 2015. Selective electrochemical reduction of CO_2 to CO with a cobalt chlorin complex adsorbed on multi-walled carbon nanotubes in water. *Chem. Commun.*, 51: 10226–10228.

[173] Zhang, X., Z. Wu, X. Zhang, L. Li, Y. Li, H. Xu, X. Li, X. Yu, Z. Zhang, Y. Liang and H. Wang. 2017. Highly selective and active CO_2 reduction electrocatalysts based on cobalt phthalocyanine/carbon nanotube hybrid structures. *Nat. Commun.*, 8: 14675.

[174] Sato, S., K. Saita, K. Sekizawa, S. Maeda and T. Morikawa. 2018. Low-energy electrocatalytic CO_2 reduction in water over Mn-complex catalyst electrode aided by a nanocarbon support and K^+ cations. *ACS Catal.*, 8: 4452–4458.

[175] Kang, P., S. Zhang, T. J. Meyer and M. Brookhart. 2014. Rapid selective electrocatalytic reduction of carbon dioxide to formate by an iridium pincer catalyst immobilized on carbon nanotube electrodes. *Angew. Chemie - Int. Ed.*, 53: 8709–8713.

[176] Choi, J., P. Wagner, R. Jalili, J. Kim, D. R. MacFarlane, G. G. Wallace and D. L. Officer. 2018. A porphyrin/graphene framework: A highly efficient and robust electrocatalyst for carbon dioxide reduction. *Adv. Energy Mater.*, 8: 1–13.

[177] Choi, J., J. Kim, P. Wagner, S. Gambhir, R. Jalili, S. Byun, S. Sayyar, Y. M. Lee, D. R. MacFarlane, G. G. Wallace and D. L. Officer. 2019. Energy efficient electrochemical reduction of CO_2 to CO using a three-dimensional porphyrin/graphene hydrogel. *Energy Environ. Sci.*, 12: 747–755.

[178] Sun, C., L. Rotundo, C. Garino, L. Nencini, S. S. Yoon, R. Gobetto and C. Nervi. 2017. Electrochemical CO_2 reduction at glassy carbon electrodes functionalized by MnI and ReI organometallic complexes. *Chem. Phys. Chem.*, 18: 3219–3229.

[179] Elgrishi, N., S. Griveau, M. B. Chambers, F. Bedioui and M. Fontecave. 2015. Versatile functionalization of carbon electrodes with a polypyridine ligand: metallation and electrocatalytic H^+ and CO_2 reduction. *Chem. Commun.* 51: 2995–2998.

[180]   Yao, S. A., R. E. Ruther, L. Zhang, R. A. Franking, R. J. Hamers and J. F. Berry. 2012. Covalent attachment of catalyst molecules to conductive diamond: CO_2 reduction using "smart" electrodes. *J. Am. Chem. Soc.*, 134: 15632–15635.

[181]   Zhanaidarova, A., C. E. Moore, M. Gembicky and C. P. Kubiak. 2018. Covalent attachment of [Ni(alkynyl-cyclam)]($^{2+}$) catalysts to glassy carbon electrodes. *Chem. Commun.*, 54: 4116–4119.

[182]   Mohamed, E. A., Z. N. Zahran and Y. Naruta. 2017. Efficient heterogeneous CO_2 to CO conversion with a phosphonic acid fabricated cofacial iron porphyrin dimer. *Chem. Mater.*, 29: 7140–7150.

[183]   Wang, Y., S. L. Marquard, D. Wang, C. Dares and T. J. Meyer. 2017. Single-site, heterogeneous electrocatalytic reduction of CO_2 in water as the solvent. *ACS Energy Lett.*, 2: 1395–1399.

[184]   Maurin, A. and M. Robert. 2016. Catalytic CO_2-to-CO conversion in water by covalently functionalized carbon nanotubes with a molecular iron catalyst. *Chem. Commun.*, 52: 12084–12087.

[185]   Zhu, M., J. Chen, L. Huang, R. Ye, J. Xu and Y. F. Han. 2019. Covalently grafting cobalt porphyrin onto carbon nanotubes for efficient CO_2 electroreduction. *Angew. Chemie. Int. Ed.*, 58: 6595–6599.

[186]   Zhu, M., J. Chen, R. Guo, J. Xu, X. Fang and Y. F. Han. 2019. Cobalt phthalocyanine coordinated to pyridine-functionalized carbon nanotubes with enhanced CO_2 electroreduction. *Appl. Catal. B Environ.*, 251: 112–118.

[187]   Su, J., J. J. Zhang, J. Chen, Y. Song, L. Huang, M. Zhu, B. I. Yakobson, B. Z. Tang and R. Ye. 2021. Building a stable cationic molecule/electrode interface for highly efficient and durable CO_2 reduction at an industrially relevant current. *Energy Environ. Sci.*, 14: 483–492.

[188]   Marianov, A. N. and Y. Jiang. 2019. Covalent ligation of Co molecular catalyst to carbon cloth for efficient electroreduction of CO_2 in water. *Appl. Catal. B Environ.*, 244: 881–888.

[189]   Lin, S., C. S. Diercks, Y. B. Zhang, N. Kornienko, E. M. Nichols, Y. Zhao, A. R. Paris, D. Kim, P. Yang, O. M. Yaghi and C. J. Chang. 2015. Covalent organic frameworks comprising cobalt porphyrins for catalytic CO_2 reduction in water. *Science*, 349: 1208–1213.

[190]   Kornienko, N., Y. Zhao, C. S. Kley, C. Zhu, D. Kim, S. Lin, C. J. Chang, O. M. Yaghi and P. Yang. 2015. Metal-organic frameworks for electrocatalytic reduction of carbon dioxide. *J. Am. Chem. Soc.*, 137: 14129–14135.

[191]   Hod, I., M. D. Sampson, P. Deria, C. P. Kubiak, O. K. Farha and J. T. Hupp. 2015. Fe-porphyrin-based metal-organic framework films as high-surface concentration, heterogeneous catalysts for electrochemical reduction of CO_2. *ACS Catal.*, 5: 6302–6309.

[192]   Diercks, C. S., S. Lin, N. Kornienko, E. A. Kapustin, E. M. Nichols, C. Zhu, Y. Zhao, C. J. Chang and O. M. Yaghi. 2018. Reticular electronic tuning of porphyrin active sites in covalent organic frameworks for electrocatalytic carbon dioxide reduction. *J. Am. Chem. Soc.*, 140: 1116–1122.

[193]   Yao, C. L., J. C. Li, W. Gao and Q. Jiang. 2018. An integrated design with new metal-functionalized covalent organic frameworks for the effective electroreduction of CO_2. *Chem. - A Eur. J.*, 24: 11051–11058.

[194]   Wu, H., M. Zeng, X. Zhu, C. Tian, B. Mei, Y. Song, X. L. Du, Z. Jiang, L. He, C. Xia and S. Dai. 2018. Defect engineering in polymeric cobalt phthalocyanine networks for enhanced electrochemical CO_2 reduction. *Chem. Electro. Chem.*, 5: 2717–2721.

[195]   Song, Y., J. J. Zhang, Z. Zhu, X. Chen, L. Huang, J. Su, Z. Xu, T. H. Ly, C. S. Lee, B. I. Yakobson, B. Z. Tang and R. Ye. 2021. Zwitterionic ultrathin covalent organic polymers for high-performance electrocatalytic carbon dioxide reduction. *Appl. Catal. B Environ.*, 284: 119750.

[196]   Tripkovic, V., M. Vanin, M. Karamad, M. E. Björketun, K. W. Jacobsen, K. S. Thygesen and J. Rossmeisl. 2013. Electrochemical CO_2 and CO reduction on metal-functionalized porphyrin-like graphene. *J. Phys. Chem. C*, 117: 9187–9195.

[197]   Boutin, E., M. Wang, J. C. Lin, M. Mesnage, D. Mendoza, B. Lassalle-Kaiser, C. Hahn, T. F. Jaramillo and M. Robert. 2019. Aqueous electrochemical reduction of carbon dioxide and carbon monoxide into methanol with cobalt phthalocyanine. *Angew. Chemie - Int. Ed.*, 58: 16172–16176.

[198]   Qiao, B., A. Wang, X. Yang, L. F. Allard, Z. Jiang, Y. Cui, J. Liu, J. Li and T. Zhang. 2011. Single-atom catalysis of CO oxidation using Pt_1/FeO_x. *Nat. Chem.*, 3: 634–641.

[199]   Peng, Y., B. Lu and S. Chen. 2018. Single atom catalysts: carbon-supported single atom catalysts for electrochemical energy conversion and storage (Adv. Mater. 48/2018). *Adv. Mater.*, 30: 1–25.

[200]   Chen, Y., S. Ji, C. Chen, Q. Peng, D. Wang and Y. Li. 2018. Single-atom catalysts: Synthetic strategies and electrochemical applications. *Joule*, 2: 1242–1264.

[201] Su, J., R. Ge, Y. Dong, F. Hao and L. Chen. 2018. Recent progress in single-atom electrocatalysts: Concept, synthesis, and applications in clean energy conversion. *J. Mater. Chem. A*, 6: 14025–14042.

[202] Li, H., H. X. Zhang, X. L. Yan, B. S. Xu and J. J. Guo. 2018. Carbon-supported metal single atom catalysts. *Xinxing Tan Cailiao/New Carbon Mater.*, 33: 1–11.

[203] Zhang, H., G. Liu, L. Shi and J. Ye. 2018. Single-atom catalysts: Emerging multifunctional materials in heterogeneous catalysis. *Adv. Energy Mater.*, 8: 1–24.

[204] Varela, A. S., W. Ju and P. Strasser. 2018. Molecular nitrogen–carbon catalysts, solid metal organic framework catalysts, and solid metal/nitrogen-doped carbon (MNC) catalysts for the electrochemical CO_2 reduction. *Adv. Energy Mater.*, 8: 1–35.

[205] Zhang, C., S. Yang, J. Wu, M. Liu, S. Yazdi, M. Ren, J. Sha, J. Zhong, K. Nie, A. S. Jalilov, Z. Li, H. Li, B. I. Yakobson, Q. Wu, E. Ringe, H. Xu, P. M. Ajayan and J. M. Tour. 2018. Electrochemical CO_2 reduction with atomic iron-dispersed on nitrogen-doped graphene. *Adv. Energy Mater.*, 8: 1–9.

[206] Pan, F., B. Li, E. Sarnello, Y. Fei, Y. Gang, X. Xiang, Z. Du, P. Zhang, G. Wang, H. T. Nguyen, T. Li, Y. H. Hu, H. C. Zhou and Y. Li. 2020. Atomically dispersed iron-nitrogen sites on hierarchically mesoporous carbon nanotube and graphene nanoribbon networks for CO_2 reduction. *ACS Nano*, 14: 5506–5516.

[207] Cheng, Q., K. Mao, L. Ma, L. Yang, L. Zou, Z. Zou, Z. Hu and H. Yang. 2018. Encapsulation of iron nitride by Fe–N–C shell enabling highly efficient electroreduction of CO_2 to CO. *ACS Energy Letters*, 3: 1205–1211.

[208] Pan, F., H. Zhang, K. Liu, D. Cullen, K. More, M. Wang, Z. Feng, G. Wang, G. Wu and Y. Li. 2018. Unveiling active sites of CO_2 reduction on nitrogen-coordinated and atomically dispersed iron and cobalt catalysts. *ACS Catal.*, 8: 3116–3122.

[209] Huan, T. N., N. Ranjbar, G. Rousse, M. Sougrati, A. Zitolo, V. Mougel, F. Jaouen and M. Fontecave. 2017. Electrochemical reduction of CO_2 catalyzed by Fe-NC materials: A structure-selectivity study. *ACS Catal.*, 7: 1520–1525.

[210] Ye, Y., F. Cai, H. Li, H. Wu, G. Wang, Y. Li, S. Miao, S. Xie, R. Si, J. Wang and X. Bao. 2017. Surface functionalization of ZIF-8 with ammonium ferric citrate toward high exposure of Fe-N active sites for efficient oxygen and carbon dioxide electroreduction. *Nano Energy*, 38: 281–289.

[211] Sheng, T. and S. G. Sun. 2017. Free energy landscape of electrocatalytic CO_2 reduction to CO on aqueous FeN_4 center embedded graphene studied by ab initio molecular dynamics simulations. *Chem. Phys. Lett.*, 688: 37–42.

[212] Varela, A. S., N. Ranjbar Sahraie, J. Steinberg, W. Ju, H.S. Oh and P. Strasser. 2015. Metal-doped nitrogenated carbon as an efficient catalyst for direct CO_2 electroreduction to CO and hydrocarbons. *Angew. Chemie - Int. Ed.*, 54: 10758–10762.

[213] Leonard, N., W. Ju, I. Sinev, J. Steinberg, F. Luo, A. S. Varela, B. Roldan Cuenya and P. Strasser. 2018. The chemical identity, state and structure of catalytically active centers during the electrochemical CO_2 reduction on porous Fe–nitrogen–carbon (Fe–N–C) materials. *Chem. Sci.*, 9: 5064–5073.

[214] Gu, J., C. S. Hsu, L. Bai, H. M. Chen and X. Hu. 2019. Atomically dispersed Fe^{3+} sites catalyze efficient CO_2 electroreduction to CO. *Science*, 364: 1091–1094.

[215] Wang, X., Z. Chen, X. Zhao, T. Yao, W. Chen, R. You, C. Zhao, G. Wu, J. Wang, W. Huang, J. Yang, X. Hong, S. Wei, Y. Wu and Y. Li. 2018. Regulation of coordination number over single Co sites: triggering the efficient electroreduction of CO_2. *Angew. Chemie - Int. Ed.*, 130: 1962–1966.

[216] Geng, Z., Y. Cao, W. Chen, X. Kong, Y. Liu, T. Yao and Y. Lin. 2019. Regulating the coordination environment of Co single atoms for achieving efficient electrocatalytic activity in CO_2 reduction. *Appl. Catal. B Environ.*, 240: 234–240.

[217] He, Q., D. Liu, J. H. Lee, Y. Liu, Z. Xie, S. Hwang, S. Kattel, L. Song and J. G. Chen. 2020. Electrochemical conversion of CO_2 to syngas with controllable CO/H_2 ratios over Co and Ni single-atom catalysts. *Angew. Chemie - Int. Ed.*, 59: 3033–3037.

[218] Pan, Y., R. Lin, Y. Chen, S. Liu, W. Zhu, X. Cao, W. Chen, K. Wu, W. C. Cheong, Y. Wang, L. Zheng, J. Luo, Y. Lin, Y. Liu, C. Liu, J. Li, Q. Lu, X. Chen, D. Wang, Q. Peng, C. Chen and Y. Li. 2018. Design of single-atom Co–N_5 catalytic site: a robust electrocatalyst for CO_2 reduction with nearly 100% CO selectivity and remarkable stability. *J. Am. Chem. Soc.*, 140: 4218–4221.

[219] Li, X., W. Bi, M. Chen, Y. Sun, H. Ju, W. Yan, J. Zhu, X. Wu, W. Chu, C. Wu and Y. Xie. 2017. Exclusive Ni–N$_4$ sites realize near-unity CO selectivity for electrochemical CO$_2$ reduction. *J. Am. Chem. Soc.*, 139: 14889–14892.

[220] Yan, C., H. Li, Y. Ye, H. Wu, F. Cai, R. Si, J. Xiao, S. Miao, S. Xie, F. Yang, Y. Li, G. Wang and X. Bao. 2018. Coordinatively unsaturated nickel–nitrogen sites towards selective and high-rate CO$_2$ electroreduction. *Energy Environ. Sci.*, 11: 1204–1210.

[221] Xiong, W., H. Li, H. Wang, J. Yi, H. You, S. Zhang, Y. Hou, M. Cao, T. Zhang and R. Cao. 2020. Hollow mesoporous carbon sphere loaded Ni–N$_4$ single-atom: Support structure study for CO$_2$ electrocatalytic reduction catalyst. *Small*, 16: 1–11.

[222] Chen, Z., X. Zhang, W. Liu, M. Jiao, K. Mou, X. Zhang and L. Liu. 2021. Amination strategy to boost the CO$_2$ electroreduction current density of M–N/C single-atom catalysts to the industrial application level. *Energy Environ. Sci.*, 14: 11–15.

[223] Zhao, C., X. Dai, T. Yao, W. Chen, X. Wang, J. Wang, J. Yang, S. Wei, Y. Wu and Y. Li. 2017. Ionic exchange of metal-organic frameworks to access single nickel sites for efficient electroreduction of CO$_2$. *J. Am. Chem. Soc.*, 139: 8078–8081.

[224] Su, P., K. Iwase, S. Nakanishi, K. Hashimoto and K. Kamiya. 2016. Nickel-nitrogen-modified graphene: An efficient electrocatalyst for the reduction of carbon dioxide to carbon monoxide. *Small*, 12: 6083–6089.

[225] Pan, F., W. Deng, C. Justiniano and Y. Li, Identification of champion transition metals centers in metal and nitrogen-codoped carbon catalysts for CO$_2$ reduction. *Appl. Catal. B Environ.*, 226: 463–472.

[226] Su, P., K. Iwase, T. Harada, K. Kamiya and S. Nakanishi. 2018. Covalent triazine framework modified with coordinatively-unsaturated Co or Ni atoms for CO$_2$ electrochemical reduction. *Chem. Sci.*, 9: 3941–3947.

[227] Jiang, K., S. Siahrostami, T. Zheng, Y. Hu, S. Hwang, E. Stavitski, Y. Peng, J. Dynes, M. Gangisetty, D. Su, K. Attenkofer and H. Wang. 2018. Isolated Ni single atoms in graphene nanosheets for high-performance CO$_2$ reduction. *Energy Environ. Sci.*, 11: 893–903.

[228] Bi, W., X. Li, R. You, M. Chen, R. Yuan, W. Huang, X. Wu, W. Chu, C. Wu and Y. Xie. 2018. Surface immobilization of transition metal ions on nitrogen-doped graphene realizing high-efficient and selective CO$_2$ reduction. *Adv. Mater.*, 30: 1–6.

[229] Cheng, Y., S. Zhao, B. Johannessen, J. P. Veder, M. Saunders, M. R. Rowles, M. Cheng, C. Liu, M. F. Chisholm, R. De Marco, H. M. Cheng, S. Z. Yang and S. P. Jiang. 2018. Atomically dispersed transition metals on carbon nanotubes with ultrahigh loading for selective electrochemical carbon dioxide reduction. *Adv. Mater.*, 30: 1–7.

[230] Ju, A. Bagger, G. P. Hao, A. S. Varela, I. Sinev, V. Bon, B. Roldan Cuenya, S. Kaskel, J. Rossmeisl and P. Strasser. 2017. Understanding activity and selectivity of metal-nitrogen-doped carbon catalysts for electrochemical reduction of CO$_2$. *Nat. Commun.*, 8: 1–9.

[231] Yang, F., X. Mao, M. Ma, C. Jiang, P. Zhang, J. Wang, Q. Deng, Z. Zeng and S. Deng. 2020. Scalable strategy to fabricate single Cu atoms coordinated carbons for efficient electroreduction of CO$_2$ to CO. *Carbon N. Y.*, 168: 528–535.

[232] Zhu, Q., X. Sun, D. Yang, J. Ma, X. Kang, L. Zheng, J. Zhang, Z. Wu and B. Han. 2019. Carbon dioxide electroreduction to C$_2$ products over copper-cuprous oxide derived from electrosynthesized copper complex. *Nat. Commun.*, 10: 1–11.

[233] Jiao, Y., Y. Zheng, P. Chen, M. Jaroniec and S. Z. Qiao. 2017. Molecular scaffolding strategy with synergistic active centers to facilitate electrocatalytic CO$_2$ reduction to hydrocarbon/alcohol. *J. Am. Chem. Soc.*, 139: 18093–18100.

[234] Wang, Y., Z. Chen, P. Han, Y. Du, Z. Gu, X. Xu and G. Zheng. 2018. Single-atomic Cu with multiple oxygen vacancies on ceria for electrocatalytic CO$_2$ reduction to CH$_4$. *ACS Catal.*, 8: 7113–7119.

[235] Chen, Z., K. Mou, S. Yao and L. Liu. 2018. Zinc-coordinated nitrogen-codoped graphene as an efficient catalyst for selective electrochemical reduction of CO$_2$ to CO. *Chem. Sus. Chem.*, 11: 2944–2952.

[236] Yang, F., P. Song, X. Liu, B. Mei, W. Xing, Z. Jiang, L. Gu and W. Xu. 2018. Highly efficient CO$_2$ electroreduction on ZnN$_4$-based single-atom catalyst. *Angew. Chemie- Int. Ed.*, 57: 12303–12307.

[237] Wang, N., Z. Liu, J. Ma, J. Liu, P. Zhou, Y. Chao, C. Ma, X. Bo, J. Liu, Y. Hei, Y. Bi, M. Sun, M. Cao, H. Zhang, F. Chang, H. L. Wang, P. Xu, Z. Hu, J. Bai, H. Sun, G. Hu and M. Zhou. 2020. Sustainability perspective-oriented synthetic strategy for zinc single-atom catalysts boosting electrocatalytic reduction of carbon dioxide and oxygen. *ACS Sustain. Chem. Eng.*, 8: 13813–13822.

[238] Zhao, Y., J. Liang, C. Wang, J. Ma and G. G. Wallace. 2018. Tunable and efficient tin modified nitrogen-doped carbon nanofibers for electrochemical reduction of aqueous carbon dioxide. *Adv. Energy Mater.*, 8: 1702524.

[239] Quan, F., G. Zhan, H. Shang, Y. Huang, F. Jia, L. Zhang and Z. Ai. 2019. Highly efficient electrochemical conversion of CO_2 and NaCl to CO and NaClO. *Green Chem.*, 21: 3256–3262.

[240] Zhao, C., Y. Wang, Z. Li, W. Chen, Q. Xu, D. He, D. Xi, Q. Zhang, T. Yuan, Y. Qu, J. Yang, F. Zhou, Z. Yang, X. Wang, J. Wang, J. Luo, Y. Li, H. Duan, Y. Wu and Y. Li. 2019. Solid-diffusion synthesis of single-atom catalysts directly from bulk metal for efficient CO_2 reduction. *Joule*, 3: 584–594.

[241] Yuan, C. Z., K. Liang, X. M. Xia, Z. K. Yang, Y. F. Jiang, T. Zhao, C. Lin, T. Y. Cheang, S. L. Zhong and A. W. Xu. 2019. Powerful CO_2 electroreduction performance with N–carbon doped with single Ni atoms. *Catal. Sci. Technol.*, 9: 3669–3674.

[242] Karapinar, D., N. T. Huan, N. Ranjbar Sahraie, J. Li, D. Wakerley, N. Touati, S. Zanna, D. Taverna, L. H. Galvão Tizei, A. Zitolo, F. Jaouen, V. Mougel and M. Fontecave. 2019. Electroreduction of CO_2 on single-site copper-nitrogen-doped carbon material: Selective formation of ethanol and reversible restructuration of the metal sites. *Angew. Chemie-Int. Ed.*, 58: 15098–15103.

[243] Han, N., P. Ding, L. He, Y. Li and Y. Li. 2020. CO_2 reduction: Promises of main group metal–based nanostructured materials for electrochemical CO_2 reduction to formate. *Adv. Energy Mater.*, 10: 1902338.

[244] Zhan, W. W., L. M. Sun, and X. G. Han. 2019. Recent progress on engineering highly efficient porous semiconductor photocatalysts derived from metal-organic frameworks. *Nano-Micro Lett.*, 11: 1.

[245] Zou, Y. and S. Wang. 2021. An investigation of active sites for electrochemical CO_2 reduction reactions: from in situ characterization to rational design. *Adv. Sci.*, 8: 2003579.

[246] Zhao, J., S. Xue, J. Barber, Y. Zhou, J. Meng and X. Ke. 2020. An overview of Cu-based heterogeneous electrocatalysts for CO_2 reduction. *J. Mater. Chem. A*, 8: 4700–4734.

[247] Yan, Y., L. Ke, Y. Ding, Y. Zhang, K. Rui, H. Lin and J. Zhu. 2021. Recent advances in Cu-based catalysts for electroreduction of carbon dioxide. *Mater. Chem. Front.*, 5: 2668–2683.

[248] Bagger, A., W. Ju, A. S. Varela, P. Strasser and J. Rossmeisl. 2017. Electrochemical CO_2 reduction: A classification problem. *Chem. Phys. Chem.*, 18: 3266–3273.

[249] Kuhl,K. P. , T. Hatsukade, E. R. Cave, D. N. Abram, J. Kibsgaard and T. F. Jaramillo. 2014. Electrocatalytic conversion of carbon dioxide to methane and methanol on transition metal surfaces. *J. Am.Chem. Soc.*, 136: 14107–14113.

[250] Zhu, W., Y. J. Zhang, H. Zhang, H. Lv, Q. Li, R. Michalsky, A. A. Peterson and S. Sun. 2014. Active and selective conversion of CO_2 to CO on ultrathin Au nanowires. *J. Am. Chem. Soc.*, 136: 16132–16135.

[251] Klinkova, A., P. De Luna, C. T. Dinh, O. Voznyy, E. M. Larin, E. Kumacheva and E. H. Sargent. 2016. Rational design of efficient palladium catalysts for electroreduction of carbon dioxide to formate. *ACS Catal.*, 6: 8115–8120.

[252] Li, F., A. Thevenon, A. Rosas-Hernández, Z. Wang, Y. Li, C. M. Gabardo, A. Ozden, C. T. Dinh, J. Li, Y. Wang, J. P. Edwards, Y. Xu, C. McCallum, L. Tao, Z. Q. Liang, M. Luo, X. Wang, H. Li, C. P. O'Brien, C. S. Tan, D. H. Nam, R. Quintero-Bermudez, T. T. Zhuang, Y. C. Li, Z. Han, R. D. Britt, D. Sinton, T. Agapie, J. C. Peters and E. H. Sargent. 2020. Molecular tuning of CO_2-to-ethylene conversion. *Nature*, 577: 509–513.

[253] Wang, H., Y. K. Tzeng, Y. Ji, Y. Li, J. Li, X. Zheng, A. Yang, Y. Liu, Y. Gong, L. Cai, Y. Li, X. Zhang, W. Chen, B. Liu, H. Lu, N. A. Melosh, Z. X. Shen, K. Chan, T. Tan, S. Chu and Y. Cui. 2020. Synergistic enhancement of electrocatalytic CO_2 reduction to C_2 oxygenates at nitrogen-doped nanodiamonds/Cu interface. *Nat. Nanotechnol.*, 15: 131–137.

[254] Zhao, K., X. Nie, H. Wang, S. Chen, X. Quan, H. Yu, W. Choi, G. Zhang, B. Kim and J. G. Chen. 2020. U1 snRNP regulates cancer cell migration and invasion *in vitro*. *Nat. Commun.*, 11: 1.

[255] Fan, K., Y. Jia, Y. Ji, P. Kuang, B. Zhu, X. Liu and J. Yu. 2020. Curved surface boosts electrochemical CO_2 reduction to formate via bismuth nanotubes in a wide potential window. *ACS Catal.*, 10: 358–364.

[256] Cao, C., D. Ma, J. Gu, X. Xie, G. Zeng, X. Li, S. Han, Q. Zhu, X. Wu and Q. Xu. 2020. Metal-organic layers leading to atomically thin bismuthene for efficient carbon dioxide electroreduction to liquid fuel. *Angew. Chemie*, 59: 15014–15020.

[257] Zheng, X., Y. Ji, J. Tang, J. Wang, B. Liu, H. G. Steinrück, K. Lim, Y. Li, M. F. Toney, K. Chan and Y. Cui. 2019. Theory-guided Sn/Cu alloying for efficient CO_2 electroreduction at low overpotentials. *Nat. Catal.*, 2: 55–61.

[258] Wen, G., D. U. Lee, B. Ren, F. M. Hassan, G. Jiang, Z. P. Cano, J. Gostick, E. Croiset, Z. Bai, L. Yang and Z. Chen. 2018. Carbon dioxide electroreduction: Orbital interactions in Bi-Sn bimetallic electrocatalysts for highly selective electrochemical CO_2 reduction toward formate production. *Adv. Energy Mater.*, 8: 1802427.

[259] Zhou, Y., R. Zhou, X. Zhu, N. Han, B. Song, T. Liu, G. Hu, Y. Li, J. Lu and Y. Li. 2020. Interface engineering in multiphase systems toward synthetic cells and organelles: from soft matter fundamentals to biomedical applications. *Adv. Mater.*, 32: 2002932.

[260] Zhong, M., K. Tran, Y. Min, C. Wang, Z. Wang, C. T. Dinh, P. De Luna, Z. Yu, A. S. Rasouli, P. Brodersen, S. Sun, O. Voznyy, C. S. Tan, M. Askerka, F. Che, M. Liu, A. Seifitokaldani, Y. Pang, S. C. Lo, A. Ip, Z. Ulissi and E. H. Sargent. 2020. Accelerated discovery of CO_2 electrocatalysts using active machine learning. *Nature*, 581: 178–183.

[261] Ma, S., M. Sadakiyo, M. Heim, R. Luo, R. T. Haasch, J. I. Gold, M. Yamauchi and P. J. A. Kenis. 2017. Electroreduction of carbon dioxide to hydrocarbons using bimetallic Cu-Pd catalysts with different mixing patterns. *J. Am. Chem. Soc.*, 139: 47–50.

7

Earth Abundant Electrocatalysts for Oxygen Evolution

Liangqi Gui,[1,2] *Beibei He*[1,2] and *Ling Zhao*[1,2,*]

7.1 Introduction

Traditional energy sources (i.e., coal, oil and natural gas) are the cornerstone of human society; however, the excessive depletion of fossil fuels causes environmental problem and ecosystem crisis (e.g., air pollution, global warming, and species loss). The dilemma of energy and environment issues is increasingly affecting the continuation of human civilization. To achieve the ambitious goal of sustainable development, the exploration of clean and renewable energy technologies is highly urgent to replace the conventional fossil energy technologies. Currently, multiple sustainable energy sources, including wind, geotherm, and solar power, are burgeoning to address energy supply. Unfortunately, the intrinsic intermittency, fluctuation, and regional dependency of these energy sources seriously limit their large-scale utilization. Besides intelligent grids, the development of energy conversion and storage technologies, such as water splitting and rechargeable metal-air betteries, is an ideal solution to efficiently use intermittent and fluctuant energy sources.[1-3] In fact, the operation of these energy conversion and storage system is extremely dependent on various fundamental electrocatalytic reactions. Particularly, electrochemical oxygen evolution reaction (OER), involving four proton-coupled electron transfer processes, is thermodynamically unfavorable and kinetically sluggish.[4] In this case, the intrinsic OER activity is far away from the requirement of practical application in those emerging energy conversion and storage devices. To handle this, the pursuit of potential catalysts to efficiently drive OER is of great significance but remains a great challenge.[5,6]

---

[1] Shenzhen Research Institute, China University of Geosciences, Shenzhen, 518057, China.

[2] Faculty of Materials Science and Chemistry, China University of Geosciences, Wuhan, 430074, China.

* Corresponding author: zhaoling@cug.edu.cn

In 1980, Sergio Trasatti revealed that iridium oxide (IrO_2) and ruthenium oxide (RuO_2) are the highly efficient catalysts for OER.[7] However, the structural stability of RuO_2 is inferior, which is originated from the unavoidable dissolution during electrochemical process.[8] On the contrary, IrO_2 exhibits good stability, whereas its activity is slightly lower than that of RuO_2. To further promote the activity and the stability of precious metal oxides, types of Ir(Ru)-based oxides have been developed. For example, Markovic and co-workers tuned the surface element concentration via a facile strategy of surface segregation to successfully synthesize the Ir-Ru bimetallic oxide, displaying an excellent stability on the basis of retaining activity.[9] Thomas F. Jaramillo's group highlighted precious metal-based $SrIrO_3$ perovskite oxide for OER.[10] As disclosed, Sr leaching readily occurred on $SrIrO_3$ perovskite surface as a formation of surface reconstructed IrO_x, ultimately inducing high catalytic activity and durability. Density functional theory (DFT) calculations indicated that the outstanding intrinsic activity was attributed to the *in situ* reconstructed IrO_x-$SrIrO_3$ hybrid. Hirai et al. reported a series of Sr-doped single phase perovskite $Ca_{1-x}Sr_xRuO_3$ as the highly active OER catalysts.[11] The introduction of Sr can tailor the strength of electron correlation in perovskite structure, thus optimizing the OER catalytic activity. Despite the significant progresses in the noble-metal based electrocatalysts, the scarce resources and expensiveness are still the essential disadvantages, which badly limit the large-scale application of Ir (Ru)-based compounds. Consequently, exploring the promising electrocatalysts with low cost and high element abundance to substitute precious-metal based materials is an urgent affair.

To this end, much attention has been paid to earth abundant materials, such as transitional metal (Mn, Fe, Co and Ni) based materials, heteroatom doped carbon nanomaterials and their hybrids.[12–14] Of note, a number of transitional metal-based alloys and oxides' electrocatalysts can't durably operate in acidic medium, and demonstrate inferior catalytic activity in neutral electrolyte. Actually, most of these alloys and oxides, for instance, Ni doped $Ba_{0.5}Sr_{0.5}Co_{0.8}Fe_{0.2}O_{3-\delta}$,[15] oxygen defect rich Co_3O_{4-x} nanosheets[16] and the composite consisting of metallic Co and $LiCoO_2$,[17] exhibit excellent OER activity and durability in alkaline media, which are superior to the benchmark IrO_2 and RuO_2 electrocatalysts. To objectively evaluate the progress of highly active OER electrocatalysts and avoid the influence of material category, we primarily pay our attention to non-noble-metal-based OER catalysts applied on alkaline electrolyte rather than acidic and neutral medium. So, for the development of transitional metal based alloys/oxides/hydroxides/oxyhydroxides/sulfides/selenides/nitrides/phosphides, heteroatom-doped carbon nanomaterials and their hybrids in alkaline media are introduced.[6,18–28]

In this chapter, we summarize the recent progress of earth abundant electrocatalysts in alkaline environment. Firstly, the evaluation criteria of the OER performance are thoroughly illustrated. Subsequently, the previouslly-reported non-precious metal based electrocatalysts toward OER in recent years is systematically introduced. Moreover, to further gain insight into underlying OER mechanism, the feasible reaction pathway and the logically thermodynamic reaction models are elaborated. Additionally, many detailed discussions of practical application for OER (i.e., electrochemical water splitting and rechargeable Zn-air batteries) are provided as well. Finally, some outlooks about the challenges and future development of earth abundant OER electrocatalysts are proposed.

7.2 Fundamental principles of OER

As illustrated in Fig. 7.1a, the catalytic performance relies heavily on electrocatalysts. In alkaline medium, the total OER process is that four hydroxide ions lose four electrons to form one oxygen molecule together with two water molecules, which can be depicted by Eq. (7.1). *M* is the active site on the surface of catalysts for catalyzing OER. This integrated OER reaction possesses the complex four-electron-transfer process, which is usually divided into four elementary reactions (Fig. 7.1b).[10,17] Specifically, the first step is that one hydroxyl ion (OH⁻) is adsorbed on the surface of catalysts and loses one electron, meanwhile the bond between the active site and OH group is built, thus generating the intermediate of *MOH* (Eq. 7.2). Subsequently, *MOH* intermediate reacts with one hydroxyl ion to form a water molecule and the second intermediate of *MO* (Eq. 7.3), along with one electron transfer. Then, the similar reaction between *MO* intermediate and hydroxyl ion reacts, which results in the formation of the third intermediate of *MOOH* with one electron transfer (Eq. 7.4). Finally, the betatopic reaction between *MOOH* intermediate and hydroxyl ion occurs, forming the resultant products of one oxygen molecule and one water molecule (Eq. 7.5). With the transfer and desorption of reaction products, the active sites are basically recycled.

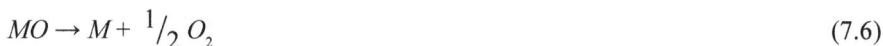

$$4OH^- \rightarrow 2H_2O + O_2 + 4e^- \tag{7.1}$$

$$M + OH^- \rightarrow MOH + e^- \tag{7.2}$$

$$MOH + OH^- \rightarrow MO + H_2O + e^- \tag{7.3}$$

$$MO + OH^- \rightarrow MOOH + e^- \tag{7.4}$$

$$MOOH + OH^- \rightarrow M + O_2 + H_2O + e^- \tag{7.5}$$

$$MO \rightarrow M + {}^1/_2\,O_2 \tag{7.6}$$

Despite the wide recognition of the reaction pathway with four elementary reactions toward OER, some researches have demonstrated another possible reaction route.[29] There is some similarity between the two pathways toward OER process, including first elementary reaction and the intermediates of *MOH* and *MO*. Evidently, the uniform elementary reaction of two route suggests that the hydroxyl ion adsorbability is crucial for OER process. The major difference between those two OER reaction routes is the generation way of oxygen gas. As shown in Fig.7.1b, one route with four elementary reactions is displayed via the yellow route, which involves the formation of *MOOH* intermediate rather than direct generation oxygen gas. On the contrary, another reaction pathway described by the purple route doesn't form the intermediate of *MOOH* and directly produce oxygen gas via the combination

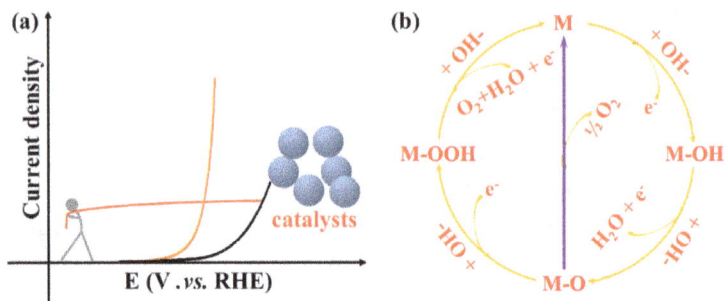

Figure 7.1. The schematic diagram of (a) the effect of catalysts and (b) OER mechanism in alkaline electrolyte.

of two *MO* intermediate (Eq. 7.6).[29] Although the different OER routes have been proposed, it is well documented that the electrochemical reaction for catalyzing OER is a heterogeneous process. The activity of totally catalytic system and the rate-determining step of OER are highly dependent on the interaction between electrocatalysts' surface and intermediates (i.e., *MOH*, *MO* and *MOOH*).

7.3 Evaluative criteria of OER

To investigate OER performances, complex and diverse electrochemical characterizations are conducted on different catalysts under various experimental conditions. In this scenario, enacting the unified evaluation criteria is the prerequisite to objectively assess the OER performances among different catalysts. Several appropriate indicators, including overpotential (η), Tafel slope, electrochemically active surface area (ECSA), mass activity (MA), electrochemical impedance spectroscopy (EIS) and durability have been widely recognized as crucial parameters for OER, which are elaborated as below.

7.3.1 Overpotential (η)

Overpotential (η), which originates from the reaction energy barrier, is one of the most important evaluation parameters for assessing OER performances. The potential gap between the OER potential at a specific current density and the thermodynamic equilibrium potential (1.23 V) is defined as the overpotential. Considering the potential of solar water splitting cells with the energy conversion efficiency of 10%, the overpotential at the current density of 10 mA cm^{-2} (denoted as η_{10}) is commonly used for assessing catalytic activity toward OER.[6,17,30] It is well accepted that the smaller value of η_{10} means better catalytic activity for OER. It is worth pointing out that the redox peaks with large current densities sometimes appear at the potential window close to 1.23 V during electrocatalytic process, corresponding to the redox of electrocatalysts (e.g., Co and Ni based OER electrocatalysts).[18,31,32] To avoid the redox peaks, the overpotentials at higher current densities (50–1000 mA cm^{-2}) can be applied for comparison. Nevertheless, the surface area and the mass loading of catalysts have great influence on the value of overpotentials, since both surface area and mass loading correlate the number of exposed active sites. Therefore, the overpotential is difficult to reflect the intrinsic electrocatalytic activity of catalysts.

7.3.2 Tafel slope and exchange current density (J_0)

The kinetics of electrocatalytic reaction is commonly assessed in virtue of Tafel slope and exchange current density (J_0). The Tafel curve originates from the linear sweep voltammetry (LSV) polarization curve via a logarithmic transformation. The Tafel slope reflects the dependency between the overpotential and current density, expressed as the following equations:

$$\eta = b \times \log (J) + a \qquad (7.7)$$

$$a = 2.303RT \times \frac{\log (J_0)}{\alpha nF} \qquad (7.8)$$

$$b = \frac{2.303RT}{\alpha nF} \qquad (7.9)$$

where η, b and J represent overpotential, Tafel slope and current density, respectively. R is the constant of perfect gas, and T denotes the absolute temperature. α is the factor of electron transfer, which is relevant to the catalysts coated on the electrode surface. F, n and J_0, respectively, represent the Faraday constant, electron transfer number during OER process and exchange current density. In the light of Eq. (7.7), η is in proportion to the value of log (J). The smaller Tafel slope denotes the faster increase rate of current density when overpotential changes, thus suggestive of a more rapid reaction kinetics. Additionally, the exchange current density (J_0), calculated via the above-mentioned Eqs. (7.7–7.9) at the overpotential of zero, can be used to evaluate the electron transfer rate between electrode and electrolyte. In general, the greater exchange current density discloses that the electron transfer occurs more easily. Similar to overpotential, Tafel slope and exchange current density are highly dependent on the exposed active sites, thus not suitable to assess the intrinsic activity of catalysts.

In practice, the capacitive current also appears in LSV process, resulting in a non-ignorable and inescapable current deviation. This non-ignorable current difference causes the inaccuracy of Tafel slope and J_0 value. Commonly, there are three effective approaches to improve the reliability for Tafel slope and J_0. Firstly, decreasing the scan rate of current in voltammetry test is a well-accepted solution to mitigate the experimental current deviation. At the low applied scan rate (normally less than 5 mV s^{-1}), the Tafel plot can be obtained via the forward sweep of voltammetry curve. If the redox peak appears, the backward sweep of voltammetry curve is adopted to better address Tafel curve. Calibrating the measured current density with iR correction is the second way, where i and R represent the observed current and the measured ohmic resistance, respectively.[33] However, the ratios of compensation (e.g., 85% iR correction and 100% iR correction) are not unified in previous researches. Thirdly, the relationship between overpotentials and logarithmic charge transfer resistance ($\log(R_{ct})$) can be also applied for plotting Tafel curve, thereby fitting the value of Tafel slope and J_0.[34] The above-mentioned R and R_{ct} are derived from electrochemical impedance spectroscopy (EIS) at the corresponding potentials.

7.3.3 Electrochemically active surface area (ECSA)

Electrochemically active surface area (ECSA), which can be estimated the amount of electrochemical active sites, can be applied to normalize the measured currents, ultimately evaluating the intrinsic activity of catalysts. Generally, the value of ECSA can be calculated by electrochemical double layer capacitances (C_{dl}), as expressed in the following equation:

$$ECSA = \frac{C_{dl}}{C_s} \qquad (7.10)$$

C_s is the specifically capacitive constant, which is highly dependent on the surface roughness of electrode. The C_{dl} value can be directly obtained via cyclic voltammogram (CV) curves at different scan rates. Notice that the applied potential window should be away from redox peaks. Furthermore, the ECSA can be accessed via other models, including integrating the redox peak area, carbon monoxide stripping, and hydrogen underpotential deposition. In principle, the larger ECSA value suggests the more exposed catalytically active sites. However, since the test condition and electrochemical process have a great impact on C_{dl} value, the ECSA is suggested to be used for comparing catalytic activity in the same work rather than in different studies.

7.3.4 Mass activity (MA), Specific activity (SA) and Turnover frequency (TOF)

To shed light on the intrinsic activity of catalysts, the mass activity (MA), specific activity (SA) and turnover frequency (TOF) are also adopted. The definition of mass activity (MA) is conducted via normalizing electrocatalytic current based on the mass loading of catalysts. Meanwhile, the MA can provide a reference for cost estimation of catalysts, making it as one of the most important indexes to evaluate the practicability, especially for precious metals-based catalysts. On the other side, the definition of specific activity (SA), as another significant parameter, is performed via normalizing electrocatalytic current based on the effective surface area of catalysts. Usually, Brunauer-Emmett-Teller (BET) surface area is employed for normalization. However, BET surface area is conducted by N_2 adsorption and desorption rather than the real electrochemically active area for OER. By contrast, the electrocatalytic current normalized via ECSA is of greater significance to reflect the inner activity of catalysts.

Turnover frequency (TOF) denotes the electron transfer ratio of single active site in unit time when electrochemical reaction occurs, which can be calculated by the following equation:

$$TOF = \frac{|J| \times A}{n \times F \times m} \qquad (7.11)$$

where J, A, and F stand for, respectively, the current density at a specific overpotential, the surface area of working electrode and Faraday constant. n represents the number of electrons involved in the electrocatalytic reaction. m is the mole number of participated active atoms or the concentration of active sites of catalysts. In principle, larger TOF value demonstrates the higher intrinsic activity with faster reaction kinetics. Indeed, it is difficult to accurately identify the number of real active sites involved in electrochemical process. According to the previous reports, active species including surface active sites, inside electrochemical active species and inert electrochemical species, are all counted, leading to an inaccurate calculation for TOF value. In order to improve the reliability of TOF, the direct measurement with high veracity is urgently required. For instance, Karthick and co-workers measured the actual number of cations involved in OER process with the aid of the change in redox peak area, thereby achieving the real value of TOF.[35] This approach is more favorable to evaluate the TOF value for transition metal based catalysts. Anantharaj et al. reported a novel test pathway for TOF in virtue of rotating ring-disk electrode (RRDE) measurement.[32] Besides, the TOF value can be identified by the valid strategies of peroxide oxidation, hydrogen underpotential deposition, and carbon monoxide stripping as well.[36]

7.3.5 Electrochemical impedance spectroscopy (EIS)

Electrochemical impedance spectroscopy (EIS) is an effective approach to analyze the polarization between electrode and electrolyte. Generally, the EIS is carried out at a specific potential with the frequency ranging from 100 kHz to 0.01 Hz and the small voltage amplitude (5 mV or 10 mV). The relationship between systemic impedance (Z) and frequency (f) can be expressed as the following equations:

$$Z(2\pi f) = Z'(2\pi f) + \hat{\imath} * Z''(2\pi f) \tag{7.12}$$

$$\hat{\imath} = \sqrt{-1} \tag{7.13}$$

where Z' and Z'', respectively, are the real-part and imaginary parts of Z. The semicircle Nyquist curves are plotted taking Z' as the X-axis and $-Z''$ as the Y-axis. The intersection point of Nyquist plots and X-axis in the high-frequency region is the ohmic resistance (R_o). R_o is mainly originated from ohmic resistance of electrolyte together with the contact resistance between electrolyte and working electrode. Correspondingly, the intersection point of Nyquist plots and X-axis in the low-frequency region is the total resistance (R_s). The difference between R_s and R_o is the electrode polarization resistance (R_{ct}, $R_{ct} = R_s - R_o$), which is usually used to evaluate the kinetics of electrochemical reaction. Normally, the smaller R_{ct} signifies a faster reaction kinetics and a higher catalytic activity.[37–39]

7.3.6 Durability

Durability is the most meaningful parameter to evaluate the real reliability of catalysts. From the point of view of practical application, the importance of durability for catalysts even precedes electrocatalytic activity. In many situations, the activity-stability trade-off issue is a formidable challenge but is still being sorted. There are several typical methods to monitor the durability of catalysts, including the continuous cycle voltammetry (CV), chronoamperometry and chronopotentiometry. In the case of CV measurement, the change of polarization curve gives a hint to evaluate the stability of catalysts. To be specific, the change of overpotential at a specific current density (i.e., 10, 50 and 100 mA cm^{-2}) in LSV curves before and after long-lasting CV cycles is recorded to determine whether the catalysts remain stable. The smaller difference between the first LSV curve and the last one denotes better OER durability for catalysts.

Chronoamperometry and chronopotentiometry are, respectively, inspected at the specific potential and the specific current density. The stable current density in chronoamperometry or the stabilized overpotential in chronopotentiometry suggest the good durability of catalysts. In particular, a high current density with a long-term operation is adopted to assess the catalysts for industrialized application. Additionally, the high potential operation readily causes the phase reconstruction on catalysts during OER process. To gain insights into the stability of catalysts, morphology, surface chemistry and structure chemistry should be also monitored.

7.4 Earth abundant OER catalysts in alkaline medium

Recently, noble-metal-free catalysts without scarcity and sumptuosity have attracted increasing attention owing to the considerable catalytic activity and sufficient durability. To replace noble-metal-based catalysts, tremendous efforts have been devoted to explore the alternative catalysts for practical application of OER, such as metals, alloys, oxides, hydroxides, oxyhydroxides, sulfides, selenides, nitrides, phosphides, heteroatom-doped carbon nanomaterials, and their hybrids. As highlighted in Fig. 7.2, the conventional elements of Mn, Fe, Co, Ni, Cu, Mo, W, B, C, N, O, F, P, S and Se, relative abundance of which are remarkably higher than that of precious metal (i.e., Pt, Ir and Ru), can be applied to synthesize non-precious-metal-based catalysts toward OER. In this part, the advance of earth abundant catalysts is intensively illustrated.

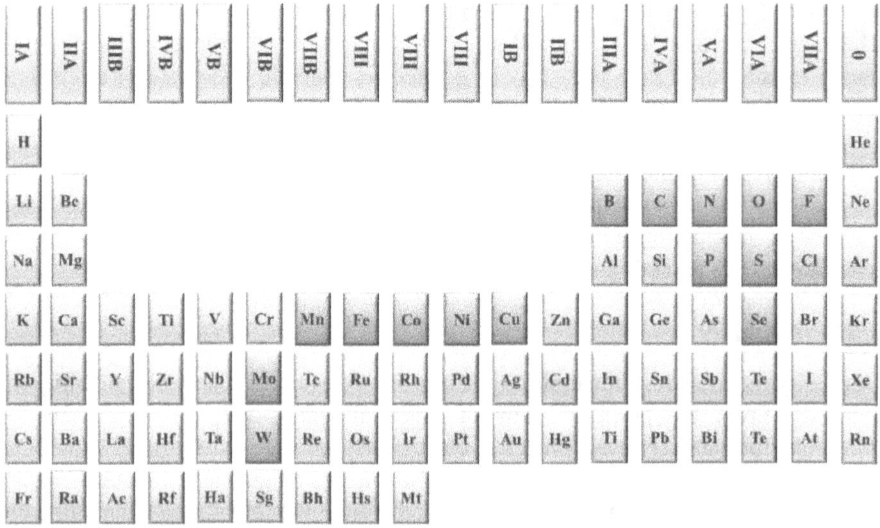

Figure 7.2. Elements used to compose earth abundant catalysts for OER.

7.4.1 Metal and alloys

Owing to the strong corrosion during OER operation process, the pristine transition metal can't operate steadily for a long time in alkaline electrolyte, thereby limiting the practical application of these catalysts. To solve this issue, hybrid catalysts, consisting of metallic nanoparticles locally or totally embedded into stable matrix (i.e., carbon nanomaterials), are designed (Table 7.1). For example, Yang's group reported Ni nanoparticles embedded in N doped carbon nanotubes as a promising

Table 7.1. Summary of OER activities of reported composite catalysts consisting of metal/alloy and carbon nanomaterials.

Catalyst	Electrode	Electrolyte	Mass loading (mg cm^{-2})	Overpotential (mV) @ 10 mA cm^{-2}	Tafel slope (mV dec^{-1})	References
Fe-NC SAC	Glassy carbon	0.1 M KOH	~ 0.20	450	114	42
Co/Co-N–C	Carbon felts	0.1 M KOH	1.0	~ 470	-	44
CoFe/N-GCT	Glassy carbon	0.1 M KOH	0.60	440	106	45
FeCoNi-NC	Glassy carbon	0.1 M KOH	-	310	20	46
Ni NPs@N-CNTs	Glassy carbon	0.1 M KOH	0.40	460	106	40
CuSA@ HNCNx	Glassy carbon	0.1 M KOH	0.20	320	64	41
NC@ Co-NGC DSNCs	Glassy carbon	0.1 M KOH	0.40	410	91	43
FeCo-NPs/ NC	Ni foam	1 M KOH	1.0	260	54.6	47
FeNi-NPs/ NC	Ni foam	1 M KOH	1.0	280	-	47
CoNi-NPs/ NC	Ni foam	1 M KOH	1.0	270	-	47
CoNi@N-DCNT	Glassy carbon	1 M KOH	0.50	300	74.9	48
Co@N-DCNT	Glassy carbon	1 M KOH	0.50	420	98.8	48
Ni@N-DCNT	Glassy carbon	1 M KOH	0.50	350	158.9	48
Co NP/NC-700	Carbon paper	1 M KOH	~ 0.18	430	-	49
Co/CNFs (1000)	Glassy carbon	1 M KOH	0.3	320	79	50
PPy/ FeTCPP/Co	Glassy carbon	1 M KOH	0.30	380	61	51

OER catalyst with high activity and good stability.[40] Wagh and co-workers synthesized single atomic Cu coordinated hollow nano-spheroids of nitrogen-deficient carbon nitride, which could efficiently drive OER.[41] The overpotential at 10 mA cm^{-2} for such catalyst is 320 mV, which is better than 370 mV for commercial IrO$_2$ and 330 mV for commercial RuO$_2$. Chen et al. developed a hybrid catalyst consisting of hierarchical porous Fe single atom and nitrogen doped carbon, which demonstrates a higher catalytic activity and durability than those of benchmark RuO$_2$.[42] Besides, Liu et al. proposed the hybrid (NC@Co-NGC DSNC) containing metallic cobalt and N doped carbon nanomaterial and investigated the reaction mechanism via DFT.[43] As illustrated in Fig. 7.3a, the overpotential at 10 mA cm^{-2} for NC@Co-NGC DSNC catalyst is 410 mV, comparable to that of commercial RuO$_2$ (370 mV). Among all

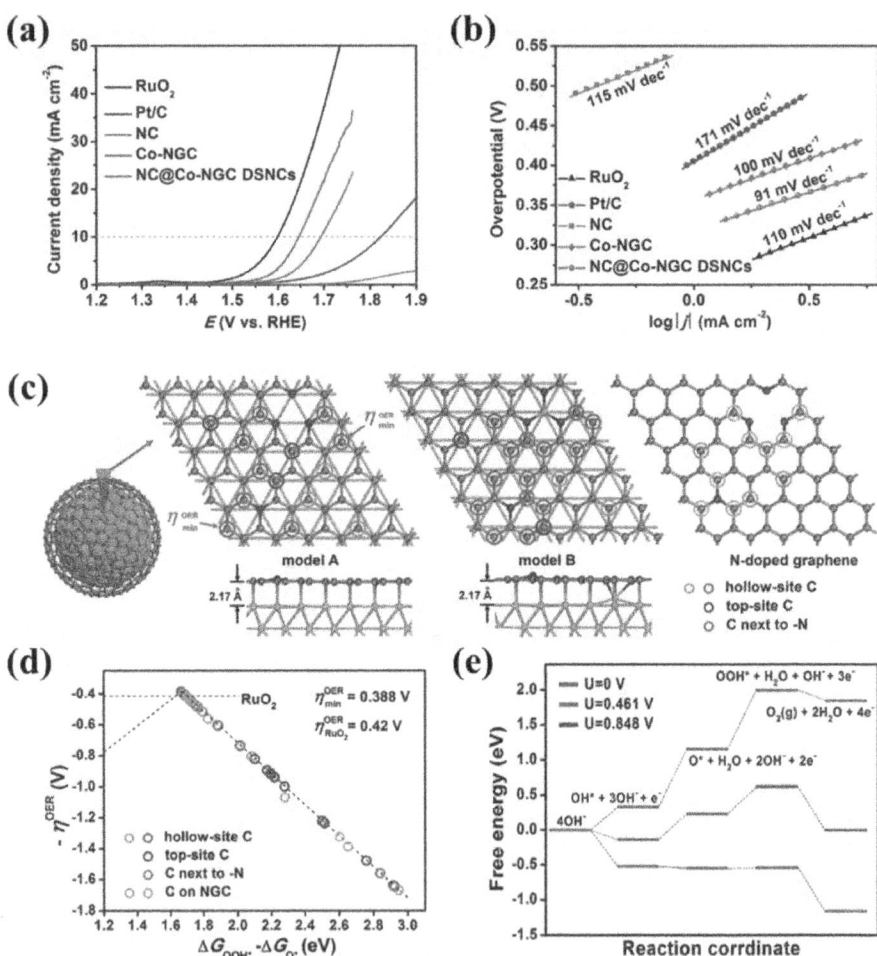

Figure 7.3. (a) LSV polarization curves and (b) Tafel plots of RuO$_2$, Pt/C, NC, Co-NGC and NC@Co-NGC DSNC catalysts measured in 0.1 M KOH solution with a scan rate of 10 mV s^{-1} at 1600 rpm. (c) Optimized structure of Co-NGC and NC@Co-NGC DSNC hybrid catalysts. (d) Overpotential volcano plots for different active sites. (e) OER routes of different catalysts at different potentials. Reproduced with permission from Ref. 43. Copyright 2017, Wiley-VCH.

the prepared catalysts, NC@Co-NGC DSNC delivers the lowest Tafel slope of 91 mV dec^{-1}, indicating the fastest kinetics of OER (Fig. 7.3b). In order to understand the underlying reaction mechanism of the NC@Co-NGC DSNC catalyst, DFT calculations are systematically performed. Figure 7.3c depicts the possible surface structures for NC@Co-NGC DSNC catalysts. The (111) plane of cubic metallic cobalt is adopted to coupling with N-doped graphene layer on account of the best compatibility between them. Gibbs free energy gap between the intermedium of O* and OOH* ($\Delta G_{OOH*} - \Delta G_{O*}$) is calculated to predict the theoretical activities of catalysts. It is clearly seen in Fig. 7.3d that NC@Co-NGC DSNC catalyst exhibits the lowest value of $\Delta G_{OOH*} - \Delta G_{O*}$, indicating the strongest adsorption of OOH* relative to O*. This offers the inner reason for the excellent activity of NC@Co-NGC DSNC hybrid catalyst. The thermodynamic process of electrocatalytic reaction is illustrated in Fig. 7.3e. Among all elementary reaction, the transformation from O* to OOH* requires the largest energy to overcome the reaction barrier at 0 V, revealing that the rate-determining step on NC@Co-NGC DSNC catalyst is the formation of OOH* (Eq. 7.4).

Rationally designing bimetal or multi-metal catalysts with a strongly synergistic effect is a feasible approach to further enhance catalytic performance. For example, the bimetallic catalyst, CoNi nanoparticles anchored inside a nitrogen-doped defective carbon nanotube network (CoNi@N-DCNT), displays a remarkable improved OER performance relative to those of single metal Ni@N-DCNT and Co@N-DCNT catalysts.[48] Peng and co-workers designed and synthesized a series of bimetallic nanoparticles confined in N-doped hollow carbon nanocubes (denoted as FeCo-NPs/NC, FeNi-NPs/NC and CoNi-NPs/NC).[47] Benefiting from the strong coupling between bimetallic alloy and N-doped carbon frameworks, all prepared catalysts demonstrated better catalytic activity than benchmark RuO_2. More importantly, it is also unveiled that tailoring the component of multi-metal (Fe-Co-Ni) based materials can optimize the adsorption and desorption energy of the reaction intermediates, thus promoting the catalytic ability for OER.

7.4.2 Oxides

Endowed with the distinctive features of flexible crystal structure, excellent physicochemical property and affordable cost, transition-metal-based oxides (TMO) are emerging as an important family of earth abundant electrocatalysts. Unfortunately, the catalytic activity of traditional TMO is relatively inferior, such as NiO, Fe_2O_3 and MnO_2. In response, several efficient strategies, including tailoring of elemental composition, phase structure, electronic structure, morphology, and constructing hybrid materials, are proposed to improve catalytic performance for catalysts. In this section, the design schemes for highly active catalysts are intensively described.

Generally, the electronic conductivities of transition metal oxides (i.e., NiO, MnO_2 et al.) are inferior. To address this, two useful strategies have been proposed to improve catalytic activity. Hetero-cation doping is one approach, which can not only increase electronic conductivity but also tune electron structure. For instance, Gao's group reported a Co doped NiO nanoflowers as an excellent oxygen electrocatalysts, displaying a better catalytic activity than the pristine NiO nanoflowers.[52] Revealed via DFT calculations, the incorporation of Co can decrease the band gap of NiO and accelerate the charge-transfer during electrochemical process, thus promoting electrocatalytic activity. Kang et al. developed a novel catalyst of Ru doped MnO_2 ($Mn_{1-x}Ru_xO_2$) nanowires.[53] The introduction of Ru results in the increasing of Mn^{3+} species and the weakening of Mn-O, contributing to catalyzing OER. When the doping amount of Ru achieves 20%, the resultant $Mn_{0.8}Ru_{0.2}O_2$ catalyst delivers a remarkable catalytic performance, which can be comparable to commercial Ir/C. Another method to boost catalytic performance is constructing heterostructure via coulping transition metal oxides with high conductive matrix. For example, Huang and co-workers elaborated and synthesized a hybrid catalyst, containing ultrafine CoO_x clusters and nitrogen-doped mesoporous carbon (denoted as CoO_x/NMC).[54] The composite catalyst, merging the advantages of each ingredient, provides an excellent catalytic activity for oxygen electrocatalysis and a prominent rechargeability for ZABs. Additionally, the formation of M-N-C (M represents transition metal element) bonds between metal oxides and carbon-based supporter is significantly instrumental in boosting catalytic activity. Liao et al. reported a series of promising composite catalysts, consisting of metal oxides (i.e., NiO, Fe_2O_3 and CuO) and polymer carbon nitride (CN).[55] The improvement in electrochemical performance of these hybrid catalysts benefits from the presence of rich M-N-C (M = Ni, Fe and Cu) bonds, which not only increases electron conductivity but also regulates intermediate adsorption during OER process.

Due to the flexible crystal structure, tunable valence state of B site cations and acceptable price, perovskite oxides (ABO_3) are widely applied for electrocatalysis. Commonly, A site of perovskite is occupied via alkaline-earth metal (i.e., Ca^{2+}, Sr^{2+} and Ba^{2+}) or rare-earth metal (i.e., La^{3+}, Pr^{3+} and Sm^{3+}), whereas the cations on the B site are transition metal ions with multi-valence states (i.e., $Fe^{2+/3+/4+}$, $Co^{2+/3+/4+}$, $Ni^{2+/3+}$ and $Mn^{2+/3+/4+}$). Cheng et al. studied the impacts of elemental composition on A site of $La_{1-x}Sr_xCoO_{3-\delta}$ during OER process.[56] The substitution of Sr can tailor electronic structure of Co and strengthen covalence of Co-O-Co bonding, thereby promoting catalytic activity. Li and co-workers investigated the activity of $CaCoO_{3-\delta}$ and $SrCoO_{3-\delta}$ for catalyzing OER in 0.1 M KOH.[57] The overpotential at 10 mA cm^{-2} of $CaCoO_{3-\delta}$ is 260 mV, which is lower than that of $SrCoO_{3-\delta}$ ($\sim$ 290 mV). Compared with the perovskite catalysts of $SrCoO_{3-\delta}$, the higher catalytic activity of $CaCoO_{3-\delta}$ originates from the shorter bond length of Co-O and the stronger covalence between O 2p and Co 3d. It is found that A site element has a significant impact on catalytic performance of perovskite oxides.

Except for engineering A site, introducing heterologous cations into B site of perovskite oxides also has been identified as the efficient way to accelerate OER. Remarkably, Yang et al. proposed a fundamental principle of perovskite oxides for OER, which guided the design of perovskite-type catalysts with high activity and durability.[27] It is disclosed that the catalytic activity of perovskites is closely related to the occupancy of e_g electron for B site metals. The e_g electron of various transition metals at different spin states is described in Fig. 7.4a.[1] In the case of perovskite oxides, the low e_g-occupancy ($e_g < 1$) denotes the strong interaction between B site metal and oxygen, indicating that the transfer of oxygenated intermediates is sluggish. On the contrary, the high e_g-occupancy ($e_g > 1$) suggests the weak bond

(a)

(b)

(c)

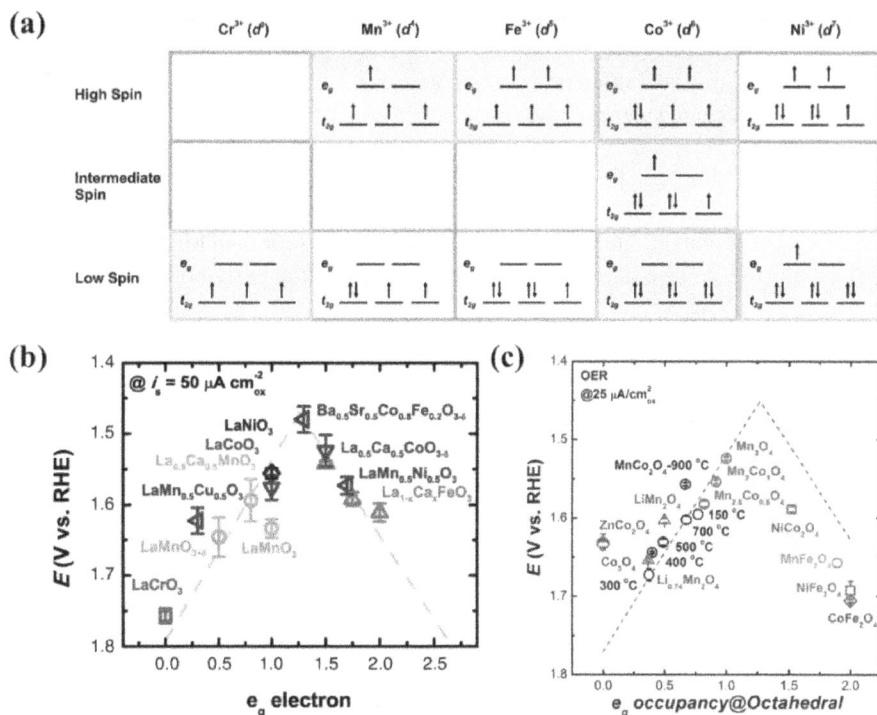

Figure 7.4. (a) Electron configurations for different transition-metal-based element (Cr, Mn, Fe, Co and Ni). The relationship between OER overpotentials and e_g electron occupancy on a variety of (b) perovskites and (c) spinel oxides. (a) Reproduced with permission from Ref. 1. Copyright 2015, Royal Society of Chemistry. (b) Reproduced with permission from Ref. 27. Copyright 2011, American Association for the Advancement of Science. (c) Reproduced with permission from Ref. 59. Copyright 2017, Wiley-VCH.

between B site metal and oxygen, demonstrating that oxygenated intermediates are uncomfortably adsorbed on catalyst surfaces. As shown in Fig. 7.4b, the optimization of e_g-occupancy is $\sim$ 1.2 to catalyze OER.[27] To achieve highly active perovskite catalysts, engineering the occupancy of e_g is an effective approach. For example, Zhu et al. reported a strategy of introducing cation deficiency to engineer the e_g-occupancy of Fe, thus greatly improving the catalytic activity of Fe-based perovskites.[38] The overpotential of $La_{0.95}FeO_{3-\delta}$ with rich La^{3+} deficiency (410 mV) is much lower than that of the pristine $LaFeO_{3-\delta}$ (510 mV). Gui et al. tailored the occupancy of e_g via cation doping in $Pr_{0.5}Ba_{0.5}Fe_{0.975}Ni_{0.025}O_{3-\delta}$, resulting in a significant improvement in catalytic performance.[58] Similarly, the occupancy of e_g electron can be tuned via anion engineering.[30] Additionally, regulating particle size is proved to be a useful method to adjust the e_g-occupancy as well. $LaCoO_{3-\delta}$ perovskite possesses an e_g-occupancy close to 1.2 and displays the enhanced catalytic activity, when the particle size is reduced to 80 nm.[39]

Similar to perovskite, manganese based spinel oxides, in which [MnO_6] octahedron is also the important part of crystalline structure, follow the correlation between e_g-occupancy of metal element in coordinate octahedron center and OER activity as well. Figure 7.4c shows the volcano shape, which evidently reflects the relation between catalytic activity and e_g filling of spinel oxides.[59] When e_g occupation reaches ~ 1.25, the valence state of manganese in spinel oxides is tervalence (Mn^{3+}), which displays the best intrinsic activity. Based on this theory, tuning the manganese element valence state close to +3 is a practicable method to enhance OER performance of manganese based spinel. Dong et al. prepared a Mn^{3+}-enriched hybrid, containing cobalt doped Mn_3O_4 nanocrystals and graphene nanosheet supporter (Co- Mn_3O_4/G).[60] The composite catalyst of Co-Mn_3O_4/G demonstrates a low overpotential of 275 mV at 10 mA cm^{-2}, which is lower than Mn_3O_4/G (338 mV) and benchmark Ir/C (282 mV). It is worth noting that the design theory is probably unsuitable for other spinel oxides, such as Co_3O_4, $ZnCo_2O_4$ and $NiCo_2O_4$.

7.4.3 Hydroxides/Oxyhydroxides

Transition metal-based hydroxides are one kind of representative electrocatalysts with high activity for catalyzing OER, which has attracted much interests. In the past few years, great progresses in designing and synthesizing hydroxide catalysts in theory and practice have been made. In this section, several typical transition metal-based hydroxide catalysts for OER have been depicted. Besides, the feasibly synthetic methods, performance-improvement strategies and reaction mechanism of these catalysts have been discussed as well.

$Ni(OH)_2$ and $Co(OH)_2$ are two basic materials, which are widely reported as highly active OER catalysts. For example, Zhou et al. introduced Fe dopants into $Ni(OH)_2$ nanosheets to obtain the single-phase $Ni_{0.83}Fe_{0.17}(OH)_2$ catalyst via a facile cation-exchange process.[18] Benefiting from the remarkable surface wettability, abundant nanopores and enhanced conductivity, $Ni_{0.83}Fe_{0.17}(OH)_2$ nanosheets deliver an ultralow overpotential of 245 mV at 10 mA cm^{-2}, which is remarkably smaller than the pristine $Ni(OH)_2$ (360 mV) and NiFe layered double hydroxide (NiFe LDH) (310 mV). Unfortunately, the highly active α-$Ni(OH)_2$ remains unstable in alkaline medium, which is readily transformed to β-$Ni(OH)_2$ with a mediocre electrocatalytic performance.[61] A similar phenomenon also occurs on $Co(OH)_2$. To address this, NH_4Cl is introduced into the reaction system during synthesis process to stabilize the phase structure of NiCo hydroxide.[62] Because of increased interlayer spacing and improved electrochemically active surface area, the optimized α-NiCo hydroxide demonstrates a lower overpotential of 285 mV at 10 mA cm^{-2} than that over β-NiCo hydroxide (362 mV). Moreover, constructing heterostructure on the surface of hydroxide is a promising method to induce high-efficiency OER electrocatalysis as well. Zhang et al. synthesized a novel heterostructured catalyst of Ag@$Co(OH)_x$ via a facile two-step method.[63] Specifically, $Co(OH)_2$ nanosheets grown on the surface of carbon cloth ($Co(OH)_2$/CC) were prepared in virtue of the typical electrodeposition technology. Subsequently, the Ag@$Co(OH)_x$ hybrid was achieved by a spontaneous redox reaction between $Co(OH)_2$ and $AgNO_3$ solution. Among all prepared electrocatalysts, Ag@$Co(OH)_x$/CC demonstrates the earliest onset potential (Fig. 7.5a), the lowest overpotential at 10 mA cm^{-2} and the smallest Tafel slope (Fig. 7.5b), together with an excellent stability (Fig. 7.5c). This indicates that the electrocatalytic ability of Ag@$Co(OH)_x$/CC catalyst outperforms those of Ag+$Co(OH)_2$/CC, $Co(OH)_2$/CC and IrO_2/CC catalysts. The home-made

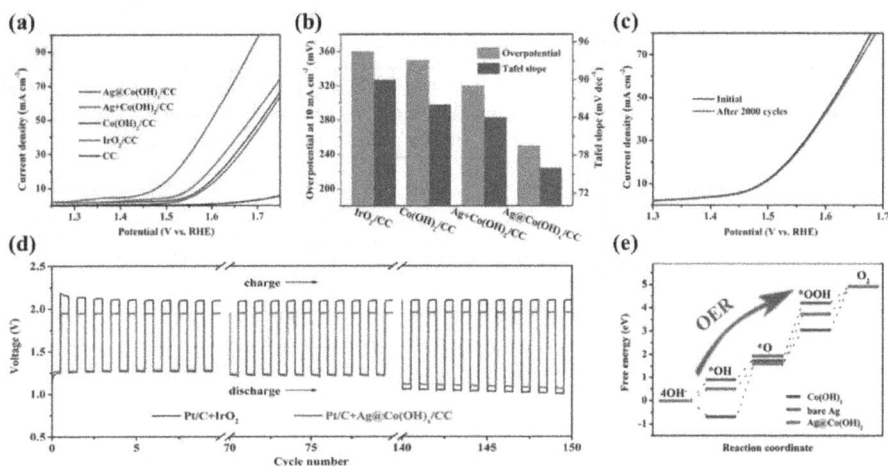

Figure 7.5. (a) LSV polarization curves, (b) overpotential at the current density of 10 mA cm⁻² of Ag@Co(OH)$_x$/CC, Ag+ Co(OH)$_2$/CC, Co(OH)$_2$/CC and IrO$_2$/CC. (c) OER polarization curves of Ag@Co(OH)x/CC before and after 2000 repeated CV cycles. (d) Charge-discharge curves of Zn-air batteries equipped with Pt/C+Ag@Co(OH)x/CC and Pt/C+IrO$_2$ air cathode. (e) Gibbs free-energy diagrams for OER pathways on Co(OH)$_2$, bare Ag and Ag@Co(OH)$_2$ catalysts. Reproduced with permission from Ref. 63. Copyright 2020, Wiley-VCH.

ZABs assembled with the Ag@Co(OH)$_x$/CC-commercial Pt/C air electrode provides a lower charging-potential and a better cycling durability than those of Pt/C-IrO$_2$ hybrid based ZABs (Fig. 7.5d), fairly agreeing with the OER performance (Fig. 7.5a). To disclose the reaction mechanism, the Gibbs free energy of four elementary reactions for OER is calculated via DFT. For all the elementary reaction, the highest Gibbs free energy of *OOH formation is observed on Ag@Co(OH)$_2$ and Ag, whereas the largest Gibbs free energy on Co(OH)$_2$ corresponds to the formation of *O (Fig. 7.5e). It is thereby concluded that the rate-determining step changes from *O formation (Eq. 2.3) to *OOH adsorption (Eq. 2.4) after introducing Ag. More importantly, the energy barrier of Ag@Co(OH)$_2$ (2.12 eV) is lower than that of Ag (2.3 eV) and Co(OH)$_2$ (2.4 eV), indicating that Ag introduction can lower energy barrier for electrochemical water oxidation. According to the experimental results and DFT calculations, it can be inferred that the significant enhancement in catalytic activity of Ag@Co(OH)$_x$/CC is mainly ascribed to the strong electronic interaction between active Co(OH)$_2$ and conductive Ag.

Notably, transition-metal-based oxyhydroxides, which have been identified as the real electrochemical active sites, are usually created on the surface of hydroxides at high potentials of OER operation.[69] Based on the occurrence of phase transformation, hydroxides are often applied for the precatalysts to synthesize oxyhydroxides, as summarized in Table 7.2. As illustrated in Fig. 7.6a, α-Co(OH)$_2$ nanosheet arrays electrodeposited on the surface of carbon fiber cloth (α-Co(OH)$_2$ NSAs/CFC) are used as a precursor for the synthesis of Fe doped CoOOH (denoted as Fe$_{0.33}$Co$_{0.67}$OOH PNSAs/CFC) via a typical anodic oxidation in 0.01 M (NH$_4$)$_2$Fe(SO$_4$)$_2$ solution.[24] During *in situ* anodic oxidation, the smooth nanosheets of hydroxides are converted to porous structure. Abundant hierarchical pore structure

Table 7.2. Summary of OER activities of reported transition-metal-based oxides, hydroxides and oxyhydroxides.

Catalyst	Electrode	Electrolyte	Mass loading (mg cm^{-2})	Overpotential (mV) @ 10 mA cm^{-2}	Tafel slope (mV dec^{-1})	References
Co-NiO NFs	Glassy carbon	1 M KOH	0.20	300	90	52
α-Mn$_{0.8}$Ru$_{0.2}$O$_2$	Glassy carbon	0.1 M KOH	0.14	~ 410	86	53
CoOx/NMC	Glassy carbon	1 M KOH	1.0	~ 280	59.8	54
NiO/CN	Glassy carbon	1 M KOH	0.285	261	58.92	55
CaCoO$_3$	Glassy carbon	0.1 M KOH	0.25	260	38	57
SrCoO$_3$	Glassy carbon	0.1 M KOH	0.25	~ 290	53	57
La$_{0.95}$FeO$_{3-δ}$	Glassy carbon	0.1 M KOH	0.232	~ 410	48	38
P-doped LaFeO$_{3-δ}$	Glassy carbon	0.1 M KOH	0.255	465	50	30
Pr$_{0.5}$Ba$_{0.5}$Fe$_{0.975}$Ni$_{0.025}$O$_{3-δ}$	Glassy carbon	0.1 M KOH	0.2264	440	73	58
Co-Mn$_3$O$_4$/G	Glassy carbon	0.1 M KOH	0.305	275	109	60
β-Ni(OH)$_2$ nanoburls	Glassy carbon	0.1 M KOH	0.205	300	43	32
Ni$_{0.83}$Fe$_{0.17}$(OH)$_2$	Glassy carbon	1 M KOH	0.204	245	61	18
α-NiCo hydroxide	Ni foam	1 M KOH	~ 3	285	63.5	62
CoFe@NiFe/NF	Ni foam	1 M KOH	-	190	45.71	64
Ag@Co(OH)$_x$	Carbon cloth	1 M KOH	1.5	250	76	63
γ-FeOOH/NF	Ni foam	1 M KOH	0.50	286	51	65
Fe$_{0.33}$Co$_{0.67}$OOH PNSAs/CFC	Carbon fiber cloth	1 M KOH	1.394	266	30	24
Cu-CoOOH/CFP	Carbon fiber paper	1 M KOH	-	234	79.1	66
L-CoOOH	Glassy carbon	1 M KOH	0.20	330	63.2	67
Au/CoOOH	-	1 M KOH	-	290	57.2	68

and strong electronic interaction between Co and Fe can facilitate charge transfer and increase electron conductivity, which is favorable for catalyzing OER. In this case, the overpotential of Fe$_{0.33}$Co$_{0.67}$OOH PNSAs/CFC at 10 mA cm^{-2} is only 266 mV, which is 65 mV and 24 mV lower than that of CoOOH PNSAs/CFC and

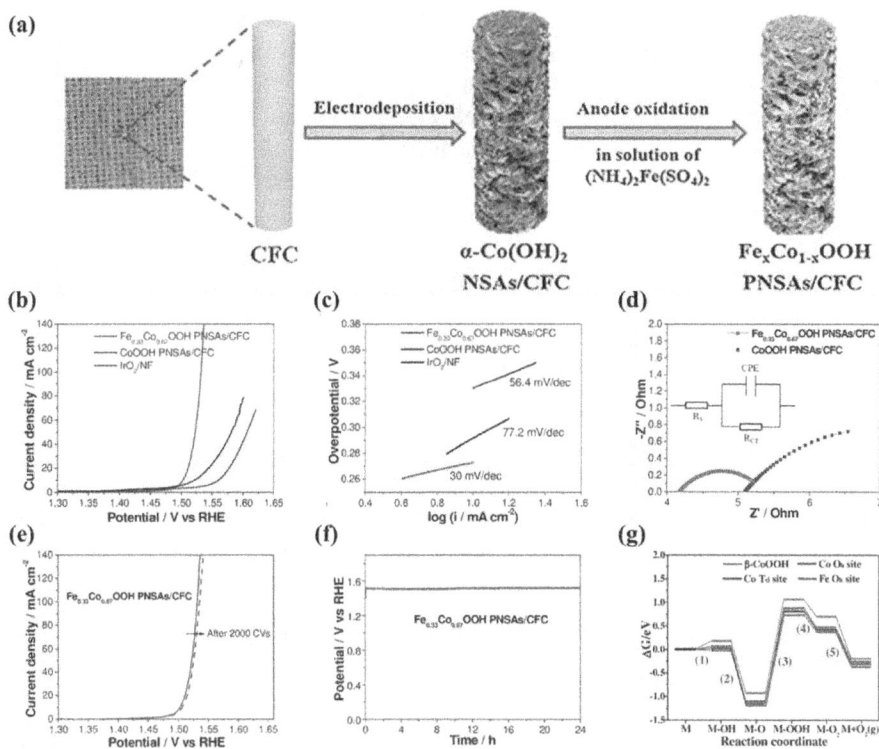

Figure 7.6. (a) Schematic illustration of synthesis for $Fe_xCo_{1-x}OOH$ PNSAs/CFC. (b) OER polarization curves and (c) Tafel plots of $Fe_{0.33}Co_{0.67}OOH$ PNSAs/CFC, CoOOH PNSAs/CFC together with IrO_2/CFC. (d) EIS Nyquist plots of $Fe_{0.33}Co_{0.67}OOH$ PNSAs/CFC and CoOOH PNSAs/CFC. (e) LSV polarization curves of $Fe_{0.33}Co_{0.67}OOH$ PNSAs/CFC before and after CV test with 2000 cycles. (f) Chronopotentiometric test at a constant current density of 10 mA cm^{-2}. (g) Gibbs free-energy diagrams for OER pathways on different possible active sites. Reproduced with permission from Ref. 24. Copyright 2018, Wiley-VCH.

IrO_2/nickel foam (NF) catalysts (Fig. 7.6b). Compared to CoOOH PNSAs/CFC, Fe incorporated $Fe_{0.33}Co_{0.67}OOH$ PNSAs/CFC displays a lower Tafel slope (Fig. 7.6c) and a smaller polarization resistance (Fig. 7.6d), indicating a faster OER kinetics process. Moreover, $Fe_{0.33}Co_{0.67}OOH$ PNSAs/CFC also demonstrates a better stability (Fig. 7.6e and f), suggesting a better feasibility in practical application. To identify the real active sites on $Fe_{0.33}Co_{0.67}OOH$ PNSAs/CFC catalyst, DFT calculation is applied to analyzing the Gibbs free energy of each elementary reaction for OER (Fig. 7.6g). The rate-controlling step on all possible active sites (containing $[FeO_6]$ (O_h), $[CoO_6]$ (O_h) and $[CoO_4]$ (T_d) of $Fe_{0.33}Co_{0.67}OOH$ together with $[CoO_6]$ (O_h) of β-CoOOH) is the formation of *M-OOH*. Among all calculated active sites, $[FeO_6]$ provides the lowest theoretical energy barrier to drive OER.

It is worth pointing out that metallic heteroatom doping into oxyhydroxides can not only introduce extra active sites, but also induce rich oxygen vacancies. Yan and co-workers reported that the highly active oxygen vacancies are generated on Cu doped CoOOH nanoplates@carbon fiber paper (Cu-CoOOH/CFP) rather than CoOOH/CFP.[66] Benefiting from the synergistic effect between Co and Cu together with rich oxygen vacancies, Cu-CoOOH/CFP shows a better OER performance than pristine CoOOH/CFP and benchmark IrO_2/CFP. These results denote that heterogeneous atom incorporation is an effective method to enhance the catalytic activity of oxyhydroxide-based electrocatalysts. Besides, the design of heterostructures is also proposed to strengthen catalytic activity of oxyhydroxides. For instance, Zhang et al. constructed the heterostructured Ni_2P/FeOOH catalyst with hierarchical structure as an efficient catalyst to catalyze OER.[22] Xu's group used Au nanoparticles to modify CoOOH (Au/CoOOH), which achieved a significantly enhanced catalytic performance.[68] Meng and co-workers synthesized oxygen vacancies enriched CoOOH nanosheets with the help of laser ablation, which demonstrated a higher electrocatalytic performance than individual CoOOH and benchmark RuO_2.[67]

7.4.4 Sulfides/Selenides

With the rapid development of noble-metal-free electrocatalysts, transition-metal-based sulfides and selenides have attracted more and more investigated interests to replace precious metal for catalyzing OER (Table 7.3). To realize the aim of large-scale application, many researchers have focused on developing sulfides and selenides with high activity and robust durability. For instance, Xu et al. highlighted an interfacial engineered nanostructure (CeO_x/CoS), consisting of hollow CoS matrix and CeO_x nanoparticles.[70] Compared with the individual CoS, CeO_x modified CoS catalysts display a remarkably improved catalytic activity, as shown in Fig. 7.7a. Of note, the surface Co^{2+}/Co^{3+} ratio of CoS is tailored by the content of CeO_x, because of the biphasic electronic interaction. When the Ce/Co molar ratio increases to 14.6%, the CeO_x/CoS hybrid catalyst has the highest ratio of Co^{2+}/Co^{3+} and exhibits the best catalytic activity with the overpotential of 269 mV to achieve 10 mA cm^{-2}, which rivals CoS, 4.8% CeO_x/CoS, 9.5% CeO_x/CoS and 19.4% CeO_x/CoS catalysts. This significant enhancement in electrocatalytic performance is attributed to the increased oxygen vacancies and the synergistic effect between CeO_x and CoS. Ding et al. proposed Fe doped NiS_2 ($Fe_xNi_{1-x}S_2$, x = 0, 0.05, 0.10 and 0.20) as the efficient electrocatalysts toward OER (Fig. 7.7b).[71] Among those studied catalysts, the pristine NiS_2 presents the worst performance for catalyzing OER owing to the inferior conductivity. Fe incorporation can induce the formation of metallic phase with high electronic conductivity in NiS_2 nanosheets, thus improving intrinsic activity. The optimized $Fe_{0.1}Ni_{0.9}S_2$ delivers a superior catalytic performance than NiS_2, $Fe_{0.05}Ni_{0.95}S_2$ $Fe_{0.2}Ni_{0.8}S_2$ and benchmark IrO_2 catalysts. Moreover, Fe substitution can protect $Fe_xNi_{1-x}S_2$ away from the formation of thick amorphous NiOOH layer, thus impeding electron-transfer kinetics. Furthermore, cooperation with conductive carbon nanomaterials to construct heterostructure can contribute to intrinsic activity of transition-metal-based sulfides as well. Dai's group reported a Co_9S_8 nanoparticles supported on graphene (Co_9S_8/G) composite catalyst for OER, which exhibited a higher electrocatalytic activity than the single Co_9S_8 (Fig. 7.7c).[72] Subsequently, the N element could be introduced into the hybrid of Co_9S_8/G via NH_3-plasma bombardment to obtain Co_9S_8/G with abundant N dopants (denoted as N-Co_9S_8/G). Benefiting from improved electronic conductivity and enriched surface

Table 7.3. Summary of OER activities of reported transition-metal-based sulfides and selenides.

Catalyst	Electrode	Electrolyte	Mass loading (mg cm^{-2})	Overpotential (mV) @ 10 mA cm^{-2}	Tafel slope (mV dec^{-1})	References
CoS@CC	Carbon cloth	1 M KOH	1.0	243	108	73
Fe$_{17.5\%}$-Ni$_3$S$_2$/NF	Ni foam	1 M KOH	~ 0.50	214	42	74
NiCo$_2$S$_4$ hollow spheres	Glassy carbon	0.1 M KOH	0.50	400	93.7	75
NiCo$_2$S$_4$/N-CNT	Glassy carbon	0.1 M KOH	~ 0.248	370	-	26
MoS$_2$/Ni$_3$S$_2$	Nickel foam	1 M KOH	-	218	88	76
(Ni, Fe)S$_2$@MoS$_2$	Carbon fiber paper	1 M KOH	-	270	43.21	31
oxygenated-CoS$_2$-MoS$_2$	Carbon fiber cloth	1 M KOH	~ 1.0	272	45	77
MoS$_2$-NiS$_2$/NGF	Glassy carbon	1 M KOH	0.286	370	-	78
Co$_3$S$_4$@MoS$_2$	Glassy carbon	1 M KOH	0.283	280	43	79
Co$_4$S$_3$/Mo$_2$C-NSC	Glassy carbon	1 M KOH	0.425	268	61.2	80
Co$_9$S$_8$/C	Glassy carbon	0.1 M KOH	-	434	119	81
NiSe$_2$	Glassy carbon	1 M KOH	0.354	332	54	82
Ni$_{0.94}$Fe$_{0.06}$Se$_2$	Glassy carbon	1 M KOH	0.354	279	39	82
NiCo$_2$Se$_4$ holey nanosheets	Glassy carbon	1 M KOH	-	300	53	83
CoFe$_2$Se$_4$/ NiCo$_2$Se$_4$ hybrid nanotubes	Glassy carbon	1 M KOH	0.28	224	48.1	84
NiCo$_2$Se$_4$/NiCoS$_4$	Glassy carbon	1 M KOH	0.714	248	98.5	85
Cu-14-Co$_3$Se$_4$/GC	Glassy carbon	1 M KOH	0.51	280	111	86
CoSe$_{2-x}$-Ni	Glassy carbon	0.1 M KOH	0.36	340	-	87
CoSe$_{2-x}$-Pt	Glassy carbon	0.1 M KOH	0.36	314	-	87
CoSe$_2$@VG-H	Carbon fiber cloth	1 M KOH	-	418	82	88
CoSe$_{1.26}$P$_{1.42}$	Carbon cloth	1 M KOH	-	255	87	89
P-CoSe$_2$/N-C	Glassy carbon	1 M KOH	-	230	36	90

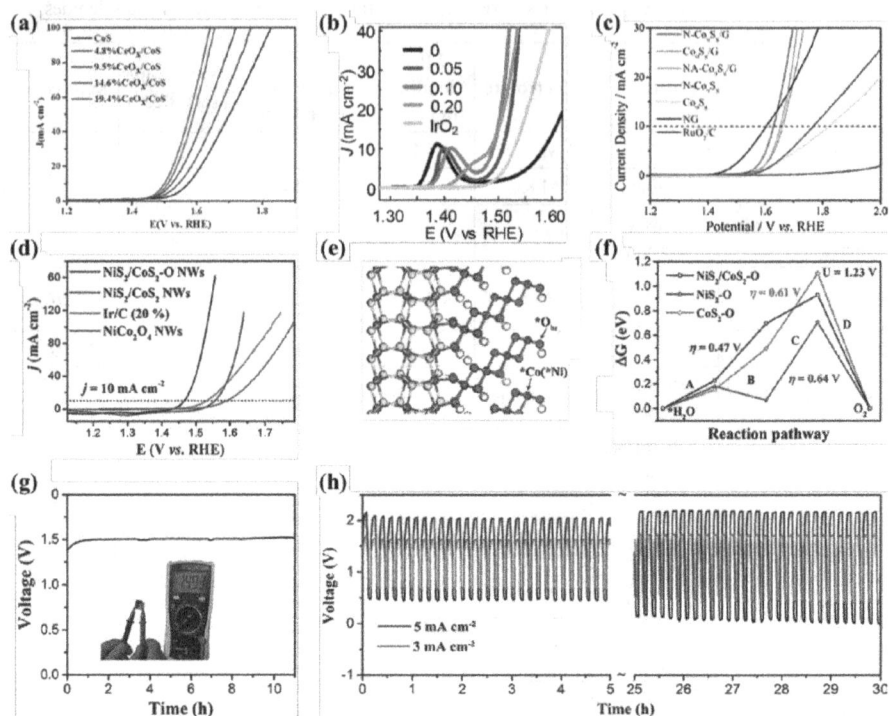

Figure 7.7. OER polarization curves of (a) CoS modified by CeO$_x$ with different content (0, 4.8%, 9.5%, 14.6% and 19.4%), (b) a series of Fe doped NiS$_2$ (Fe$_x$Ni$_{1-x}$S$_2$, x = 0, 0.05, 0.10 and 0.20) and IrO$_2$ catalysts, (c) RuO$_2$/C, NG, Co$_9$S$_8$, Co$_9$S$_8$/G, N-Co$_9$S$_8$, NA-Co$_9$S$_8$/G and N-Co$_9$S$_8$/G, (d) NiS$_2$/CoS$_2$-O NWs, NiS$_2$/CoS$_2$ NWs, NiCo$_2$O$_4$ NWs and commercial Ir/C catalysts. (e) Atomic structure and (f) OER energetic profiles on CoS$_2$-O, NiS$_2$-O and NiS$_2$/CoS$_2$-O. (g) The open circuit voltage of NiS$_2$/CoS$_2$-O NWs-based portable Zn-air battery. (h) Charge-discharge curves of Zn-air batteries equipped with NiS$_2$/CoS$_2$-O NWs air cathode at 3 and 5 mA cm^{-2}. (a) Reproduced with permission from Ref. 70. Copyright 2018, Wiley-VCH. (b) Reproduced with permission from Ref. 71. Copyright 2019, The Royal Society of Chemistry. (c) Reproduced with permission from Ref. 72. Copyright 2016, The Royal Society of Chemistry. (d–h) Reproduced with permission from Ref. 91. Copyright 2017, Wiley-VCH.

active sites, N-Co$_9$S$_8$/G demonstrates a much lower overpotential of 409 mV at 10 mA cm^{-2} than 441 mV for Co$_9$S$_8$/G and 602 mV for Co$_9$S$_8$.

Besides, engineering oxygen vacancies also can improve catalytic activity of sulfides. Yin et al. elaborated a NiS$_2$@CoS$_2$ hybrid nanowires with rich oxygen vacancies (NiS$_2$@CoS$_2$-O NWs) for accelerating OER electrocatalysis.[91] As can be seen in Fig. 7.7d, the overpotential of NiS$_2$@CoS$_2$-O NWs catalyst at 10 mA cm^{-2} is as low as 235 mV, smaller than 320 mV for NiS$_2$/CoS$_2$ NWs, 360 mV for NiCo$_2$O$_4$ NWs and 300 mV for commercial Ir/C. To understand the inner reason of the remarkable enhancement, DFT calculation is used to analyze thermodynamics process of each elementary reaction for OER process. The optimized structure model of NiS$_2$/CoS$_2$-O, NiS$_2$-O and CoS$_2$-O is shown in Fig. 7.7e. At 1.23 V, the highest Gibbs free energy on NiS$_2$/CoS$_2$-O is the formation of *O, whereas the process of transformation from *O to *OOH over NiS$_2$-O and CoS$_2$-O catalysts require the highest energy barriers (Fig. 7.7f). Therefore, the rate-determining-step of

NiS_2/CoS_2-O hybrid for catalyzing OER is different from that of single phase NiS_2-O and CoS_2-O catalysts. Moreover, the Gibbs free energy of rate-determining-step on different catalysts is followed in the order of NiS_2/CoS_2-O < CoS_2-O < NiS_2-O < CoS_2 < NiS_2. To explore the practicability of the material, the solid-state rechargeable ZABs equipped with NiS_2/CoS_2-O NWs air cathode catalyst was assembled. The homemade ZABs using NiS_2/CoS_2-O NWs as air cathode displays a high open-circuit voltage (Fig. 7.7g) and an excellent cycling stability (Fig. 7.7h), illustrative of the great feasibility in electrochemical devices.

Similar to sulfides, the transition-metal-based selenides have been proven as the alternative electrocatalysts. For example, Du et al. synthesized hierarchical Ni_3Se_4 ultrathin nanosheets with Fe dopant (denoted as $(Ni,Fe)_3Se_4$) via a facile solvothermal method.[21] When the OER current density reaches 10 mA cm^{-2}, the overpotential of 250 mV for $(Ni,Fe)_3Se_4$ is lower than 320 mV for Ni_3Se_4, indicating that Fe substitution can improve the catalytic activity of selenides (Fig. 7.8a). The decreased overpotential can be given the credit to increased surface area and enriched active sites together with optimized electronic environment of Fe, Ni bi-metal. Moreover, it has been also identified that introducing cation dopant benefits the electronic conductivity of selenides, thereby boosting the transfer of electron during OER process. Zeng's group elaborated a series of manganese substituted cobalt selenide nanosheets $((CoMn)Se_2)$ with orthorhombic phase for catalyzing OER.[92] When the mole ratio of Co/Mn is 4 : 1, the $(Co_4Mn_1)Se_2$ catalyst shows excellent catalytic activity with a lower overpotential of 274 mV at the current density of 10 mA cm^{-2} than 317 mV for $CoSe_2$ and 347 mV for IrO_2 (Fig. 7.8b). This increased activity mainly originates from reduced electrical resistivity (Fig. 7.8c), optimized electronic structure and tailored entropy of cations. Regulating anion deficiency into selenides is an effective pathway to further improve catalytic activity. In virtue of selenic-acid-assisted etching strategy, Zhang and co-workers prepared a carbon coated $Co_{0.85}Se_{1-x}$ nanocages $(Co_{0.85}Se_{1-x}@C)$ with abundant Se vacancies derived from metal-organic framework (ZIF-67).[93] The resultant $Co_{0.85}Se_{1-x}@C$ nanocages display a remarkable catalytic activity towards OER. For instance, the overpotential of $Co_{0.85}Se_{1-x}@C$ at 10 mA cm^{-2} is as low as 231 mV, which is smaller than that of $Co_{0.85}Se@C$ (291 mV) (Fig. 7.8d) and comparable to other reported transition-metal-based selenides catalysts (Table 3). Moreover, the turnover frequency (TOF) of 1.46 s^{-1} for $Co_{0.85}Se_{1-x}@C$ at an overpotential of 300 mV is higher than 0.89 s^{-1} for $Co_{0.85}Se@C$ (Fig. 7.8e), suggestive of a higher intrinsic activity. The improvement of catalytic activity is mainly attributed to increased concentration of Se vacancies, as revealed by electron paramagnetic resonance (EPR) spectra (Fig. 7.8f). To obtain insights into the OER mechanism induced via Se vacancies, the electric density of state (DOS) and Gibbs free-energy of elementary reaction for $Co_{0.85}Se@C$ and $Co_{0.85}Se_{1-x}@C$ are analyzed by DFT calculation. As seen in Fig. 7.8g, $Co_{0.85}Se@C$ and $Co_{0.85}Se_{1-x}@C$ have no bandgap at Fermi level, implying that both of them are conductive. It is notable that the electron density of $Co_{0.85}Se_{1-x}@C$ (12.7 eV) on Fermi level is slightly higher than that of $Co_{0.85}Se@C$ (11.4 eV). The simulative OER process on the major $Co_{0.85}Se$ surface of (010) plane is shown in Fig. 7.8h. At 1.23 V, the process of conversion from *O to *OOH possesses the smallest Gibbs free energy on (010) plane of both $Co_{0.85}Se@C$ and $Co_{0.85}Se_{1-x}@C$ (Fig. 7.8i). Notably, the activation

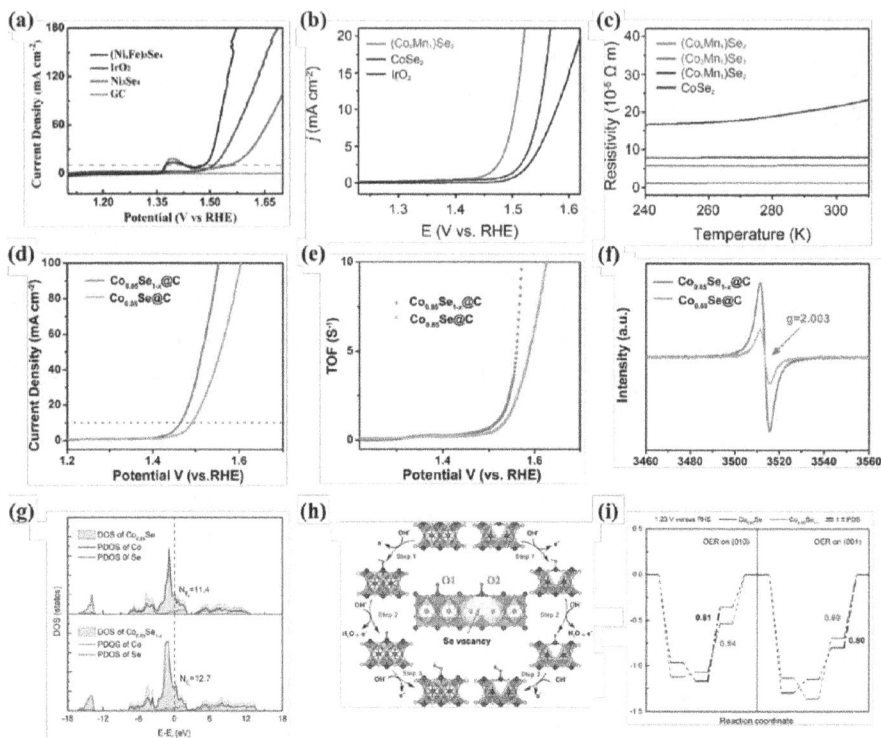

Figure 7.8. OER polarization curves of (a) (Ni,Fe)$_3$Se$_4$, Ni$_3$Se$_4$, IrO$_2$ and glassy carbon (GC) electrode and (b) (Co$_4$Mn$_1$)Se$_2$, CoSe$_2$ and commercial IrO$_2$. (c) Electrical resistivity of (Co$_4$Mn$_1$)Se$_2$, (Co$_2$Mn$_1$) Se$_2$, (Co$_1$Mn$_1$)Se$_2$ and CoSe$_2$ nanosheets. (d) LSV polarization curves, (e) TOF plots and (f) EPR spectra of Co$_{0.85}$Se$_{1-x}$@C and Co$_{0.85}$Se@C. (g) Densities of states for Co$_{0.85}$Se$_{1-x}$@C and Co$_{0.85}$Se@C. Simulation of the OER configurations (h) with the corresponding free energies (i). (a) Reproduced with permission from Ref. 21. Copyright 2018, The Royal Society of Chemistry. (b, c) Reproduced with permission from Ref. 92. Copyright 2018, Elsevier. (d–i) Reproduced with permission from Ref. 93. Copyright 2021, Wiley-VCH.

free energy of *OOH formation on Co$_{0.85}$Se$_{1-x}$@C (1.77 eV) is lower than that on Co$_{0.85}$Se@C (2.04 eV), suggesting a faster kinetics rate. In the case of (001) plane, the rate-determining-step for these catalysts is the transformation from *OOH to O$_2$ rather than *OOH formation. The energy barrier of 1.92 eV for O$_2$ desorption on Co$_{0.85}$Se$_{1-x}$@C (001) plane is smaller than 2.03 eV on Co$_{0.85}$Se@C (001) plane. By contrast, the (010) plane demonstrates a lower energy barrier of rate-determining-step, indicating that (010) plane is more active for catalyzing OER. Briefly, the abundant Se vacancies can effectively increase electronic conductivity and decrease potential barrier of rate-determining-step, resulting in the significant improvement of catalytic activity. Besides Co$_{0.85}$Se$_{1-x}$@C, introducing Se vacancies can improve the catalytic activity of Ni and Fe-based selenides (NiSe$_{2-x}$@C, FeSe$_{2-x}$@C) as well.

7.4.5 Nitrides/Phosphides

Transition-metal-based nitrides are the potential electrocatalysts toward OER. The nitrides are commonly prepared via nitridation process, where ammonia (NH$_3$) or

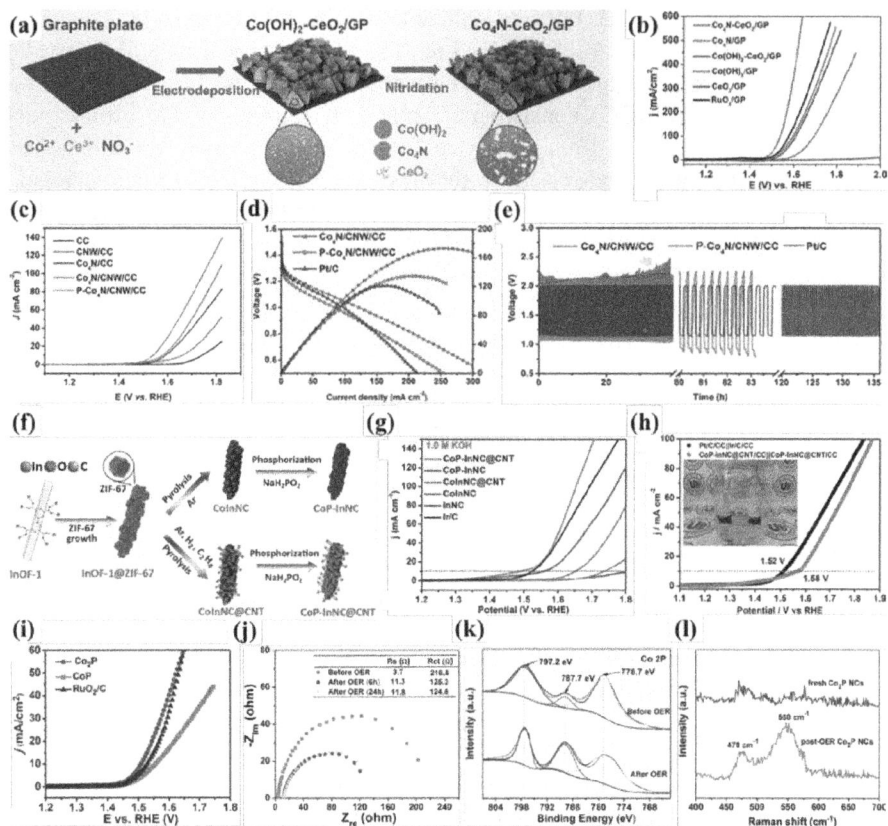

Figure 7.9. (a) Schematic illustration of synthesis for Co_4N-CeO_2/GP. (b) LSV polarization curves of Co_4N-CeO_2/GP, Co_4N/GP, $Co(OH)_2$-CeO_2/GP, $Co(OH)_2$/GP, CeO_2/GP and commercial RuO_2/GP. (c) OER polarization curves of Co_4N/CNW/CC, P-Co_4N/CNW/CC, Co_4N/CC, CNW/CC and CC. (d) Battery voltage and power density of Zn-air batteries with Co_4N/CNW/CC, P-Co_4N/CNW/CC and Pt/C as cathodes. (e) Galvanostatic discharge-charge cycling curves at 10 mA cm^{-2} of rechargeable Zn-air batteries with the electrode of Co_4N/CNW/CC, P-Co4N/CNW/CC and Pt/C catalyst on carbon cloth, respectively. (f) Schematic illustration of synthesis for sugar-gourd-like CoP-InNC and CoP-InNC@CNT composites. (g) OER polarization curves of CoP-InNC@CNT, CoP-InNC, CoInNC@CNT, CoInNC, InNC and benchmark Ir/C. (h) LSV curves of an overall water splitting in a two-electrode configuration. (i) LSV polarization curves of Co_2P, CoP and RuO_2/C. (j) Nyquist plots of fresh and post-OER Co_2P NCs. (k) XPS spectra of Co 2p core level of Co_2P NCs before and after OER at an overpotential of 300 mV for 6 h. (l) Raman spectra for Co_2P NCs before and after OER for 6 h in 1 M KOH. (a, b) Reproduced with permission from Ref. 94. Copyright 2020, WILEY-VCH. (c–e) Reproduced with permission from Ref. 95. Copyright 2016, ACS. (f–h) Reproduced with permission from Ref. 110. Copyright 2020, WILEY-VCH. (i–l) Reproduced with permission from Ref. 111. Copyright 2018, WILEY-VCH.

polymer with abundant nitrogen serves as nitrogen source. The hybrid of Co_4N-CeO_2 grown on the surface of graphite plate (Co_4N-CeO_2/GP) was reported by Sun et al. via a typical thermochemical treatment on the precursor of $Co(OH)_2$-CeO_2 nanosheet electrodeposited on graphite supporter (Fig. 7.9a).[94] Benefiting from outstanding hydrophilia, interfacial synergistic effect of Co_4N-CeO_2 together with strong electronic interaction between CeO_2 and Co_3O_4 *in situ* formed on the surface of Co_4N, Co_4N-CeO_2/GP catalyst displays a low overpotential of 239 mV at 10 mA cm^{-2}, smaller than Co_4N /GP (263 mV) and benchmark RuO_2/GP (265 mV) (Fig. 7.9b). As proposed by Meng and co-workers, the composite electrocatalyst (denoted as Co_4N/CNW/CC), consisting of Co_4N derived from ZIF-67, carbon fibers network (CNW) originated from polypyrrole (PPy) nanofibers network and carbon cloth (CC) matrix was prepared via a pyrolysis method.[95] Due to the synergistic effect between Co_4N and Co-N-C structure together with the stability and electronic conductivity of 3D carbon fibers network, Co_4N/CNW/CC exhibits a remarkable electrochemical performance. For instance, the overpotential of Co_4N/CNW/CC hybrid at 10 mA cm^{-2} is only 310 mV, which is lower than that of P-Co_4N/CNW/CC (350 mV), Co_4N/CC (360 mV) and CNW/CC (400 mV) (Fig. 7.9c). Furthermore, the ZABs using Co_4N/CNW/CC catalyst as air cathode displays a higher power density (174 mW cm^{-2}) than the ZABs equipped with P-Co_4N/CNW/CC (135 mW cm^{-2}) and Pt/C (122 mW cm^{-2}) cathode (Fig. 7.9d). The cycling durability of Co_4N/CNW/CC cathode is superior to Co_4N/CNW/CC and Pt/C as well (Fig. 7.9e). Except for constructing composite materials, introducing nitrogen vacancies into nitrides is also an effective way to improve catalytic activity. He's group reported a nitrogen deficient Co_4N_x as highly active catalyst for OER.[96] According to experimental study and DFT calculation, nitrogen vacancies can tailor the number of e_g electrons of Co cation and decrease energy barrier of *O formation, thereby contributing to electrocatalytic activity of nitrides.

Similar to nitrides, non-noble-metal-based phosphides demonstrate superior catalytic activity relative to precious metal-based electrocatalysts as well (Table 7.4). For instance, the CoP-InNC@CNT catalyst (cobalt phosphide NPs embedded in carbon nanotubes and nitrogen-doped carbon) derived from the composite of indium-organic framework (InOF-1) and ZIF-67 was prepared by Huang's group, following the combination synthesis method of pyrolysis and thermochemical treatment (Fig. 7.9f).[110] In the alkaline electrolyte of 1 M KOH solution, the overpotential of CoP-InNC@CNT hybrid at 10 mA cm^{-2} is 270 mV, which is lower than that of CoP-InNC (330 mV) and benchmark Ir/C (295 mV) (Fig. 7.9g). The excellent OER performance of CoP-InNC@CNT can be attributed to the uniquely porous nanostructure, highly electronic conductivity, abundant active species (e.g., CoP nanoparticles, Co-N-C structure and N doped carbon) together with the strong synergetic effect between these electrochemical active sites. Moreover, the overall water splitting assembled with CoP-InNC@CNT as both anode and cathode displays an outstanding device performance comparable to Pt/C (cathode) and Ir/C (anode) based electrolytic cell (Fig. 7.9 h). Except for rationally designing composite structure, introducing heteroatom into phosphides can effectively enhance catalytic activity as well. Xu et al. systematically investigated the electrocatalytic performance of phosphides with different cation composition and found that the catalytic activity of transition-metal-based phosphides followed the order of FeP < NiP < CoP < FeNiP < FeCoP < CoNiP < FeCoNiP.[19] The improvement of electrocatalytic activity for phosphides with dopants is benefited from the increased electronegativity of metal-P binding together with the high-valence-state of cations. Besides, the formation of electrochemically active CoOOH often occurs on the surface of phosphides during

Table 7.4. Summary of OER activities of reported transition-metal-based nitrides and phosphides.

Catalyst	Electrode	Electrolyte	Mass loading (mg cm^{-2})	Overpotential (mV) @ 10 mA cm^{-2}	Tafel slope (mV dec^{-1})	References
Co$_4$N	Glassy carbon	1 M KOH	~ 0.714	~ 260	86	97
Ni doped Co$_4$N	Glassy carbon	1 M KOH	~ 0.714	233	61	97
Co$_3$Fe$_1$N/ graphene	Glassy carbon	1 M KOH	0.20	266	32	98
Co$_4$N @ mesoporous nitrogen-doped carbon	Glassy carbon	1 M KOH	~ 0.285	257	58	99
Co$_4$N @ nitrogen doped carbon box	Glassy carbon	0.1 M KOH	0.30	290	67.89	100
Fe-Co$_4$N@N-C nanosheet array	Carbon cloth	0.1 M KOH	~ 1.0	~ 310	62	23
NC-Co/CoN$_x$ nanoarrays	Glassy carbon	1 M KOH	-	289	-	101
FeNi$_3$N/NF	Ni foam	1 M KOH	-	202	40	102
FeP nanorods	Carbon fiber paper	1 M KOH	0.70	350	64	103
FeP @ graphitic carbon	Glassy carbon	1 M KOH	~ 0.35	278	87	104
Ni$_2$P/Ni/NF	Ni foam	1 M KOH	-	200	-	105
Ni$_2$P @ N/P co-doped carbon tubes	Glassy carbon	1 M KOH	~ 0.40	440	110	106
NiFeP @ N/P co-doped carbon tubes	Glassy carbon	1 M KOH	~ 0.40	350	78	106
CoP nanoframes	Glassy carbon	1 M KOH	~ 0.265	323	49.6	107
CoP/reduced graphene oxide	Glassy carbon	1 M KOH	~ 0.28	340	66	108
NiCoP	Ni foam	1 M KOH	~ 1.60	280	87	109

OER process. Based on surface reconstruction, Li and co-workers reported two kinds of cobalt phosphides (Co$_2$P and CoP) with different ability of phase conversion and observed the catalysts change on the process of catalyzing OER.[111] The overpotential of Co$_2$P (280 mV) at 10 mA cm^{-2} is lower than that of CoP (330 mV), suggesting a better catalytic activity (Fig. 7.9i). With reaction progress, the polarization resistance of Co$_2$P during OER process gradually decays until the test time reaches 24 h (Fig. 7.9j). As revealed, the improved catalytic activity originates from the presence of CoOOH (Fig. 7.9k and l).

7.4.6 Heteroatom-doped carbon nanomaterials

Besides, heteroatom-doped carbon nanomaterials have been proven to be another type of earth-abundant electrocatalysts for OER. Based on abundant resource, low cost, and expectable performance, carbon-based electrocatalysts have been intensively investigated. Lai et al. designed three nitrogen doped carbon nanomaterials with different microstructure, including 1D nitrogen-doped carbon nanotubes (N-CNT), 2D nitrogen-doped reduced graphene oxide (N-rGO) and 3D nitrogen-doped carbon nanocages (N-CNC-900) (Fig. 7.10a).[112] Benefiting from the large specific surface area and enriched micropores, porous N-CNC-900 catalyst shows a remarkable catalytic activity, comparable to commercial RuO$_2$. When the OER current density reaches 10 mA cm^{-2}, the overpotential of N-CNC-900 is 291 mV, which is much lower than 375 mV for N-CNT and 481 mV for N-rGO (Fig. 7.10b). Moreover, it has been identified that pyridinic N can induce electron-transfer from near carbon

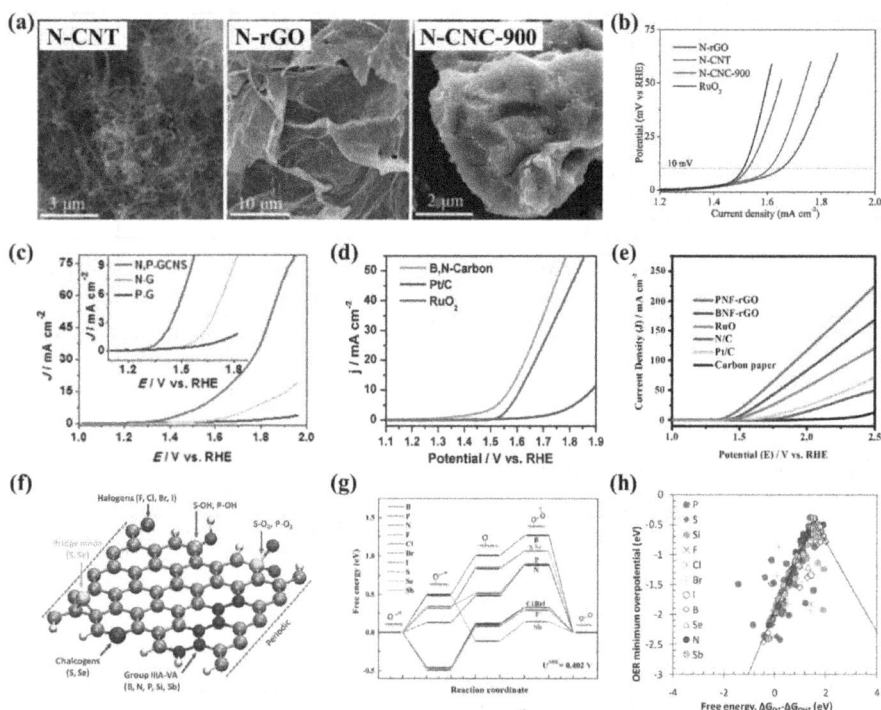

Figure 7.10. (a) SEM images of N-CNT, N-rGO and N-CNC-900. OER polarization curves of (b) N-CNT, N-rGO, N-CNC-900 and commercial RuO$_2$, (c) N, P-GCNS, N-G and P-G, (d) B, N-carbon, Pt/C and RuO$_2$, (e) PNF-rGO, PNF-rGO, RuO$_2$, N/C, Pt/C and carbon paper. (f) The illustration of possible sites for each dopant in crystalline structure of graphene nanoribbons. (g) Gibbs free-energy diagrams for OER pathways and (h) relationship between overpotential and adsorption free energy $\Delta G_{O^*} - \Delta G_{OH^*}$ on different possible active sites of various dopants. (a, b) Reproduced with permission from Ref. 112. Copyright 2021, Elsevier. (c) Reproduced with permission from Ref. 113. Copyright 2015, ACS. (d) Reproduced with permission from Ref. 114. Copyright 2018, WILEY-VCH. (e) Reproduced with permission from Ref. 115. Copyright 2019, Elsevier. (f–h) Reproduced with permission from Ref. 116. Copyright 2015, WILEY-VCH.

atom to nitrogen atom with p-type doping, leading to electronic deficiency of carbon atom (δ^+). The electron deficient carbon atom is instrumental to the adsorption of reaction intermedium, thereby accelerating kinetics rate of OER. To further improve the catalytic activity, multiple elements have been introduced into carbon materials. For instance, Li and co-workers developed a carbon-based mixture (denoted as N, P-GCNS), consisting of nitrogen, phosphorus co-doped graphene and nitrogen, phosphorus co-doped carbon nanosheets, as the highly active catalyst for OER.[113] Due to the synergetic effect between N and P elements, N, P-GCNS demonstrates an overpotential at 10 mA cm^{-2} of 340 mV, which is lower than single-element-doped materials of N doped graphene (N-G) and P doped graphene (P-G) (Fig. 7.10c). Su's group reported a B, N dual-doped nanocarbon with rich carbon deficiency (B, N-Carbon), the electrocatalytic activity of which is superior to that of benchmark RuO$_2$ and Pt/C (Fig. 7.10d).[114] P, N, F tri-doped 3D graphitic carbon (PNF-rGO) and B, N, F tri-doped 3D graphitic carbon (BNF-rGO) were proposed as highly active OER catalysts as well (Fig. 7.10e).[115] Both PNF-rGO and BNF-RGO catalysts exhibit a much higher catalytic activity than commercial RuO$_2$, N/C together with Pt/C. The above results indicate that introducing heteroatom dopants into nanocarbon materials is a desirable solution to obtain electrocatalysts with high activity and acceptable prices. In order to provide a guidance for designing and evaluating metal-free catalysts, Xia's group systematically investigated the reaction process of carbon nanomaterials with different dopants (i.e., B, N, P, F, Cl, Br, I, S, Se and Sb) for catalyzing OER with the aid of DFT conclusion.[116] The possible sites for each dopant in crystalline structure of graphene nanoribbon is optimized (Fig. 7.10f), which reflects the impact of doping element. For all mentioned heteroatom doped graphene, Gibbs free energy of every elementary reaction for OER is calculated (Fig. 7.10g). It is evident that the dopants play a great influence on tailoring energy barriers of elementary reaction during OER process. Figure 7.10h presents a volcano relation between $\Delta G_{O^*} - \Delta G_{OH^*}$ and doping position. As known, the value of $\Delta G_{O^*} - \Delta G_{OH^*}$ can be used as the descriptor to assess the catalytic activity of carbon-based electrocatalysts. When the value of $\Delta G_{O^*} - \Delta G_{OH^*}$ is adjusted to ~ 1.5 eV, the corresponding doping structure displays the lowest overpotential, indicating the best activity for catalyzing OER. As a result, rational designing of the doped position and tuning the type of dopants are beneficial to improve the catalytic activity of metal-free electrocatalysts.

7.5 Conclusion and perspectives

To achieve the ambitious goal of carbon-neutral energy roadmap, the development of clean and renewable energy technologies is highly needed to replace the conventional fossil energy technologies. As known, electrochemical OER is one of the essential reactions for the burgeoning energy technologies. Consequently, there is an urgent need for the design of alternative OER electrocatalysts, to meet the actual demands like having high performance, being stable, and cost-efficient. In this chapter, we have outlined evaluation criteria, reaction mechanisms, and practical applications of earth-abundant OER electrocatalysts in alkaline media, to build a fundamental understanding of reaction mechanism and the innovation and development of OER.

Despite the delightful scientific achievements, considerable challenges remain to be overcome, as illustrated in the following aspects:

1. Objectively, the experimental conditions significantly affect the measured OER performance. Meanwhile, the unified evaluation criteria of OER are lacking. To compare various electrocatalysts prepared from disparate groups in a relatively fair manner, a third-party evaluation system is highly desirable.

2. To reveal the underlying OER mechanism in the atomic scale, more *in situ/ operando* characterization techniques and DFT calculations are encouraged. Furthermore, the application of machine learning and/or artificial intelligence is more conducive to building a fundamental and comprehensive understanding in OER process. This basic principle can in turn provide the explicit guidance in the rational design of promising electrocatalysts.

3. Actually, the real active sites and the intrinsic activities of numerous electrocatalysts are still controversial. It is unclear whether electrocatalysts themselves or surface reconstructed species possess the real OER activity. Currently, the scientific research on the dynamic reconstruction of electrocatalysts for OER is obviously inadequate. In-depth studies on dynamic electrocatalysts during OER operation are still required. The identification of real electrochemically active sites provides a new thinking of the construction of pre-electrocatalysts or electrocatalysts.

4. To achieve the large-scale application, the high current density and the long-term durability of electrocatalysts are indispensable. As of now, most of the reported OER electrocatalysts are investigated at a low current density and a short-term stability. The operation at a high current density for a long-lasting time will be another challenge. Moreover, from the view of industrial application, another pivotal issue worth thinking about is the electrocatalysts against impurities contamination, such as chlorine oxidation.

In conclusion, according to these advances and challenges, we believe that earth abundant electrocatalysts would overcome the activity and stability limitation of OER and finally achieve the commercial applications in the future.

References

[1] Hong, W. T., M. Risch, K. A. Stoerzinger, A. Grimaud, J. Suntivich and Y. Shao-Horn. 2015. Toward the rational design of non-precious transition metal oxides for oxygen electrocatalysis. *Energy Environ. Sci.*, 8: 1404–1427.

[2] Jiang, W. J., T. Tang, Y. Zhang and J. S. Hu. 2020. Synergistic modulation of non-precious-metal electrocatalysts for advanced water splitting. *Acc. Chem. Res.*, 53: 1111–1123.

[3] Xu, H., S. Ci, Y. Ding, G. Wang and Z. Wen. 2019. Recent advances in precious metal-free bifunctional catalysts for electrochemical conversion systems. *J. Mater. Chem. A*, 7: 8006–8029.

[4] Wu, Z. P., X. F. Lu, S. Q. Zang and X. W. Lou. 2020. Non-noble-metal-based electrocatalysts toward the oxygen evolution reaction. *Adv. Funct. Mate.*, 30: 1910274.

[5] Lyu, F., Q. Wang, S. M. Choi and Y. Yin. 2019. Noble-metal-free electrocatalysts for oxygen evolution. *Small*, 15: 1804201.

[6] Osgood, H., S. V. Devaguptapu, H. Xu, J. Cho and G. Wu. 2016. Transition metal (Fe, Co, Ni, and Mn) oxides for oxygen reduction and evolution bifunctional catalysts in alkaline media. *Nano Today*, 11: 601–625.

[7]   TRASATTI, S. 1980. Electrocatalysis by oxides-attempt at a unifying approach. *J. Electroanal. Chem.*, 111: 125–131.

[8]   Klyukin, K., A. Zagalskaya and V. Alexandrov. 2019. Role of dissolution intermediates in promoting oxygen evolution reaction at RuO_2 (110) surface. *J. Phys. Chem. C*, 123: 22151–22157.

[9]   Danilovic, N., R. Subbaraman, K. C. Chang, S. H. Chang, Y. Kang, J. Snyder, A. P. Paulikas, D. Strmcnik, Y. T. Kim, D. Myers, V. R. Stamenkovic and N. M. Markovic. 2014. Using surface segregation to design stable Ru-Ir oxides for the oxygen evolution reaction in acidic environments. *Angew. Chem. Int. Ed.*, 53: 14016–14021.

[10]   Linsey, C. F. D., C. Seitz, Kazunori Nishio, Yasuyuki Hikita, Joseph Montoya, Andrew Doyle, Charlotte Kirk, Aleksandra Vojvodic, Harold Y. Hwang, Jens K. Norskov, Thomas F. Jaramillo. 2016. A highly active and stable $IrO_x/SrIrO_3$ catalyst for the oxygen evolution reaction. *Science*, 353: 1011–1014.

[11]   Hirai, S., T. Ohno, R. Uemura, T. Maruyama, M. Furunaka, R. Fukunaga, W.-T. Chen, H. Suzuki, T. Matsuda and S. Yagi. 2019. $Ca_{1-x}Sr_xRuO_3$ perovskite at the metal-insulator boundary as a highly active oxygen evolution catalyst. *J. Mater. Chem. A*, 7: 15387–15394.

[12]   Han, L., S. Dong and E. Wang. 2016. Transition-metal (Co, Ni, and Fe)-based electrocatalysts for the water oxidation reaction. *Adv. Mater.*, 28: 9266–9291.

[13]   Dong, Y. and S. Komarneni. 2020. Strategies to develop earth-abundant heterogeneous oxygen evolution reaction catalysts for pH-neutral or pH-near-neutral electrolytes. *Small Methods*, 5: 2000719.

[14]   Wang, J., Y. Gao, H. Kong, J. Kim, S. Choi, F. Ciucci, Y. Hao, S. Yang, Z. Shao and J. Lim. 2020. Non-precious-metal catalysts for alkaline water electrolysis: operando characterizations, theoretical calculations, and recent advances. *Chem. Soc. Rev.*, 49: 9154–9196.

[15]   Dong, F., L. Li, Z. Kong, X. Xu, Y. Zhang, Z. Gao, B. Dongyang, M. Ni, Q. Liu and Z. Lin. 2021. Materials engineering in perovskite for optimized oxygen evolution electrocatalysis in alkaline condition. *Small*, 17: 2006638.

[16]   Li, M., F. Luo, Q. Zhang, Z. Yang and Z. Xu. 2020. Atomic layer Co_3O_{4-x} nanosheets as efficient and stable electrocatalyst for rechargeable zinc-air batteries. *Journal of Catalysis*, 381: 395–401.

[17]   Gui, L., Y. Liu, J. Zhang, B. He, Q. Wang and L. Zhao. 2020. *In situ* exsolved Co nanoparticles coupled on $LiCoO_2$ nanofibers to induce oxygen electrocatalysis for rechargeable Zn-air batteries. *J. Mater. Chem. A*, 8: 19946–19953.

[18]   Zhou, Q., Y. Chen, G. Zhao, Y. Lin, Z. Yu, X. Xu, X. Wang, H. K. Liu, W. Sun and S. X. Dou. 2018. Active-site-enriched iron-doped nickel/cobalt hydroxide nanosheets for enhanced oxygen evolution reaction. *ACS Catal.*, 8: 5382–5390.

[19]   Xu, J., J. Li, D. Xiong, B. Zhang, Y. Liu, K. H. Wu, I. Amorim, W. Li and L. Liu. 2018. Trends in activity for the oxygen evolution reaction on transition metal (M = Fe, Co, Ni) phosphide pre-catalysts. *Chem. Sci.*, 9: 3470–3476.

[20]   Chen, S., L. Zhao, J. Ma, Y. Wang, L. Dai and J. Zhang. 2019. Edge-doping modulation of N, P-codoped porous carbon spheres for high performance rechargeable Zn-air batteries. *Nano Energy*, 60: 536–544.

[21]   Du, J., Z. Zou, C. Liu and C. Xu. 2018. Hierarchical Fe-doped Ni_3Se_4 ultrathin nanosheets as an efficient electrocatalyst for oxygen evolution reaction. *Nanoscale*, 10: 5163–5170.

[22]   Zhang, Y., L. You, Q. Liu, Y. Li, T. Li, Z. Xue and G. Li. 2021. Interfacial charge transfer in a hierarchical $Ni_2P/FeOOH$ heterojunction facilitates electrocatalytic oxygen evolution. *ACS Appl. Mater. Interfaces*, 13: 2765–2771.

[23]   Xu, Q., H. Jiang, Y. Li, D. Liang, Y. Hu and C. Li. 2019. *In-situ* enriching active sites on co-doped Fe-$Co_4N@N$-C nanosheet array as air cathode for flexible rechargeable Zn-air batteries. *Appl. Catal. B: Environ.*, 256: 117893.

[24]   Ye, S. H., Z. X. Shi, J. X. Feng, Y. X. Tong and G. R. Li. 2018. Activating CoOOH porous nanosheet arrays by partial Iron substitution for efficient oxygen evolution reaction. *Angew. Chem. Int. Ed.*, 57: 2672–2676.

[25]   Han, X., X. Ling, Y. Wang, T. Ma, C. Zhong, W. Hu and Y. Deng. 2019. Generation of nanoparticle, atomic-cluster, and single-atom cobalt catalysts from zeolitic imidazole frameworks by spatial isolation and their use in zinc-air batteries. *Angew. Chem. Int. Ed.*, 58: 5359–5364.

[26]   Han, X., X. Wu, C. Zhong, Y. Deng, N. Zhao and W. Hu. 2017. NiCo$_2$S$_4$ nanocrystals anchored on nitrogen-doped carbon nanotubes as a highly efficient bifunctional electrocatalyst for rechargeable zinc-air batteries. *Nano Energy*, 31: 541–550.

[27]   Suntivich, J., K. J. May, H. A. Gasteiger, J. B. Goodenough, Y. Shao-Horn. 2011. A perovskite oxide optimized for oxygen evolution catalysis from molecular orbital principles. *Science*, 334: 1383–1385.

[28]   Qu, K., Y. Zheng, S. Dai and S. Z. Qiao. 2016. Graphene oxide-polydopamine derived N, S-codoped carbon nanosheets as superior bifunctional electrocatalysts for oxygen reduction and evolution. *Nano Energy*, 19: 373–381.

[29]   Suen, N. T., S. F. Hung, Q. Quan, N. Zhang, Y. J. Xu and H. M. Chen. 2017. Electrocatalysis for the oxygen evolution reaction: recent development and future perspectives. *Chem. Soc. Rev.*, 46: 337–365.

[30]   Li, Z., L. Lv, J. Wang, X. Ao, Y. Ruan, D. Zha, G. Hong, Q. Wu, Y. Lan, C. Wang, J. Jiang and M. Liu. 2018. Engineering phosphorus-doped LaFeO$_{3-\delta}$ perovskite oxide as robust bifunctional oxygen electrocatalysts in alkaline solutions. *Nano Energy*, 47: 199–209.

[31]   Liu, Y., S. Jiang, S. Li, L. Zhou, Z. Li, J. Li and M. Shao. 2019. Interface engineering of (Ni, Fe) S$_2$@MoS$_2$ heterostructures for synergetic electrochemical water splitting. *Appl. Catal. B: Environ.*, 247: 107–114.

[32]   Anantharaj, S., P. E. Karthik and S. Kundu. 2017. Petal-like hierarchical array of ultrathin Ni(OH)$_2$ nanosheets decorated with Ni(OH)$_2$ nanoburls: A highly efficient OER electrocatalyst. *Catal. Sci. Technol.*, 7: 882–893.

[33]   Anantharaj, S., S. R. Ede, K. Karthick, S. Sam Sankar, K. Sangeetha, P. E. Karthik and S. Kundu. 2018. Precision and correctness in the evaluation of electrocatalytic water splitting: Revisiting activity parameters with a critical assessment. *Energy Environ. Sci.*, 11: 744–771.

[34]   Vrubel, H., T. Moehl, M. Gratzel and X. Hu. 2013. Revealing and accelerating slow electron transport in amorphous molybdenum sulphide particles for hydrogen evolution reaction. *Chem. Commun.*, 49: 8985–8987.

[35]   Karthick, K., S. Anantharaj, P. E. Karthik, B. Subramanian and S. Kundu. 2017. Self-assembled molecular hybrids of CoS-DNA for enhanced water oxidation with low cobalt content. *Inorg. Chem.*, 56: 6734–6745.

[36]   Anantharaj, S., P. E. Karthik, B. Subramanian and S. Kundu. 2016. Pt nanoparticle anchored molecular self-assemblies of DNA: An extremely stable and efficient HER electrocatalyst with ultralow Pt content. *ACS Catal.*, 6: 4660–4672.

[37]   Gui, L., Y. Xu, Q. Tang, X. Shi, J. Zhang, B. He and L. Zhao. 2021. Silver decorated cobalt carbonate to enable high bifunctional activity for oxygen electrocatalysis and rechargeable Zn-air batteries. *J. Colloid Interface Sci.*, 603: 252–258.

[38]   Zhu, Y., W. Zhou, J. Yu, Y. Chen, M. Liu and Z. Shao. 2016. Enhancing electrocatalytic activity of perovskite oxides by tuning cation deficiency for oxygen reduction and evolution reactions. *Chem. Mater.*, 28: 1691–1697.

[39]   Zhou, S., X. Miao, X. Zhao, C. Ma, Y. Qiu, Z. Hu, J. Zhao, L. Shi and J. Zeng. 2016. Engineering electrocatalytic activity in nanosized perovskite cobaltite through surface spin-state transition. *Nat. Commun.*, 7: 11510.

[40]   Han, H., S. Chao, X. Yang, X. Wang, K. Wang, Z. Bai and L. Yang. 2017. Ni nanoparticles embedded in N doped carbon nanotubes derived from a metal organic framework with improved performance for oxygen evolution reaction. *Int. J. Hydrog. Energy*, 42: 16149–16156.

[41]   Wagh, N. K., S. S. Shinde, C. H. Lee, J.-Y. Jung, D.-H. Kim, S.-H. Kim, C. Lin, S. U. Lee and J.-H. Lee. 2020. Densely colonized isolated Cu-N single sites for efficient bifunctional electrocatalysts and rechargeable advanced Zn-air batteries. *Appl. Catal. B: Environ.*, 268: 118746.

[42]   Du, C., Y. Gao, J. Wang and W. Chen. 2020. A new strategy for engineering a hierarchical porous carbon-anchored Fe single-atom electrocatalyst and the insights into its bifunctional catalysis for flexible rechargeable Zn-air batteries. *J. Mater. Chem. A*, 8: 9981–9990.

[43]   Liu, S., Z. Wang, S. Zhou, F. Yu, M. Yu, C. Y. Chiang, W. Zhou, J. Zhao and J. Qiu. 2017. Metal-organic-framework-derived hybrid carbon nanocages as a bifunctional electrocatalyst for oxygen reduction and evolution. *Adv. Mater.*, 29: 1700874.

[44] Yu, P., L. Wang, F. Sun, Y. Xie, X. Liu, J. Ma, X. Wang, C. Tian, J. Li and H. Fu. 2019. Co nanoislands rooted on Co-N-C nanosheets as efficient oxygen electrocatalyst for Zn-Air Batteries. *Adv. Mater.*, 31: 1901666.

[45] Liu, X., L. Wang, P. Yu, C. Tian, F. Sun, J. Ma, W. Li and H. Fu. 2018. A stable bifunctional catalyst for rechargeable zinc-air batteries: iron-cobalt nanoparticles embedded in a nitrogen-doped 3D carbon matrix. *Angew. Chem. Int. Ed.*, 57: 16166–16170.

[46] Tang, X., R. Cao, L. Li, B. Huang, W. Zhai, K. Yuan and Y. Chen. 2020. Engineering efficient bifunctional electrocatalysts for rechargeable zinc-air batteries by confining Fe-Co-Ni nanoalloys in nitrogen-doped carbon nanotube@nanosheet frameworks. *J. Mater. Chem. A*, 8: 25919–25930.

[47] Xie, D., D. Yu, Y. Hao, S. Han, G. Li, X. Wu, F. Hu, L. Li, H. Y. Chen, Y. F. Liao and S. Peng. 2021. Dual-active sites engineering of N-doped hollow carbon nanocubes confining bimetal alloys as bifunctional oxygen electrocatalysts for flexible metal-air batteries. *Small*, 17: 2007239.

[48] Zhu, C., W. Yang, J. Di, S. Zeng, J. Qiao, X. Wang, B. Lv and Q. Li. 2021. CoNi nanoparticles anchored inside carbon nanotube networks by transient heating: Low loading and high activity for oxygen reduction and evolution. *J. Energy Chem.*, 54: 63–71.

[49] Zhao, J., F. Rong, Y. Yao, W. Fan, M. Li and Q. Yang. 2018. Co NP/NC hollow nanoparticles derived from yolk-shell structured ZIFs@polydopamine as bifunctional electrocatalysts for water oxidation and oxygen reduction reactions. *J. Energy Chem.*, 27: 1261–1267.

[50] Yang, Z., C. Zhao, Y. Qu, H. Zhou, F. Zhou, J. Wang, Y. Wu and Y. Li. 2019. Trifunctional self-supporting cobalt-embedded carbon nanotube films for ORR, OER, and HER triggered by solid siffusion from bulk metal. *Adv. Mater.*, 31: 1808043.

[51] Yang, J., X. Wang, B. Li, L. Ma, L. Shi, Y. Xiong and H. Xu. 2017. Novel iron/cobalt-containing polypyrrole hydrogel-derived trifunctional electrocatalyst for self-powered overall water splitting. *Adv. Funct. Mater.*, 27: 1606497.

[52] Qian, J., X. Guo, T. Wang, P. Liu, H. Zhang and D. Gao. 2019. Bifunctional porous Co-doped NiO nanoflowers electrocatalysts for rechargeable zinc-air batteries. *Appl. Catal. B: Environ.*, 250: 71–77.

[53] Kang, B., X. Jin, S. M. Oh, S. B. Patil, M. G. Kim, S. H. Kim and S.-J. Hwang. 2018. An effective way to improve bifunctional electrocatalyst Activity of manganese oxide via control of bond competition. *Appl. Catal. B: Environ.*, 236: 107–116.

[54] Huang, K., R. Wang, S. Zhao, P. Du, H. Wang, H. Wei, Y. Long, B. Deng, M. Lei, B. Ge, H. Gou, R. Zhang and H. Wu. 2020. Atomic species derived CoO_x clusters on nitrogen doped mesoporous carbon as advanced bifunctional electro-catalysts for Zn-air battery. *Energy Stor. Mater.*, 29: 156–162.

[55] Liao, C., B. Yang, N. Zhang, M. Liu, G. Chen, X. Jiang, G. Chen, J. Yang, X. Liu, T. S. Chan, Y. J. Lu, R. Ma and W. Zhou. 2019. Constructing conductive interfaces between nickel oxide nanocrystals and polymer carbon nitride for efficient electrocatalytic oxygen evolution reaction. *Adv. Funct. Mater.*, 29: 1904020.

[56] Cheng, X., E. Fabbri, Y. Yamashita, I. E. Castelli, B. Kim, M. Uchida, R. Haumont, I. Puente-Orench and T. J. Schmidt. 2018. Oxygen evolution reaction on perovskites: A multieffect descriptor study combining experimental and theoretical methods. *ACS Catal.*, 8: 9567–9578.

[57] Li, X., Z. Cui, Y. Li, S. Xin, J. Zhou, Y. Long, C. Jin, John B. Goodenough. 2019. Exceptional oxygen evolution reactivities on $CaCoO_3$ and $SrCoO_3$. *Sci. Adv.*, 5: eaav6262.

[58] Gui, L., Z. Huang, G. Li, Q. Wang, B. He and L. Zhao. 2019. Insights into Ni-Fe couple in perovskite electrocatalysts for highly efficient electrochemical oxygen evolution. *Electrochim. Acta*, 293: 240–246.

[59] Wei, C., Z. Feng, G. G. Scherer, J. Barber, Y. Shao-Horn and Z. J. Xu. 2017. Cations in octahedral sites: a descriptor for oxygen electrocatalysis on transition-metal spinels. *Adv. Mater.*, 29: 1606800.

[60] Dong, M., X. Liu, L. Jiang, Z. Zhu, Y. Shu, S. Chen, Y. Dou, P. Liu, H. Yin and H. Zhao. 2020. Cobalt-doped Mn_3O_4 nanocrystals embedded in graphene nanosheets as a high-performance bifunctional oxygen electrocatalyst for rechargeable Zn-Air batteries. *Green Energy Environ.*, 5: 499–505.

[61] Kim, M. and K. Kim. 1998. A Study on the phase transformation of electrochemically precipitated nickel hydroxides using an electrochemical quartz crystal microbalance. *J. Electrochem. Soc.*, 145: 507–511.

[62] Wei, M., J. Li, W. Chu and N. Wang. 2019. Phase control of 2D binary hydroxides nanosheets via controlling-release strategy for enhanced oxygen evolution reaction and supercapacitor performances. *J. Energy Chem.*, 38: 26–33.

[63] Zhang, Z., X. Li, C. Zhong, N. Zhao, Y. Deng, X. Han and W. Hu. 2020. Spontaneous synthesis of silver-nanoparticle-decorated transition-metal hydroxides for enhanced oxygen evolution reaction. *Angew. Chem. Int. Ed.*, 59: 7245–7250.

[64] Yang, R., Y. Zhou, Y. Xing, D. Li, D. Jiang, M. Chen, W. Shi and S. Yuan. 2019. Synergistic coupling of CoFe-LDH arrays with NiFe-LDH nanosheet for highly efficient overall water splitting in alkaline media. *Appl. Catal. B: Environ.*, 253: 131–139.

[65] Wang, K., H. Du, S. He, L. Liu, K. Yang, J. Sun, Y. Liu, Z. Du, L. Xie, W. Ai and W. Huang. 2021. Kinetically controlled, scalable synthesis of γ-FeOOH nanosheet arrays on nickel foam toward efficient oxygen evolution: the key role of in-situ-generated γ-NiOOH. *Adv. Mater.*, 33: 2005587.

[66] Yan, L., B. Zhang, Z. Liu and J. Zhu. 2021. Synergy of copper doping and oxygen vacancies in porous CoOOH nanoplates for efficient water oxidation. *Chem. Eng. J.*, 405: 126198.

[67] Meng, C., M. Lin, X. Sun, X. Chen, X. Chen, X. Du and Y. Zhou. 2019. Laser synthesis of oxygen vacancy-modified CoOOH for highly efficient oxygen evolution. *Chem. Commun.*, 55: 2904–2907.

[68] Feng, S., L. Yang, Z. Zhang, Q. Li and D. Xu. 2020. Au-decorated CoOOH nanoplate hierarchical hollow structure for plasmon-enhanced electrocatalytic water oxidation. *ACS Appl. Energy Mater.*, 3: 943–950.

[69] Lee, S.-Y., I.-S. Kim, H.-S. Cho, C.-H. Kim and Y.-K. Lee. 2021. Resolving potential-dependent degradation of electrodeposited Ni(OH)₂ catalysts in alkaline oxygen evolution reaction (OER): in situ XANES studies. *Appl. Catal. B: Environ.*, 284: 119729.

[70] Xu, H., J. Cao, C. Shan, B. Wang, P. Xi, W. Liu and Y. Tang. 2018. MOF-derived hollow CoS decorated with CeOₓ nanoparticles for boosting oxygen evolution reaction electrocatalysis. *Angew. Chem. Int. Ed.*, 57: 8654–8658.

[71] Ding, X., W. Li, H. Kuang, M. Qu, M. Cui, C. Zhao, D. C. Qi, F. E. Oropeza and K. H. L. Zhang. 2019. An Fe stabilized metallic phase of NiS₂ for the highly efficient oxygen evolution reaction. *Nanoscale*, 11: 23217–23225.

[72] Dou, S., L. Tao, J. Huo, S. Wang and L. Dai. 2016. Etched and doped Co₉S₈/graphene hybrid for oxygen electrocatalysis. *Energy Environ. Sci.*, 9: 1320–1326.

[73] Surendran, S., S. Shanmugapriya, H. Ramasamy, G. Janani, D. Kalpana, Y. S. Lee, U. Sim and R. K. Selvan. 2019. Hydrothermal deposition of CoS nanostructures and its multifunctional applications in supercapattery and water electrolyzer. *Appl. Surf. Sci.*, 494: 916–928.

[74] Zhang, G., Y.-S. Feng, W.-T. Lu, D. He, C.-Y. Wang, Y.-K. Li, X.-Y. Wang and F.-F. Cao. 2018. Enhanced catalysis of electrochemical overall water splitting in alkaline media by Fe doping in Ni₃S₂ nanosheet arrays. *ACS Catal.*, 8: 5431–5441.

[75] Feng, X., Q. Jiao, H. Cui, M. Yin, Q. Li, Y. Zhao, H. Li, W. Zhou and C. Feng. 2018 One-pot synthesis of NiCo₂S₄ hollow spheres via sequential ion-exchange as an enhanced oxygen bifunctional electrocatalyst in alkaline solution. *ACS Appl. Mater. Interfaces*, 10: 29521–29531.

[76] Zhang, J., T. Wang, D. Pohl, B. Rellinghaus, R. Dong, S. Liu, X. Zhuang and X. Feng. 2016. Interface engineering of MoS₂/Ni₃S₂ heterostructures for highly enhanced electrochemical overall-water-splitting activity. *Angew. Chem. Int. Ed.*, 55: 6702–6707.

[77] Hou, J., B. Zhang, Z. Li, S. Cao, Y. Sun, Y. Wu, Z. Gao and L. Sun. 2018. Vertically aligned oxygenated-CoS₂-MoS₂ heteronanosheet architecture from polyoxometalate for efficient and stable overall water splitting. *ACS Catal.*, 8: 4612–4621.

[78] Kuang, P., M. He, H. Zou, J. Yu and K. Fan. 2019. 0D/3D MoS₂-NiS₂/N-doped graphene foam composite for efficient overall water splitting. *Appl. Catal. B: Environ.*, 254: 15–25.

[79] Guo, Y., J. Tang, Z. Wang, Y.-M. Kang, Y. Bando and Y. Yamauchi. 2018. Elaborately assembled core-shell structured metal sulfides as a bifunctional catalyst for highly efficient electrochemical overall water splitting. *Nano Energy*, 47: 494–502.

[80] Liu, Y., X. Luo, C. Zhou, S. Du, D. Zhen, B. Chen, J. Li, Q. Wu, Y. Iru and D. Chen. 2020. A modulated electronic state strategy designed to integrate active HER and OER components as hybrid heterostructures for efficient overall water splitting. *Appl. Catal. B: Environ.*, 260: 118197.

[81] Li, L., L. Song, H. Guo, W. Xia, C. Jiang, B. Gao, C. Wu, T. Wang and J. He. 2019. N-Doped porous carbon nanosheets decorated with graphitized carbon layer encapsulated Co_9S_8 nanoparticles: an efficient bifunctional electrocatalyst for the OER and ORR. *Nanoscale*, 11: 901–907.

[82] Zhou, J., L. Yuan, J. Wang, L. Song, Y. You, R. Zhou, J. Zhang and J. Xu. 2020. Combinational modulations of $NiSe_2$ nanodendrites by phase engineering and iron-doping towards an efficient oxygen evolution reaction. *J. Mater. Chem. A*, 8: 8113–8120.

[83] Fang, Z., L. Peng, H. Lv, Y. Zhu, C. Yan, S. Wang, P. Kalyani, X. Wu and G. Yu. 2017. Metallic transition metal selenide holey nanosheets for efficient oxygen evolution electrocatalysis. *ACS Nano*, 11: 9550–9557.

[84] Wang, H., Z. Sun, X. Zou, J. Ren and C. Y. Zhang. 2021. Controllable synthesis of $CoFe_2Se_4$/$NiCo_2Se_4$ hybrid nanotubes with heterointerfaces and improved oxygen evolution reaction performance. *Nanoscale*, 13: 6241–6247.

[85] Wang, K., Z. Lin, Y. Tang, Z. Tang, C.-L. Tao, D.-D. Qin and Y. Tian. 2021. Selenide/sulfide heterostructured $NiCo_2Se_4$/$NiCoS_4$ for oxygen evolution reaction, hydrogen evolution reaction, water splitting and Zn-air batteries. *Electrochim. Acta*, 368: 137584.

[86] Dai, J., D. Zhao, W. Sun, X. Zhu, L.-J. Ma, Z. Wu, C. Yang, Z. Cui, L. Li and S. Chen. 2019. Cu(II) ions induced structural transformation of cobalt selenides for remarkable enhancement in oxygen/hydrogen electrocatalysis. *ACS Catals.*, 9: 10761–10772.

[87] Zhuang, L., Y. Jia, H. Liu, X. Wang, R. K. Hocking, H. Liu, J. Chen, L. Ge, L. Zhang, M. Li, C. L. Dong, Y. C. Huang, S. Shen, D. Yang, Z. Zhu and X. Yao. 2019. Defect-induced Pt-Co-Se coordinated sites with highly asymmetrical electronic distribution for boosting oxygen-involving electrocatalysis. *Adv. Mater.*, 31: 1805581.

[88] Xia, Z., H. Sun, X. He, Z. Sun, C. Lu, J. Li, Y. Peng, S. Dou, J. Sun and Z. Liu. 2019. *In situ* construction of $CoSe_2$@vertical-oriented graphene arrays as selfsupporting electrodes for sodium-ion capacitors and electrocatalytic oxygen evolution. *Nano Energy*, 60: 385–393.

[89] Zhu, Y., H.-C. Chen, C.-S. Hsu, T.-S. Lin, C.-J. Chang, S.-C. Chang, L.-D. Tsai and H. M. Chen. 2019. Operando unraveling of the structural and chemical stability of P-substituted $CoSe_2$ electrocatalysts toward hydrogen and oxygen evolution reactions in alkaline electrolyte. *ACS Energy Lett.*, 4: 987–994.

[90] Zhang, H., T. Wang, A. Sumboja, W. Zang, J. Xie, D. Gao, S. J. Pennycook, Z. Liu, C. Guan and J. Wang. 2018. Integrated hierarchical carbon flake arrays with hollow P-doped $CoSe_2$ nanoclusters as an advanced bifunctional catalyst for Zn-air batteries. *Adv. Funct. Mater.*, 28: 1804846.

[91] Yin, J., Y. Li, F. Lv, M. Lu, K. Sun, W. Wang, L. Wang, F. Cheng, Y. Li, P. Xi and S. Guo. 2017. Oxygen vacancies dominated NiS_2/CoS_2 interface porous nanowires for portable Zn-air batteries driven water splitting devices. *Adv. Mater.*, 29: 1704681.

[92] Zhao, X., X. Li, Y. Yan, Y. Xing, S. Lu, L. Zhao, S. Zhou, Z. Peng and J. Zeng. 2018. Electrical and structural engineering of cobalt selenide nanosheets by Mn modulation for efficient oxygen evolution. *Appl. Catal. B: Environ.*, 236: 569–575.

[93] Zhang, L., C. Lu, F. Ye, R. Pang, Y. Liu, Z. Wu, Z. Shao, Z. Sun and L. Hu. 2021. Selenic acid etching assisted vacancy engineering for designing highly active electrocatalysts toward the oxygen evolution reaction. *Adv. Mater.*, 33: 2007523.

[94] Sun, H., C. Tian, G. Fan, J. Qi, Z. Liu, Z. Yan, F. Cheng, J. Chen, C. P. Li and M. Du. 2020. Boosting activity on Co_4N porous nanosheet by coupling CeO_2 for efficient electrochemical overall water splitting at high current densities. *Adv. Funct. Mater.*, 30: 1910596.

[95] Meng, F., H. Zhong, D. Bao, J. Yan and X. Zhang. 2016. *In situ* coupling of strung Co_4N and intertwined N-C fibers toward free-standing bifunctional cathode for robust, efficient, and flexible Zn-air batteries. *J. Am. Chem. Soc.*, 138: 10226–10231.

[96] Liu, H., J. Lei, S. Yang, F. Qin, L. Cui, Y. Kong, X. Zheng, T. Duan, W. Zhu and R. He. 2021. Boosting the oxygen evolution activity over cobalt nitride nanosheets through optimizing the electronic configuration. *Appl. Catal. B: Environ.*, 286: 119894.

[97] Li, D., W. Zhang, J. Zeng, B. Gao, Y. Tang and Q. Gao. 2021. Nickel-doped Co_4N nanowire bundles as efficient electrocatalysts for oxygen evolution reaction. *Sci. China Mater.*, 64: 1889–1899.

[98] Liu, H., X. Lu, Y. Hu, R. Chen, P. Zhao, L. Wang, G. Zhu, L. Ma and Z. Jin. 2019. Co_xFe_yN nanoparticles decorated on graphene sheets as high-performance electrocatalysts for the oxygen evolution reaction. *J. Mater. Chem. A*, 7: 12489–12497.

[99]   Yuan, W., S. Wang, Y. Ma, Y. Qiu, Y. An and L. Cheng. 2020. Interfacial engineering of cobalt nitrides and mesoporous nitrogen-doped carbon: toward efficient overall water-splitting activity with enhanced charge-transfer efficiency. *ACS Energy Lett.*, 5: 692–700.

[100]   Ge, H., G. Li, J. Shen, W. Ma, X. Meng and L. Xu. 2020. Co_4N nanoparticles encapsulated in N-doped carbon box as tri-functional catalyst for Zn-air battery and overall water splitting. *Appl. Catal. B: Environ.*, 275: 119104.

[101]   Guan, C., A. Sumboja, W. Zang, Y. Qian, H. Zhang, X. Liu, Z. Liu, D. Zhao, S. J. Pennycook and J. Wang. 2019. Decorating Co/CoN_x nanoparticles in nitrogen-doped carbon nanoarrays for flexible and rechargeable zinc-air batteries. *Energy Stor. Mater.*, 16: 243–250.

[102]   Zhang, B., C. Xiao, S. Xie, J. Liang, X. Chen and Y. Tang. 2016. Iron-nickel nitride nanostructures in situ grown on surface-redox-etching nickel foam: efficient and ultrasustainable electrocatalysts for overall water splitting. *Chem. Mater.*, 28: 6934–6941.

[103]   Xiong, D., X. Wang, W. Li and L. Liu. 2016. Facile synthesis of iron phosphide nanorods for efficient and durable electrochemical oxygen evolution. *Chem. Commun.*, 52: 8711–8714.

[104]   Yao, Y., N. Mahmood, L. Pan, G. Shen, R. Zhang, R. Gao, F. E. Aleem, X. Yuan, X. Zhang and J. J. Zou. 2018. Iron phosphide encapsulated in P-doped graphitic carbon as efficient and stable electrocatalyst for hydrogen and oxygen evolution reactions. *Nanoscale*, 10: 21327–21334.

[105]   You, B., N. Jiang, M. Sheng, M. W. Bhushan and Y. Sun. 2016. Hierarchically porous urchin-like Ni_2P superstructures supported on nickel foam as efficient bifunctional electrocatalysts for overall water splitting. *ACS Catal.*, 6: 714–721.

[106]   Wang, J. and F. Ciucci. 2019. *In-situ* synthesis of bimetallic phosphide with carbon tubes as an active electrocatalyst for oxygen evolution reaction. *Appl. Catal. B: Environ.*, 254: 292–299.

[107]   Ji, L., J. Wang, X. Teng, T. J. Meyer and Z. Chen. 2020. CoP nanoframes as bifunctional electrocatalysts for efficient overall water splitting. *ACS Catal.*, 10: 412–419.

[108]   Jiao, L., Y. X. Zhou and H. L. Jiang. 2016. Metal-organic framework-based CoP/reduced graphene oxide: high-performance bifunctional electrocatalyst for overall water splitting. *Chem. Sci.*, 7: 1690–1695.

[109]   Liang, H., A. N. Gandi, D. H. Anjum, X. Wang, U. Schwingenschlogl and H. N. Alshareef. 2016. Plasma-assisted synthesis of NiCoP for efficient overall water splitting. *Nano Lett.*, 16: 7718–7725.

[110]   Chai, L., Z. Hu, X. Wang, Y. Xu, L. Zhang, T. T. Li, Y. Hu, J. Qian and S. Huang. 2020. Stringing bimetallic metal-organic framework-derived cobalt phosphide composite for high-efficiency overall water splitting. *Adv. Sci.*, 7: 1903195.

[111]   Li, H., Q. Li, P. Wen, T. B. Williams, S. Adhikari, C. Dun, C. Lu, D. Itanze, L. Jiang, D. L. Carroll, G. L. Donati, P. M. Lundin, Y. Qiu and S. M. Geyer. 2018. Colloidal cobalt phosphide nanocrystals as trifunctional electrocatalysts for overall water splitting powered by a zinc-air battery. *Adv. Mater.*, 30: 1705796.

[112]   Lai, C., X. Liu, C. Cao, Y. Wang, Y. Yin, T. Liang and D. D. Dionysiou. 2021. Structural regulation of N-doped carbon nanocages as high-performance bifunctional electrocatalysts for rechargeable Zn-air batteries. *Carbon*, 173: 715–723.

[113]   Li, R., Z. Wei and X. Gou. 2015. Nitrogen and phosphorus dual-doped graphene/carbon nanosheets as bifunctional electrocatalysts for oxygen reduction and evolution. *ACS Catal.*, 5: 4133–4142.

[114]   Sun, T., J. Wang, C. Qiu, X. Ling, B. Tian, W. Chen and C. Su. 2018. B, N codoped and defect-rich nanocarbon material as a metal-free bifunctional electrocatalyst for oxygen reduction and evolution reactions. *Adv. Sci.*, 5: 1800036.

[115]   Murugesan, B., N. Pandiyan, M. Arumugam, M. Veerasingam, J. Sonamuthu, A. R. Jeyaraman, S. Samayanan and S. Mahalingam. 2019. Two dimensional graphene oxides converted to three dimensional P, N, F and B, N, F tri-doped graphene by ionic liquid for efficient catalytic performance. *Carbon*, 151: 53–67.

[116]   Zhao, Z., M. Li, L. Zhang, L. Dai and Z. Xia. 2015. Design principles for heteroatom-doped carbon nanomaterials as highly efficient catalysts for fuel cells and metal-air batteries. *Adv. Mater.*, 27: 6834–6840.

8

Self-Support Nanoarrays as Electrocatalysts for Energy Conversion Applications

Xiaojun Shi and *Huanwen Wang**

8.1 Introduction

The over-exploitation of fossil fuel energy gives rise to energy shortage and environmental problems such as polluting gas emission and climate warming. To that end, carbon-neutral[1–3] policies have recently been put on the agenda. For the sustainable development of society, it is urgent to develop clean and renewable new energy technology. As a highly admirable method of conversion, electrochemical processes, which could convert small molecules (oxygen, water, carbon dioxide, nitrogen, and urea) into valuable fuels and chemicals (hydrogen, hydrocarbons, ammonia and so on).[4–7] In general, these conversion techniques involve some electrochemical reactions, such as the oxygen reduction reaction (ORR),[8–10] oxygen evolution reaction (OER),[11–14] hydrogen evolution reaction (HER),[15–18] carbon dioxide reduction reaction (CRR),[19–21] nitrogen reduction reaction (NRR),[22–25] and urea oxidation reaction (UOR).[26–28] However, the disadvantages of high overpotential and slow kinetics restrict the further application and development of these conversion techniques.[6,29] Therefore, the exploration of high activity and strong durability of electrocatalyst has become a top priority.

Currently, the state-of-the-art electrocatalysts are still based on expensive and rare precious metals (Pt, IrO_2, and RuO_2) with minimized overpotential, which to some extent restricts large-scale sustainable energy conversion.[30–33] Therefore, a large number of researchers set their sights on transition metal-based catalysts with abundant reserves, low cost, flexible electronic configuration, excellent electrochemical activity and stability.[6,9,34] In particular, with the recent development of nanotechnology and

Faculty of Material and Chemistry, China University of Geosciences, Wuhan, 430074, China.
* Corresponding author: wanghw@cug.edu.cn

material science, the derivatization and development of nanoscale catalysts have been promoted. Nano-catalyst is provided with the advantages of abundant active sites, fast material transfer and limited volume change, and have been recognized as an effective strategy to improve catalyst activity.[35,36] Despite these achievements, conventional powdery catalyst electrodes require binders and conductors to carry out a time-consuming and tanglesome preparation process, which sacrifice some of the active sites and thus diminish the positive effects of nanomaterials.[4,37] Moreover, such an electrode structure introduces some additional random interface, which is not conducive to the direct transfer of electrons and ions and may induce side reactions (Fig. 8.1a).[38,39] Secondly, the high surface energy of nanomaterials can lead to agglomeration of transition metals, thus reducing the effective utilization rate of the catalyst.[40,41] Finally, the active nanomaterials physically adhere to the collector surface and are prone to powder shedding, especially in catalytic reactions with gas generation.[42] In addition, its poor bending and folding ability limits its application in some flexible electronic devices.

In contrast to the conventional powder catalysts, a self-supported electrode material is here defined as an electrode without binder and conductive carbon black where the active materials are grown directly on the conductive substrate. As self-supported and unbonded electrodes, enormous superiority is obtained by integrating nanostructured electrocatalysts into a variety of conductive substrates.[43,44] By virtue of *in situ* growth strategies, rational design of nanoarray architectures is able to boost electrocatalysis performance based on several features[38,39,45–48] (Fig. 8.1b): (1) the *in situ* growth strategy ensures the tight contact of the interface, thus making the structure and properties of the material durable, and avoiding the structure collapse and volume expansion during the reaction process; (2) the array structure is conducive to the electrolyte entering the active site and expanding the specific surface area, thus significantly increasing the effective active site; (3) the ordered hierarchical structure can not only provide sufficient channels for material transport, but also timely

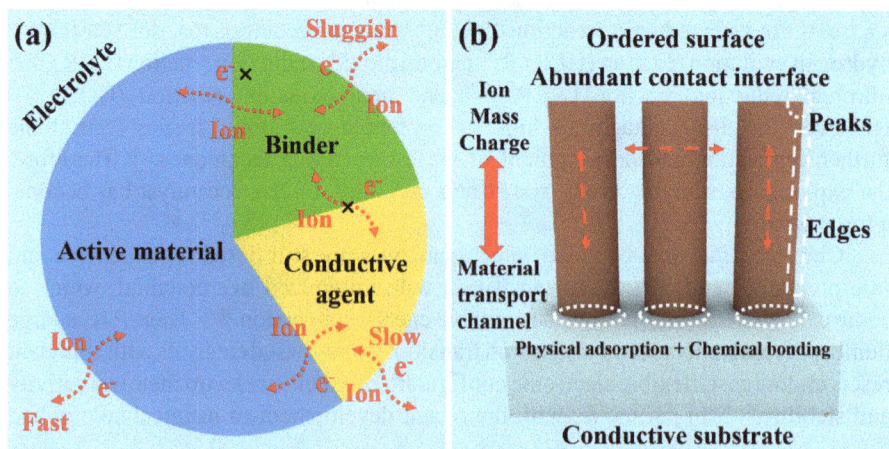

Figure 8.1. (a) Electron and ion exchange between electrodes of conventional powder catalysts and between them and electrolyte. (b) Preponderances of self-supported electrocatalysts in terms of mass transport and charge transfer.

release the gas generated in the state of high current density to promote the catalytic reaction; (4) the peaks and edges of nanoarray architecture own acknowledged unexceptionable catalytic advantage; (5) the synergistic effect between matrix and nanoarray structure can be developed by selecting suitable substrates to generate catalysts, thus improving the electrocatalytic activity. Based on these advantages, a lot of self-supported earth-abundant nanoarrays in the literature have presented significant catalytic efficiency and stability toward the ORR, OER, HER, UOR, NRR, CRR and so on,[49–51] and promising prospects are presented by nanoarrays' structure even better than precious metal catalysts. Furthermore, some concrete instances exhibit excellent bifunctional properties in practical energy storage and conversion application such as fuel cells, rechargeable metal-air batteries and water-alkaline electrolyzers.[4,43,52] From the perspective of scalability and sustainability, it will be essential to summarize the development, application and future prospect of nanoarray architecture catalysts.

Here, we conclude the types of array catalysts, synthesis methods and the applications of a large number of low-cost, earth-abundant, high-performance and stable electrocatalysts in the field of catalysis. In terms of the design and synthesis of array architectures, this paper reviews the concepts of building array manufacturing and general synthesis strategies, providing promising insights for the development of excellent performance of transition metal array architectures. Promising strategies for improving electrocatalytic activity and stability, such as heterostructure regulation, interface and defect engineering, and inhibition of metal clusters, are highlighted. Through theoretical simulation and *in situ* identification, the internal mechanisms of electronic structure optimization, intermediate adsorption promotion and coordination environment improvement were analyzed. Finally, we present the challenges and opportunities in valuable directions and promising paths towards performance excellence in transition metal array architectures. This portion not only provides guidelines for us to fully grasp the design and regulation strategies of architectural arrays, but also provides inspiration for us to carefully manufacture ideal electrocatalysts for future energy conversion systems.

8.2 Synthetic methods of nanoarrays

The composition, morphology, size and microstructure of array architecture catalysts can be controlled by different synthesis strategies. In turn, the relationship between physicochemical properties and the resulting electrocatalytic behavior can provide feedback for optimizing the synthesis process. To this end, we classified the array catalysts on the basis of the reported literature. According to the different chemical composition of the array architecture catalyst, it can be classified as metal foil/ plate/ foam, carbon paper, carbon cloth and conductive glass, etc. According to the different morphology of array catalyst, the free-standing arrays can be categorized as single metal catalysts (SACs),[53–55] metal oxides,[56,57] chalcogenides,[10,58,59] metal carbides/ nitrides,[52,60,61] hydroxides[62–64] and metal organic framework (MOF)[65–67] and carbon-based composite materials.[49,68,69] In addition, the nanorods,[70–73] nanoneedles,[74–77] nanosheets,[78,79] nanoparticles,[80–83] hollow nanoarrays,[84,85] nanoarray on array,[86–88] and nanoarray on hollow arrays'[89,90] composite structures are presented in Fig. 8.2a,b according to the different manifested microstructure.

The construction of nanostructured array electrocatalysts often involves individual or integrated approaches and can be achieved through a flexible combination of different synthesis tactics. At the surface of substrate and arrays, complicated physical and chemical reactions are often involved in *in situ* growth process. Herein, we will focus on several typical preparation methods of nanoarray architecture catalysts in Fig. 8.3, including chemical bath deposition,[91–93] hydro/solvothermal

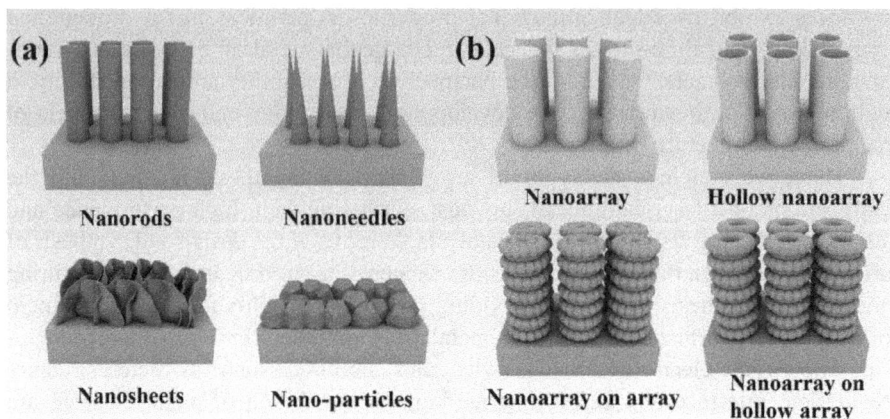

Figure 8.2. (a), (b) Nanoarray categories with different micromorphology and composite structure.

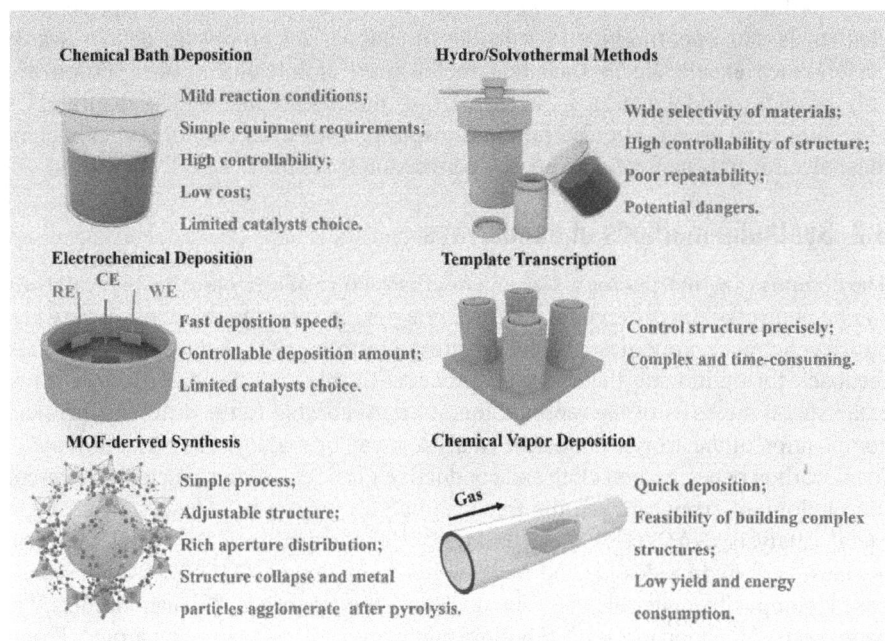

Figure 8.3. Schematics of the widely used synthesis methods for nanoarray electrocatalysts and their respective strength and weakness.

methods,[94-96] electrochemical deposition,[97-99] template transcription,[100-102] MOF-derived synthesis[103-105] and chemical vapor deposition.[106-108] Each method possesses its own advantages and disadvantages, and reasonable utilization can maximize the excellent performance of self-supported nanoarray catalyst.

8.2.1 Chemical Bath Deposition (CBD)

CBD is a sort of preparation method of array catalyst with mild reaction conditions, simple equipment requirements, high controllability and low cost.[91-93] CBD involves a complete process of continuous nucleation on a conductive substrate in a solution rich in precursor ions, with only standing or stirring required. Of course, it is on this basis that the possibilities of material selection are restricted. At present, only some ordinary metal oxide, hydroxide, sulfide, and selenide-based array materials are acquirable. For example, CuS nanosheet arrays were prepared by *in situ* growth strategy of CBD on copper sheets and served as efficient electrode materials for CO_2 reduction.[109] Another example is where nanosheets on nanorods type composite with interconnected hierarchical nanoarray structure ($FeOOH/TiO_2/ZnO$)[110] were successfully synthesized by two-step CBD methods. As CBD is still an array catalyst synthesis method under research and development, its preparation process is affected by a variety of factors, such as temperature, deposition time, pH value, precipitation, doping, etc. On the other hand, by regulating these parameters, array catalysts with different morphology, structure and uniformity can be obtained. It is advantageous compared with other array catalyst preparation methods, which make it manifest a hopeful development prospect.

8.2.2 Hydro/Solvothermal methods

Solvothermal is a further development of hydrothermal method, which differs from hydrothermal method is that the used solvent is organic rather than water. Hydrothermal method and solvent method are one of the most common methods for the synthesis of self-supported array catalysts, which is provided with preponderances of wide selectivity of materials and high controllability of structure. They are both in an aqueous or non-aqueous solution under a closed system, based on a certain temperature and pressure; the original mixture reacts to nucleate and grow on a substrate.[94-96] In most cases, the added substrate induces non-uniform nucleation, thus guaranteeing uniform deposition of the nanoarray. The absence of substrates may result in severe agglomeration due to rapid nucleation and crystal growth. In general, the micro morphology and chemical structure of nanoarrays can be effectively regulated by changing the type of solvent, solution concentration, pH, temperature, time and other parameters. Among them, temperature is the main parameter affecting the morphology and microstructure. A large number of advanced array catalyst materials can be obtained by hydro/solvothermal methods. At present, hydro/solvothermal methods have been widely employed in the preparation of metal-based nanoarray electrocatalysts, such as oxides, hydroxides, phosphates, chalcogenides and so on. For instance, a new type of electrocatalyst with core-shell carbon/$NiCo_2O_4$[111] double microtubes architecture was successfully synthesized through a hydrothermal method combined with the calcination process. Abundant defects and pore distribution in inner and outer tubes endowed it excellent OER catalytic performance. In addition, Yang et al.[112] reported a highly efficient HER catalyst for nano-VS_4 directional decoration of Ni_3S_2-on-NF by simple solvothermal methods. Jin et al.[113] synthesized nitrogen-doped carbon layer to encapsulate Co and FeCo nanoparticles mixed nanowire array (Co-FeCo/N-G) by hydrothermal method and plasma-assistance strategy, and prepared a foldable and high-performance self-supported air electrode. Although hydro/solvothermal methods are simple and rapid, and possess favorable prospects for industrialization, there are poor repeatability problems and potential dangers caused by difficult control of the reaction process. Therefore, the device modification of hydro/solvothermal method to adapt to complex parameter changes is also a key issue for its further development.

8.2.3 Electrochemical deposition

Electrochemical deposition synthesis is a process in which coherent films are generated on electrodes by electrochemically induced reactions in precursor solution.[97–99] The electrode potential, as a general-purpose oxidation-reducing agent, is sufficient to drive most chemical reactions to deposit arrays of metal compounds on the substrate, including metal hydroxides, oxides, sulfides, phosphates, selenides, etc. The process of deposition is carried out under certain electrolyte and operating conditions. The difficulty of metal electrodeposition and the morphology of the sediment are related to the properties of the deposited metal, and also depend on the composition of electrolyte, pH value, temperature, current density and other factors. On this basis, electrochemical deposition demonstrates the advantages of fast deposition speed and highly controllable deposition amount, which is another efficient method for preparing array materials. For example, the Ag-doped CoOOH nanosheets array[114] produced by electrochemical deposition method can not only realize the uniform distribution of Ag atoms in the nanosheet array, but also promote the transformation of $Co(OH)_2$ into CoOOH active phase. Li et al.[115] triumphantly synthesized 3D heterogeneous NiFe layered double hydroxides with strong interfacial interaction through electrochemical deposition and electrochemical dealloying methods. Co_3O_4 with excellent bi-functional catalytic activity was also deposited on carbon cloth and metal substrate as the electrode material of efficient and stable Zn-air battery. Of course, the generation of crystal nucleus and the growth process of crystal are not easy to control. Therefore, the influence of parameters such as current magnitude, precursor solution composition and kinetic process of ions on morphology and structure after deposition becomes the focus of research.

8.2.4 Template transcription

Template method is a general and reliable technique for directional synthesis of 1D nanoarray materials by depositing related substances in the holes or surface of hard or soft templates.[100–102] At the same time, this means that the template synthesis method must be combined with other array preparation techniques to achieve the corresponding purpose. After removing the template, a porous or hollow 1D nanoarray structure is usually obtained, which will help catalyze material transport during the process. As an effective method to prepare nanomaterials, the main characteristic of template method is that no matter which chemical reaction takes place in liquid phase or gas phase, the reaction is carried out in an effectively controlled area, which is the main difference between template method from ordinary method. The template method can be divided into soft template and hard template according to the characteristics of the template itself and the different ability of limiting the domain. In the soft template method, the template is usually an ordered polymer formed by emulsion droplets, surfactant molecules and polymer micelles. Weak intermolecular or intramolecular interactions are the forces that hold the template together, thus forming aggregates of different spatial structural features. Hard templates are mainly rigid templates maintained by covalent bonds, such as polymers with different spatial structures, anodic alumina films, porous silicon, metal templates, etc. Compared with the soft template, the hard template exhibits higher stability and better narrow zone limit effect, and can strictly control the size and morphology of nanomaterials. The morphology of nanomaterials prepared with hard template usually shows minor changes because of the ordinary structure of hard template.

SiO_2, ZnO, anodic Al_2O_3, MgO and various metal substrates were commonly served as hard templates for the synthesis of array self-supported electrode materials. $Ni_3ZnC_{0.7}$/NCNT[116] arrays with ZnO nanorods as template and Zn source were synthesized by one-step template strategy. $Ni_3ZnC_{0.7}$/NCNT-700 arrays displayed excellent catalytic activity for hydrogen evolution and oxygen evolution at low overpotential due to the introduction of $Ni_3ZnC_{0.7}$ nanodots and nitrogen doping of carbon nanotubes. Gao et al.[117] employed Co_3O_4 nanowire arrays prefabricated on Ni foam as templates and metal sources to integrate with organic ligands to form MOF layers. The optimized CO_3O_4@MOF-74 structure presented enormously enhanced catalytic activity for OER. Ni-Mo phosphide nanotube arrays (O-NiMoP/NF)[118] doped with oxygen were synthesized on nickel foam (NF) by electrodeposition and *in situ* template etching. The self-supported O-NiMOP/NF electrode exhibited attractive bi-functional catalytic activity for HER and UOR due to the O-NiMOP modulated electron structure and nanotube array structure.

Soft templates are usually ordered polymers formed by amphiphilic molecules. Tao et al.[119] reported a two-step synthesis method for a 3D ordered mesoporous Cu sphere array by templating mesoporous structural colloidal crystals in conventional pores using a two-template method using polymethyl methacrylate (PMMA) and a non-ionic surfactant Brij 58. Therefore, the well-organized 3D interconnect dual-continuous mesoporous structure revealed the superiorities of a large number of exposed catalytic active sites, efficient mass transfer and high conductivity, resulting in excellent electrocatalytic CRR performance. An independent mesoporous decorated Palladium nanotube array (P-PDNTA)[120] was created using a dual-template electrodeposition method as a bifunctional electrocatalyst. The soft template based on phytotriol could alter the surface chemistry of the inner surface of aluminum anodic oxide, thus promoting the controlled electrodeposition of highly stable Pd nanotubes. Sacrificing the soft template resulted in a large number of mesopores on the nanotube, which increased the electrochemically active surface area by 189% compared to the PdNTA baseline without mesopores. In general, the template method usually has the advantage of using the template as the carrier to accurately control the size, shape, structure and properties of nanomaterials, but it also has the disadvantage of complex and time-consuming preparation process. Therefore, the template method, as a preparation technique for array materials for sustainable new energy system development, offers promising features with significant advantages and limitations.

8.2.5 MOF-derived synthesis

Metal Organic Frameworks (MOFs) are crystalline porous materials with periodic network structure formed by self-assembly of transition metal ions and organic ligands.[103–105] It has the characteristics of high specific surface area, unsaturated metal site structure and functional diversity. MOF is generally prepared by hydrothermal or solvothermal methods, and the derived materials are rich in variety and widely applied. Of course, MOF precursors can also be derived from a variety of 1D, 2D or 3D porous array materials, including metal hydroxide, oxide, sulfide, phosphide, selenide and their composites with carbon. MOF is a common precursor material in the synthesis of array materials, and the whole process is provided with the advantages of flexible and adjustable morphology and structure. Metal-organic frameworks (MOF) with inherently porous structures and well-dispersed metal sites are promising candidates for electrocatalysis. However, the catalytic efficiency of most MOFs is significantly limited by the adsorption/desorption energy and very low conductivity of the intermediates they form during electrocatalysis. Therefore, the regulation and synergy between metal sites in the polymetallic organic framework are generally used to promote the electrocatalytic process. For example, Co was introduced into a conductive copper catecholate (Cu-CAT)[121] nanorods array that grows directly on a flexible carbon cloth for HER. NiIr MOF nanosheet arrays were grown *in situ* on NF (NiIr-MOF/NF)[122] and supplied for a bifunctional self-supported electrocatalyst to oxidize organic small molecules of methanol to value-added chemicals formates while effectively promoting hydrogen production. Of course, it was well known that arrays of porous carbon-based transition metal compounds derived from MOF can be afforded as promising electrocatalyst materials. For example, carbon-limited $NiCo@NiCoO_2$ core-shell nanoparticles ($NiCo@NiCoO_2$/C PMRAs)[90] were produced by reductive carbonization of bimetallic (Ni, Co) metal-organic skeleton microrod arrays. The synthesized material combined a number of preponderances required for electrocatalytic reactions, including large specific surface area, high conductivity and multiple electrocatalytic active sites of OER. Zhong et al.[123] proposed an air cathode structure design strategy called MOF-on-MOF and creatively developed a highly efficient oxygen catalyst consisting of multistage Co_3O_4 nanoparticles anchored on a flexible carbon cloth with nitrogen doped carbon nanoarrays (Co_3O_4@NCNMAs/CC). This layered and independent structure ensured a high catalyst load on the air cathode with multiple electrocatalytic active sites, which undoubtedly improves the reaction kinetics and energy density of the all-solid Zn-air battery. However, MOF-derived array materials are prone to structural collapse and agglomeration of metal particles at high temperature.

8.2.6 Chemical Vapor Deposition (CVD)

In addition to common methods such as CBD, hydrothermal/solvothermal synthesis, electrochemical deposition, template synthesis, and MOF-derived synthesis, chemical vapor deposition (CVD) is a deposition method that utilizes evaporative precursor materials to produce high-quality solid materials on a specific substrate.[106–108] The corresponding advantages that CVD occupied are rapid deposition, easy access to raw materials, and suitable for synthesis of complex array structures. The deposition process in the reaction chamber is frequently complex and changeable. The pressure, temperature, gas flow rate, the ratio between gases and the external energy inflicted in the reaction chamber will affect the final deposition effect. Zhou et al.[83] reported a novel hybrid nanostructure (MoP/Co_2P), which was anchored on carbon fiber paper (CFP) by a simple hydrothermal method. Thus, $Co(CO_3)_{0.35}Cl_{0.20}(OH)_{1.10}$ nanowire array (NA)/CFP was obtained, followed by the deposition of MoO_2 nanoparticles via CVD methods and subsequent *in situ* phosphatizing process. In addition, synthesized perpendicular-aligned Co-doped MoS_2 nanosheets on graphite foil substrate performs[124] by the reaction between dripping $CoCl_2$-$MoCl_5$ precursor film and sulfur vapor released from elemental sulfur powder through a simple CVD method. The unique structure of Co-doped MoS_2 provided reaction sites with more active centers and ensured electron transport almost along the high electron mobility base plane of MoS_2. This array grown *in situ* on a graphite-foil substrate can be furnished as an effective electrocatalyst for HER. Although there are variety of advantages of CVD method, its equipment and operating conditions limit the development of large-scale nanoarray synthesis technology.

8.3 Energy conversion applications as electrocatalysts

Profiting from rich active sites, luxuriant diffusion channels and highly ordered structure, self-supported nanomaterials show fascinating catalytic properties in a wide range of energy storage and conversion reactions. The diversity of nanoarrays on earth and the development of preparation technology provide many possibilities and prospects for their wide application. Especially in recent years, the application in the field of electrocatalysis, including ORR, OER, HER, CRR, NRR, URR as well as the application of bifunction efficient and stable self-supporting electrode materials in metal air battery and water electrolysis device, has gradually approached or even exceeded the performance of commercial noble metal catalyst. In this section, we will introduce the research status of nanomaterials as self-supported electrocatalysts in recent years.

8.3.1 Oxygen Reduction Reaction (ORR)

ORR is an important positive electrode reaction in various metal air batteries and fuel cells. It refers to the reaction process in which oxygen absorbs electrons and is reduced to H_2O (acid electrolyte), OH^- (alkaline electrolyte), and O_2^{2-} (proton solution) on the surface of the electrode.[125–127] In fact, the electrochemical reduction of oxygen is a multi-electron reaction, which is composed of several successive and parallel reaction steps. In alkaline electrolyte, ORR takes the form of two reaction pathways, one is four-electron pathway and the other is two-electron pathway. The four-electron reaction is a direct reduction from O_2 to OH^-, without involving intermediate by-product peroxide. In the two-electron reduction pathway, O_2 is first reduced to HO_2^-, in which hydrogen peroxide intermediates are formed, and then further reduced to OH^-. For ORR under ideal condition, the reaction potential should be near the equilibrium electrode potential; however, in the process of real reaction, because of the $O = O$ bond rupture, and limited oxygen solubility in the electrolyte, the reaction overpotential (offset the potential difference in response to the actual equilibrium electrode potential) will increase, leading to the slow reaction kinetics. The slow reaction kinetics is the most important reason for the high reaction overpotential. In fact, Pt is considered the benchmark for ORR catalysts because of its optimal catalytic activity. However, scarce earth storage, high cost, extremely poor stability (particle agglomeration, dissolution, stripping, etc.) and low fault tolerance for various impurities (such as methanol and CO) seriously restrict the wide application of Pt-based ORR electrocatalysts. Therefore, to find and develop catalysts with high activity and strong durability to speed up the oxygen reduction reaction process is the core direction of current catalytic battery research (Table 8.1).

Since 2014, single atom catalysts (SACs) have rapidly become the research frontier in the field of catalysis due to their superiorities of homogeneity, high atomic utilization rate and admirable catalytic stability. With the continuous progress

Table 8.1. Performance of recent nanoarray electrocatalysts based on the ORR.

Catalyst	Substrate	Electrolyte	$E_{1/2}$ (V)	Limited current density (mA cm^{-2})	Tafel slope (mV dec^{-1})	Year	Ref.
Co(OH)$_2$@NC.	CC	1 M KOH	0.83	≈ 13	63	2021	128
D-ZIF	NF	0.1 M KOH	0.6	2.3	120	2021	129
MoCoZn/NCNTA	MoO$_3$ NW	0.1 M KOH	0.914	3.8	60	2021	130
CuCo$_2$S$_4$/NF	NF	0.1 M KOH	0.81	5.9	/	2020	131
CoFe@NCNT/CFC	CF	0.1 M KOH	0.873	8.5	125	2020	132
Mn-CoN-1.5	CC	0.1 M KOH	0.65	15.9	/	2020	133
Co-CoO$_x$/N-C	CC	0.1 M KOH	0.86	4.98	64.4	2020	134
CF@CuCoNC	Cu foam	1 M KOH	0.84	21	/	2020	135
SS-Co-SAC NSAs	CC	0.1 M KOH	0.81	59.1	/	2019	136
NF@Co$_{3-x}$Ni$_x$O$_4$	NF	1 M KOH	E_{on} = 0.91	/	/	2019	137
Fe-Co$_4$N@N-C	CC	0.1 M KOH	0.83	5	58	2019	138
Co-FeCo/N-G	CC	0.1 M KOH	0.82	2.28 @ 0.5 V	/	2019	113
Co$_3$O$_4$@NiFe LDH	NF	0.1 M KOH	0.75	/	58.7	2019	87
ZnO@Zn/Co-ZIF	CC	1 M KOH	0.87	14.02	/	2019	84
CoNCNTF/CNF	CNF	0.1 M KOH	0.857	/	59	2018	139
2D-MCo$_3$O$_4$-NCNAs	CF	0.1 M KOH	0.74	6.89	/	2018	140
CoNi@NCNT/NF	NF	1 M KOH	E_{on} = 0.97	20	/	2018	141

of advanced characterization techniques (extended X-ray absorption fine structure, spherical aberration electron microscopy), the development of SAC level on the atomic scale to clarify the catalyst of structure-activity relationship has become a reality, as well as links to heterogeneous catalysis and homogeneous catalysis provides opportunity, particularly on the catalytic mechanism research, which provides a feasible operation platform. Zhang et al.[142] developed a Co single atom anchored in a nitrogen-doped porous carbon nanosheet array, which was synthesized from a Co-MOF precursor and then etched unwanted Co clusters. The well-dispersed Co single atoms were connected to the carbon network by N-Co bonds, in which additional porosity and active surface area were generated by removing the Co metal clusters. Interestingly, individual Co atoms exhibited a lower oxygen release reaction (OER) overpotential and a higher oxygen reduction reaction (ORR) saturation current than those containing excess Co nanoparticles, suggesting that cobalt clusters are redundant in driving OER and ORR. A self-supporting carbon nanosheet array composed of SS-Co-SAC[136] was derived by *in situ* growth of metal imidazole framework on carbon cloth (Fig. 8.4a). Atomic resolution High-Angle

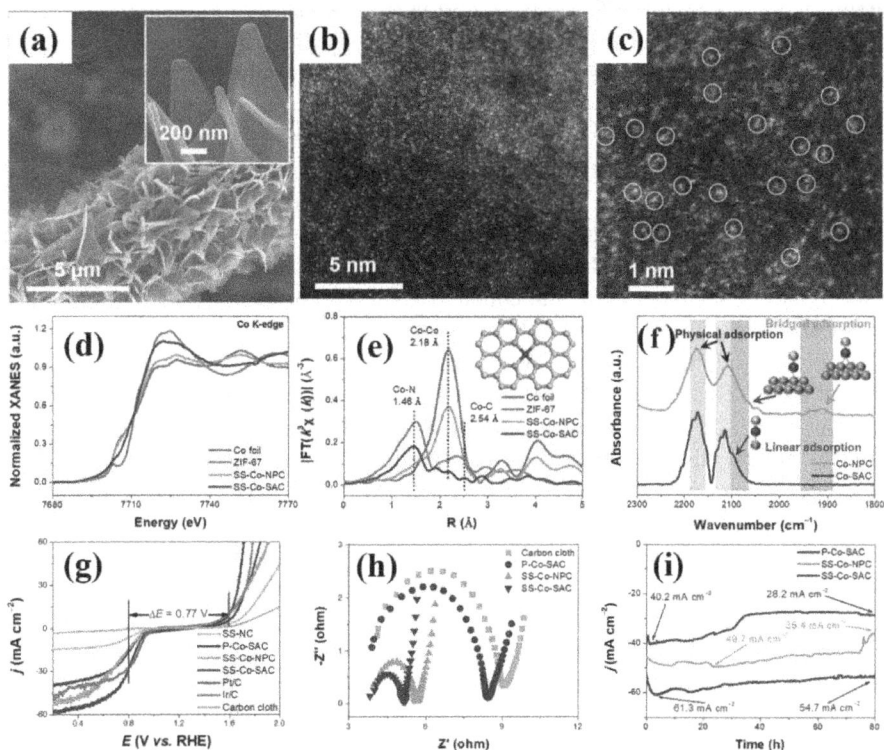

Figure 8.4. (a) AFM, and (b, c) HAADF-STEM images of SS-Co-SAC NSAs; (d) Co K-edge XANES; (e) the corresponding Fourier transforms EXAFS spectrums of Co foil, ZIF-67, SS-Co-NPC, SS-Co-SAC; (f) *in situ* FT-IR spectra of CO adsorption of SS-Co-SAC and SS-Co-NPC NSAs; (g) overall ORR and OER polarization curves of SS-Co-SAC NSAs, SS-Co-NPC NSAs, SS-NC NSAs, P-Co-SAC, carbon cloth, commercial Pt/C, and Ir/C; (b) EIS spectra; (c) I-t curves of SS-Co-SAC NSAs, SS-Co-NPC NSAs, and P-Co-SAC. Reproduced with permission from Ref. 136. Copyright 2019, Wiley-VCH.

Annular Dark-Field Scanning Transmission Electron Microscope (HAADF-STEM) images in Fig. 8.4b and c clearly show that individual Co atoms (white dots) are individually dispersed in nanosheets at very high densities, ranging in size from 0.1 to 0.2 nm. Both XAS and FT-IR spectra confirm that SS-Co-SAC NSAs contain positively charged Co single atoms in Fig. 8.4d–f, while SS-Co-NPC NSAs were composed of single atoms and Co nanoparticles. Impressively, the synthesis of SS-Co-SAC significantly increased the availability of active sites ($\approx$ 22.3%@2.3wt.%) due to the monoatomic dispersion of Co in a single nanosheet array architecture. The SS-Co-SAC electrode exhibited excellent electrocatalytic performance for both ORR and OER in Fig. 8.4g–i, which was benefiting from the high utilization of the active site. In addition, noble metal SACs could effectively solve the problem of poor stability caused by noble metal agglomeration and structural damage. For example, in cetylpyridine chloride and potassium chloride solutions, H_2PtCl6 and Na_2PdCl_4 were reduced with hematoic acid to prepare Pd@Pt core-shell nanodot arrays (Pd@PtNAs[143]). The platinum nanodot arrays with a diameter of about 3 nm were uniformly dispersed on the surface of the Pd decahedron. This ultra-small platinum spot array can effectively increase the proportion of low coordination platinum atoms. Compared with Pt/C catalysts sold in the market, the catalytic performance of Pd@PtNAs for oxygen reduction reaction was significantly improved. The ORR mass activity of Pd@PtNAs (36.9 wt.% Pt) was 2.2 times higher than that of Pt/C catalyst (0.09 A/mg Pt).

In addition to SAC arrays, transition metal oxides, hydroxides and sulfides array structures also show characteristic performance and application in ORR. Co_3O_4 and some derived spinel oxides have attracted more and more attention as a promising ORR catalyst due to their unique preponderances of low cost and surperb catalytic activity. However, the ORR activity of the Co_3O_4 catalyst is still much lower than expected due to its poor electrical conductivity due to its semiconductor properties. Yu et al.[144] significantly improved the ORR catalytic activity of Co_3O_4 by using a controlled N-doping strategy. The results of DFT calculation showed that the N dopant enhanced the electronic conductivity, increased the adsorption strength of O_2 and improved the reaction kinetics through a combination of a series of factors. A highly efficient ORR electrocatalyst for ultra-thin $Co(OH)_2$ nanosheets supported on a nitrogen-doped carbon nanosheet array (named $Co(OH)_2$@NC)[128] performed a high half-wave potential of 0.83 V, and a 3D array structure catalyst consisting of $Co_{1-x}S$[145] nanoneedles coated with nitrogen-doped carbon (NC) was grown *in situ* on a carbon fiber paper substrate. The investigative results showed that metastable hexagonal $Co_{1-x}S$ reveals attractive electrocatalytic activity. In addition, $Co_{1-x}S$@NC showed excellent ORR catalytic activity and long-term stability, due to the wrapping and protection effect of NC derived from polydopamine on $Co_{1-x}S$, the strong electronic coupling between $Co_{1-x}S$ and NC, and the 3D array structure, which can be compared to precious metals and other state-of-the-art bifunctional oxygen electrocatalysts reported.

8.3.2 Oxygen Evolution Reaction (OER)

OER is the main reaction of water anode electrolysis and also an indispensable reaction in the charging process of metal air battery.[11,146,147] It is an electrochemical reaction process in which H_2O or OH^- in the electrolyte loses electrons and then oxidizes to obtain O_2. A large overpotential is required on the oxygen electrode to overcome the complex and dynamically slow four-electron transfer process, which hinders the OER reaction efficiency. At present, precious metal Ir and Ru based materials are considered as reference materials for OER, and their high cost greatly hinders their further commercialization in large-scale applications.

In recent years, transition metal oxides, hydroxides, sulfides, phosphates and selenides have become potential candidate materials for OER due to their unique electrochemical activity and low cost (Table 8.2). By optimizing the electronic structure of the catalyst, the intrinsic catalytic activity can be optimized, the rapid electron transfer can be realized, the local environmental dynamics of the catalyst surface might be changed, and the catalytic performance can be further improved. Of course, SACs usually show only selective OER or ORR activity due to their specific affinity for oxygen reaction intermediates (*O, *OH, *OOH). In this case, it is necessary to explore multicomponent and hierarchical structures. Hydroxides exhibit good corrosion resistance and stability in alkaline environments, and MOOH (M = Ni, Fe, Co, Mn, Cu and so on)[52,178,179] has been reported to have satisfactory intrinsic catalytic activity as a true catalytic species. In particular, nickel-based hydroxides and oxides are equipped with excellent stability. In addition, it was found that the presence of iron ions in nickel hydride oxides and oxides can significantly improve the catalytic activity by reducing OER overpotential. Therefore, NiFe hydroxides and oxides are one of the most promising OER catalysts.[178,179] Li et al. [151] came up with a time-saving and energy-saving method for direct growth of NiFe layered double hydroxide (NiFe-LDH) nanosheets on nickel foam at ambient temperature and pressure (Fig. 8.5a). The NiFe-LDH nanosheets were vertically rooted in the NF and cross each other, forming highly porous arrays that result in a large number of exposed active sites, reducing charge/mass transfer resistance and enhancing mechanical stability. As a self-supported electrocatalyst, the representative sample (NF@NiFe-LDH-1.5-4) showed excellent high current density catalytic activity for OER in alkaline electrolyte, with a Tafel slope of 38.1 mV dec^{-1} (Fig. 8.5b–d). In addition, Wen et al.[180] proposed a Schottky heterojunction nanosheet array consisting of dispersed NiFe hydroxide nanoparticles and ultra-thin NiS nanosheets (NiFe LDH/NiS) to synergically regulate mass transfer and electron structure to trigger oxygen evolution reaction (OER) activity at high currents. In catalytic systems, the rich porosity of NiS nanosheet arrays assisted to provide rich catalytic sites and good electrolyte penetration for rapid mass transfer. Furthermore, theoretical calculations showed that the coupling of NiFe LDH with NiS could adjust the binding strength of the D-band center and the oxygen intermediate of NiFe atoms to obtain favorable OER dynamics. Moreover, MOF nanosheet arrays were provided

Table 8.2. Performance of recent nanoarray electrocatalysts based on the OER.

Catalyst	Substrate	Electrolyte	$\eta_{j=10}$ (mV)	Tafel slope (mV dec^{-1})	Year	References
WS$_2$@Co$_3$S$_4$ NW/CC	CC	1M KOH	280	52	2021	148
Cu@Cu NWs@LDH	Cu foam	1M KOH	212	68.9	2021	149
IrCo NRAs.	CC	1M KOH	257.3	70.1	2021	150
NF@NiFe-LDH-1.5-4	NF	1M KOH	η_{100} = 190	38.1	2021	151
FN@VSB-5/NF	NF	1M KOH	η_{100} = 273	43.1	2021	152
F-CoMoO$_{4-x}$-2@GF	Graphite felt	1M KOH	256	64.4	2021	153
NiFe-LDH/CC	CC	1M KOH	226	/	2021	154
MoS$_2$-CoS$_2$@PCMT	Carbon textile	1M KOH	360	93	2021	155
FeNi$_2$P/CC	CC	1M KOH	210	47.8	2021	156
Ni/Ni-M NRAs	Si wafer	1M KOH	250	54	2021	157
CoP-CoO/CC	CC	1M KOH	210	90	2021	158
CeO$_2$/C	CF	1M KOH	297	46	2021	159
Ni-Fe-S	NF	1M KOH	201	36.2	2021	160
W-NiCoP/NF	NF	1M KOH	η_{100} = 291	97.2	2021	161
Fe-S-NiMoO$_4$/MoO$_3$	NF	1M KOH	η_{500} = 271	41	2021	162
CoFe MLDH/Ti$_3$C$_2$	Ti$_3$C$_2$ MXene	1M KOH	170	31.5	2021	163
so-Fe-Ni(OH)$_2$	CFC	1M KOH	226	35.9	2021	164
Fe,Rh-Ni$_2$P/NF.	NF	1M KOH	η_{30} = 226	52.7	2021	165
Co(OH)$_2$@NC	CF	1M KOH	285	108	2021	128
Ru-MoS$_2$-Mo$_2$C/TiN	CC	1M KOH	280	202	2021	166
NiMP/NF	NF	1M KOH	η_{100} = 279	34	2021	167
NiCoZn/NC	CC	1M KOH	228	/	2021	168
CoSAs-MoS$_2$/TiN	Titanium nitride	1M KOH	340.6	73.5	2021	88
NiFe-LDH@CoS$_x$/NF	NF	1M KOH	206	62	2021	169
Fe-Ni$_2$P@PC/Cu$_x$S	Cu foam	1M KOH	η_{50}=330	140	2021	170
NiFe/NF	NF	1M KOH	190	39	2020	171
cMOF/LDH	CC	1M KOH	η_{50}=216	34.1	2020	172
Fe-NiO-Ni CHNAs	CFC	1M KOH	245	43.4	2020	173
Fe$_x$Ni$_{3-x}$S$_2$ @ NF	NF	1M KOH	η_{100}=252	64	2020	174
NiCo$_2$S$_4$/FeOOH	CC	1M KOH	200	71	2020	175
CoMoNiS-NF-31	NF	1M KOH	166	71	2019	176
CoSe$_2$@VG/CC	CC	1M KOH	328	82	2019	177

as structural guidance templates to synthesize edge-enriched NiFe LDH nanoarrays with rich coordination unsaturated sites. Impressively, the obtained NiFe-LDH[181] nanosheet array catalysts on nickel foam emerged captivating electrocatalytic

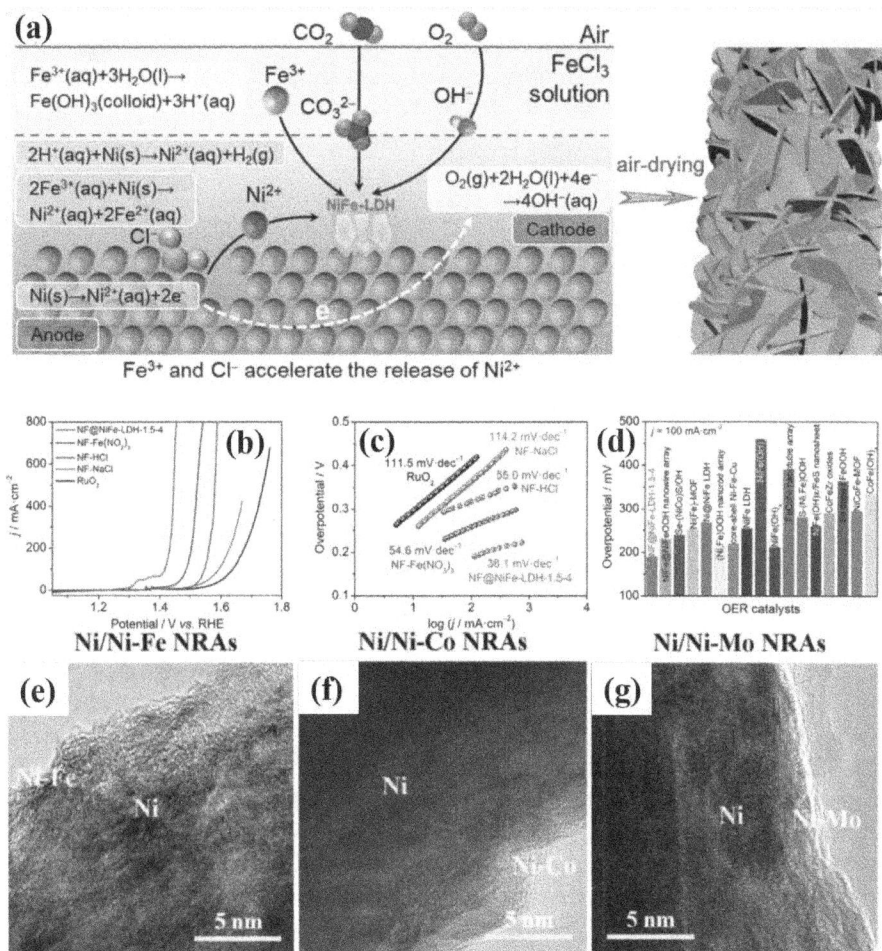

Figure 8.5. (a) The possible formation mechanism of NF@NiFe-LDH-1.5-4; (b) OER polarization curves and (c) Tafel slopes of NF@NiFe-LDH-1.5-4, NF-Fe(NO₃)₃, NF-HCl, NF-NaCl and RuO₂; (d) overpotentials required to achieve a current density of 100 mA cm⁻² on NF@NiFe-LDH-1.5-4 and other related OER catalysts. Reproduced with permission from Ref. 151. Copyright 2021, Wiley-VCH. The HR-TEM images (lower panel) of (e) Ni/Ni-Fe; (f) Ni/Ni-Co; and (g) Ni/Ni-Mo NRAs. Reproduced with permission from Ref. 157. Copyright 2021, Elsevier.

activity in OER, with an overpotential of only 205 mV (@10 mA cm⁻²), which was superior to all reported NiFe-LDH nanostructures. X-ray absorption spectra (XRS) and DFT calculations showed that the exposed edge of NiFe-LDH is rich in iron and oxygen vacancies, which optimizes the electronic state of the NiFe-LDH and enhances the adsorption of oxygen-containing intermediates, leading to outstanding OER catalytic activity.

Moreover, transition metal oxides, sulfides and other array materials are also widely studied in OER. An independent Ni/Ni-M (M = Fe, Co, Mo) bimetallic oxide core/shell nanorods (Ni/Ni-M NRA)[157] were prepared by electrodeposition of transition metals on black nickel sheets (Fig. 8.5e-f). All three types of Ni/Ni-M NRA showed enhanced electrocatalytic activity for oxygen evolution reaction (OER). Self-supported $Ni_3S_2/FeNi_2S_4$ nanosheets (Ni-Fe-S)[160] were synthesized and *in situ* Raman measurements were performed to monitor phase transition under realistic conditions. Combined with a series of transfer techniques, it can be concluded that Ni-Fe-S nanosheets act as pre-catalysts for dynamic structure reconstruction during electrolysis. In the 1 M KOH medium, the active substances were identified as γ-NiOOH for OER and Ni0 for HER. Therefore, in order to understand and explore the real active species in oxygen electrocatalyst, focusing on the conversion and transfer of substances in catalytic reaction can help us to better design excellent electrocatalytic materials.

8.3.3 Hydrogen Evolution Reaction (HER)

Hydrogen is a clean and sustainable energy source and is considered as one of the alternatives to fossil fuels. HER is a hydrogen production process with promising application prospects. The reduction reaction of water molecules in alkaline electrolyte to generate H_2 is HER, which is the cathode reaction in the process of water electrolysis.[182–184] In alkaline media, the first Volmer reaction occurs when H_2O diffuses to the electrode surface to combine electrons, forming OH⁻, at which time H atoms are adsorbed on the metal surface. The reaction of H_2 generation can still be divided into two reaction processes. One is Tafel reaction, in which H atoms adsorbed on the metal surface directly combine to form H_2. In the second Heyrovsky reaction, water molecules are reduced to OH⁻ and H_2 by combining H atoms adsorbed on the surface. Although there are not numerous reactants involved in HER, the reaction steps are more complicated, so it requires higher energy and slower chemical kinetics. Pt and Pt-based materials are the most effective catalysts for HER, but their scarcity and cost restrict their application in HER. The HER activity of Pt in alkaline medium is usually 2 ~ 3 orders of magnitude lower than that in acidic solution, while the efficiency of water electrolytic cell in alkaline electrolyte is higher than that in acidic medium. Therefore, it is of great significance to explore efficient and abundant replacement catalysts for alkaline HER. In view of the strong points of self-supported electrocatalysts, various nanoarrays based on transition metal hydroxides, chalcogenides and phosphating compounds have attracted special attention in HER applications due to their attractive properties (Table 8.3).

Based on the fact that Ni metal is another active hydrogen cathode material besides precious metal, it brings their own advantages of abundant reserves and low cost, and a lot of scientific research has been put into it (Ni alloys, oxides, hydroxides, sulfides, phosphates, selenides, etc.). Naturally, nickel foam (NF) has turned into the most familiarly used metal substrate in HER, which can not only serve as conductive support, but also provide catalytic activity as metal source. Other transition metals such as Fe, Co and Cu have also been proved to be high-efficiency catalyst materials for HER. For example, a novel CoFe LDH was developed by simple hydrothermal and electrodeposition methods coupled with an array of NiFe-LDH nanosheets supported on a nickel foam (denoted CoFe@NiFe/NF).[206] It was noteworthy that the obtained CoFe@NiFe/NF/NF catalysts exhibited excellent electrocatalytic activity and stability for HER due to the strong synergistic effect between CoFe-LDH

Table 8.3. Performance of recent nanoarray electrocatalysts based on the HER.

Catalyst	Substrate	Electrolyte	$E_{j=10}$ (mV)	Tafel slope (mV dec^{-1})	Year	References
FeP@PPy/CTs	CT	1 M KOH	103.1	49.2	2021	185
M-NiCo	Ti foam	1 M KOH	109.2	110.3	2021	186
MoO$_3$-Co(OH)$_2$@Ag	Ag NW	1 M KOH	η_{100} = 290	60.9	2021	187
Co@CoO/NF	NF	1 M KOH	76	81	2021	188
Cr-CoP-NR/CC	CC	1 M KOH	10	44.8	2021	189
O-NiMoP/NF	NF	1 M KOH	97	49	2021	118
NiMP	NF	1 M KOH	68	87	2021	167
NiCoZn/NC	CFC	1 M KOH	74	60.12	2021	168
Co-TiO$_2$@Ti(H$_2$)	Ti plate	1 M KOH	78	67.8	2021	190
CoSAs-MoS$_2$/TiN NRs	TiN	1 M KOH	131.9	52.9	2021	88
NiFe-LDH@CoS$_x$	NF	1 M KOH	136	73	2021	169
Fe-Ni$_2$P@PC/Cu$_x$S	PC	1 M KOH	112.9	76	2021	170
Pt-Ni NTAs	Ti foil	1 M KOH	23	38	2021	191
V-Ni$_3$N/NF	NF	1 M KOH	83	45	2021	192
Ru-Co$_2$P/N-C/NF	NF	1 M KOH	65	65	2021	81
Mo-CoP/NC/TF	Ti foil	1 M KOH	78	51.2	2021	193
a-CoMoP$_x$/CF	CF	1 M KOH	59	55	2020	194
NiCoN|Ni$_x$P|NiCoN	NF	1 M KOH	165	139.2	2020	195
(Ru-Co)O$_x$	CC	1 M KOH	44.1	23.5	2020	196
ReS$_2$/ReO$_2$	Graphite film	1 M KOH	150	65	2020	197
MoSe$_2$-NiSe$_2$-CoSe$_2$	NF	1 M KOH	38	35	2020	72
Ni/NiFeMoO$_x$/NF.	NF	1 M KOH	20	/	2019	198
CoFe-PBA	Ti foam	1 M KOH	48	66	2019	199
CS@CNC NAs/CC	CC	1 M KOH	84	38	2019	78
NS-horn/NF	NF	1 M KOH	177	139	2019	200
CoMoNiS-NF-31	NF	1 M KOH	113	85	2019	201
1 T-MoS$_2$/CC	CC	1 M KOH	151	55	2019	202
1T-MoSe$_2$/NiSe NS/NW	NiSe	1 M KOH	120	86	2019	203
Fe/P-CoS$_2$ PCNW	CC	1 M KOH	80	56	2019	204
MoX$_n$ NWAs/CFP	CFP	1 M KOH	52	40	2019	205

and NiFe-LDH and the unique structural characteristics. A self-supporting 3D NiFe-LDH@CoS$_x$/NF[169] bifunctional electrocatalyst was fabricated by electrodeposition of amorphous CoS$_x$ on 2D NiFe-LDH nanosheets supported by porous nickel foam (NF) in Fig. 8.6a. The integration of CoS$_x$ HER-active electrocatalyst with the outstanding NiFe-LDH OER catalyst guaranteed bifunction electrocatalytic activity. In addition, the formation of the interface between the

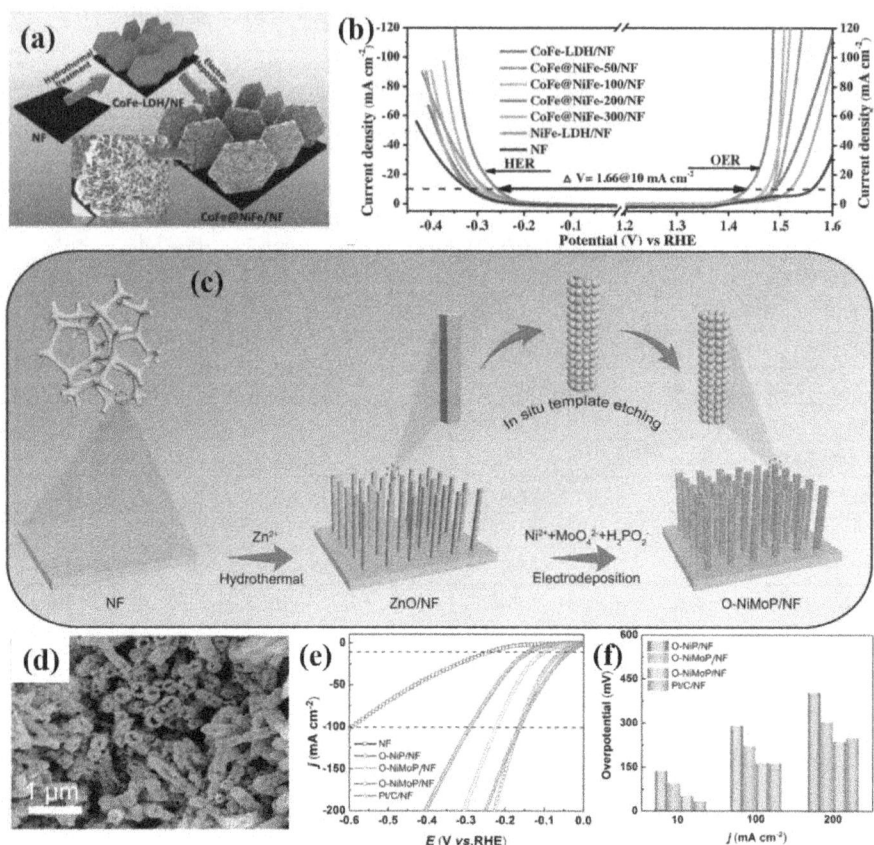

Figure 8.6. (a) Schematic diagram for the fabrication of CoFe@NiFe/NF architecture; (b) Steady-state polarization curves of CoFe-LDH/NF, CoFe@NiFe-50/NF, CoFe@NiFe-100/NF, CoFe@NiFe-200/NF, NiFe-300/NF, NiFe-LDH/NF and NF in 1.0 M KOH for HER and OER; Reproduced with permission from Ref. 206. Copyright 2019, Elsevier. (c) Schematic illustration of the preparation process of O-NiMoP/NF; (d) SEM images of O-NiMoP/NF; (e) HER polarization curves of NF, O-NiP/NF, O-NiMoP$_t$/NF, O-NiMoP/NF, and Pt/C/NF in 1.0 M KOH with a scan rate of 5 mV s^{-1}; (f) Comparison of overpotentials (η) of different catalysts at selected current densities. Reproduced with permission from Ref. 118. Copyright 2021, Wiley-VCH.

amorphous CoS$_x$ structure and the NiFe-LDH nanosheet synergistically promotes electron transfer. Thus, this novel NiFe-LDH@CoS$_x$/NF fractional assembly required ultra-low overpotentials of 136 mV for HER and 206 mV for OER to provide a current density of 10 mA cm^{-2} in an alkaline medium (1 M KOH, Fig. 8.6b).

A general urea regulated hydrothermal phosphating strategy was used to directly grow layered CoMnP/Ni$_2$P[207] nanosheet-based microplate arrays on nickel foam (NF). Due to the unique hierarchical structure of CoMnP/Ni$_2$P/NF array, CoP-MnP synergy and superhydrophobic properties, the optimized CoMnP/Ni$_2$P/NF showed high activity for HER and OER with low overpotential at 0.5 M H$_2$SO$_4$ (HER: 84 mV; OER: 165 mV) and 1 M KOH (HER: 108 mV; OER: 209 mV), and performed better than most reported electrocatalysts at a current density of 10 mA cm^{-2}. Furthermore, combining HER with urea oxidation reaction (UOR) in alkaline electrolyte for energy saving hydrogen production assisted by urea is a promising method to reduce energy consumption for H_2 production. Ni-Mo phosphide nanotube arrays doped with oxygen were synthesized on NF (O-NiMo/NF)[118] by electrodeposition and *in situ* template etching (Fig. 8.6c). The self-supported O-NiMoP/NF electrode exhibited inviting bi-functional catalytic activity for HER and UOR due to the O-NiMoP modulated electron structure and nanotube array structure (Fig. 8.6d). Especially in its HER and UOR (HER| |UOR) coupled system for H_2 production, the current density received significantly reduced battery voltage of 1.55 at 50 mA cm^{-2}, which is lower than the traditional electrolysis of about 300 mV (Fig. 8.6e, f). DFT calculations showed that significant HER and UOR activity originates from the Ni site, and the modulated electron environment induced by Mo, P and O atoms promotes hydrolytic dissociation during HER and balances the adsorption/desorption of intermediates during UOR. Ni base nanoarray as its HER| |OER or HER| |UOR double functions in the system of electric catalyst developed for energy saving H_2 production provides a new method.

8.3.4 CO₂ Reduction Reaction (CRR)

The continuous emission and accumulation of CO_2 and other greenhouse gases pose a great threat to the Earth's ecology. In order to alleviate the accumulation of CO_2, electrocatalytic CO_2 reduction is a promising strategy for promoting global carbon balance and tackling global climate change by converting CO_2 into clean energy using renewable energy.[20,21,208] Electrocatalytic CO_2 reduction is easy to be scaled up in industry, no additional H_2 supply is needed in the process of use, and the reaction conditions are mild and controllable. Also, it is one of the most promising reduction methods at present. CRR is a complicated process involving multiple proton-electron transfers, especially the competition of HER. Traditional metal catalysts can be divided into three types according to their main CRR products: (1) Au, Ag and Zn for CO production; (2) Sn, Hg, Pb for HCOOH production; (3) catalysts such as Cu for the production of hydrocarbons and alcohol. Therefore, how to effectively control the reaction direction and promote the combination of electrons and intermediates in the reaction path is the key to improve the performance of CRR.

SACs have broad prospects for electrochemical CRR, but the low density of active sites, the poor conductivity and mass transfer of single atom electrodes greatly restrict their manifestation. A prepared nickel single-atom electrode consisting of isolated, high-density and low-valent nickel(I) sites anchored on a self-standing N-doped carbon nanotube array with nickel–copper alloy encapsulation on a carbon-fiber paper (Fig. 8.7a,b).[209] The combination of Ni SAC and self-supporting array structure produced excellent electrocatalytic CRR performance in Fig. 8.7c,d. The introduction of Cu could adjust the electron configuration of d-band and enhance the adsorption of hydrogen, thus hindering the HER. The single Ni atom electrode exhibits a specific current density of −32.87 mA·cm^{-2} and turnover frequency of 1962 h^{-1} at a mild overpotential of 620 mV for CO formation with 97% Faradic efficiency. Under well-controlled conditions, large areas of vertically aligned bismuth nanosheet arrays were grown on copper substrates of various shapes and sizes (Fig. 8.7e).[210] The product presented an ultrathin sheet thickness of two to three

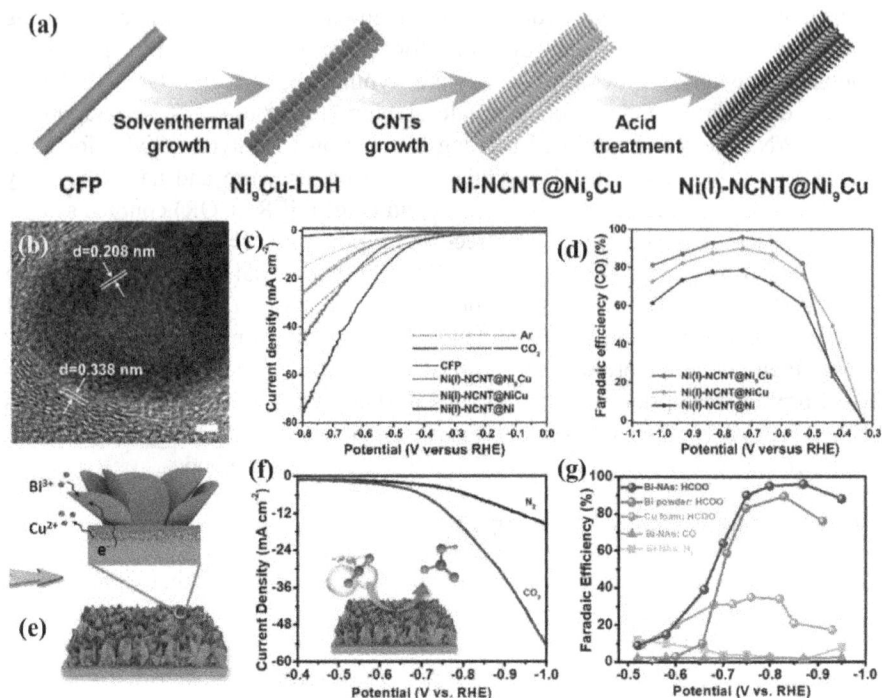

Figure 8.7. (a) Structural characterization of Ni(I)-NCNT@Ni$_9$Cu array on CFP; (b) HRTEM image of individual Ni(I)-NCNT@Ni$_9$Cu tip area; (c) LSV curves acquired in Ar and CO2-saturated 0.5 M KHCO$_3$ aqueous solution at a scan rate of 5 mV s^{-1}; (d) CO Faradaic efficiency at various applied potentials. Reproduced with permission from Ref. 209. Copyright 2020, Wiley-VCH. (e) Schematic diagram structure of Bi-NAs; (f) Polarization curves of Bi-NAs in 0.5 M KHCO$_3$ solution; (g) potential-dependent Faradaic efficiency for formate, CO, and H$_2$ measured on Bi-NAs, commercial Bi powders, and the bare Cu foam. Reproduced with permission from Ref. 210. Copyright 2021, Wiley-VCH.

atomic layers, a large surface area and abundant porosity between the sheets. Most notably, bismuth nanosheet arrays grown on Cu foam can achieve high CO$_2$ Faraday efficiency > 90%, current density up to 50 mA cm^{-2}, and desired stability (Fig. 8.7f,g). The CRR is converted to ethanol by TiO$_2$/MoS$_2$[211] nanosheet arrays synthesized by atomic layer deposition (ALD) TiO$_2$ on the surface of MoS$_2$ nanosheet arrays. The results showed that 50% faradaic efficiency (FE) for ethanol was achieved over the obtained electrocatalyst at only –0.60 V (vs. RHE) in CO$_2$-saturated 0.5 M KHCO$_3$. The experimental results and theoretical calculations showed that the Mo and Ti active sites formed at the interface of TiO$_2$ and MoS$_2$ can regulate the binding energy of CO and promote the Co-Co coupling reaction and its subsequent transformation.

8.3.5 Nitrogen Reduction Reaction (NRR)

NH$_3$ is an important chemical in synthetic fertilizers, and the Haber-Bosch process for nitrogen fixation was one of the greatest inventions of the 20th century, converting atmospheric N$_2$ into NH$_3$. However, the Haber-Bosch method not only leeches on to harsh reaction conditions, but also causes environmental pollution and greenhouse gas emissions. Electrocatalytic nitrogen reduction shows a cleaner, low-cost and sustainable carbon-free strategy for NH_3 production.[212-214] Electrocatalytic NRR for NH_3 synthesis is considered as a promising alternative to the Haber-Bosch process due to its mild reaction conditions, the use of renewable energy generation and distributed production. At present, there are some problems in NRR that need to be addressed urgently, such as low NH_3 yield, low Faradic efficiency and poor selectivity. How to achieve efficient electrocatalytic NH_3 synthesis is a challenging problem in this field. The key to solve this problem is to construct a reasonable nitrogen reduction reaction system, develop efficient NRR electrocatalyst, optimize the selectivity of electrocatalyst, and improve the electron utilization ratio, which is also the focus of the current research on NRR. Through rational design and adjustment of the active components, crystal structure and surface properties of the electrocatalyst, the catalyst can preferentially bind nitrogen atoms rather than hydrogen atoms, so as to effectively inhibit HER occurrence and further improve the efficiency of NRR. In addition, in the electrocatalytic process, the catalyst is prone to deactivation and decomposition, which requires that the designed catalyst should have good stability and long service life. In order to obtain efficient and stable NRR catalysts, a series of developmental design strategies, such as alloying, heterojunction, size effect, interface effect and defect engineering, have been provided to effectively promote the NRR activity of the catalysts and accelerate the industrialization process of electrocatalytic NH_3 production.

A bifunction Ni-Fe nanoarray electrocatalyst has been developed with excellent structural characteristics suitable for good NRR and OER processes, including highly exposed active sites derived from subnanoscale nanomesh, layered pores generated by arrays of nanolayers, and active sites for 2D Ni and Fe.[215] In Fig. 8.8a–c, the electrode showed excellent NRR activity with an NH_3 yield of 16.89 μg h^{-1} mg^{-1} at −350 mV (vs RHE) and a Faraday efficiency (FE) of 12.50%, in addition to OER activity of a negligible overpotential (191 mV at 10 mA cm^{-2}). Further mechanism studies by DFT calculation showed that NRR takes place through the associated distal path, and the N_2^* → NNH* intermediate step in NRR and the free energy of OH* + H_2O → O* + H_2 regulate the electronic structure of the catalyst by utilizing the strong synergy between Ni and Fe. The sisal like porous MoO_2 nanosheet arrays (MoO_2/NF)[216] supported by NF were synthesized by simple hydrothermal and subsequent annealing processes (Fig. 8.8d,e). The analysis results showed that Mo^{4+} in MoO_2/NF was more favorable than Mo^{6+} in MoO_3/NF during NRR, resulting in impressive NH_3 electrosynthesis activity (NH_3 yield: 9.56 μg h^{-1} cm^{-2}; Faraday efficiency: 5.14%) in 0.1 M KOH (Fig. 8.8g, h). In addition, the sisal like porous nanosheet array structure not only provided a high surface area for abundant accessible active sites, but also limited the coalescence of bubbles, enabling the active sites to be better exposed to nitrogen by rapid removal of small bubbles.

In addition, photoelectrochemical (PEC) nitrogen reduction reactions (NRR) produce NH_3 under environmental conditions by combining the advantages of electrocatalysis and photocatalysis represent attractive prospects for nitrogen fixation. Due to the limitation of suitable P-type semiconductor, nitrogen fixation of PEC should be realized by cathode NRR integrated with photoanode rather than direct photocathode NRR. Even with this approach, it is still hazy how to design the catalytic active sites on the dark cathode to achieve the high activity and selectivity

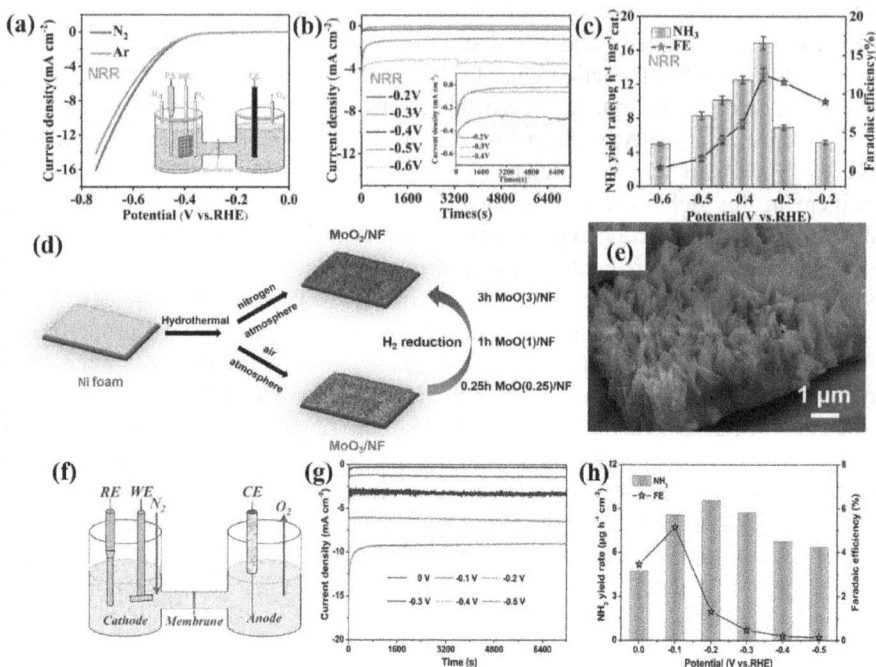

Figure 8.8. (a) Linear sweep voltammetry curves of NiFe-nanomesh array electrode recorded with a scan rate of 5 mV s⁻¹ in N₂-saturated and Ar-saturated 0.1 M Na₂SO₄ solution; (b) I-t curves from –0.2 to –0.6 V; (c) Faradaic efficiency (FE) and NH₃ yields at different applied potentials. Reproduced with permission from Ref. 215. Copyright 2020, American Chemical Society. (d) Illustration of the fabrication of MoO₂/NF, MoO₃/NF and MoO(x)/NF catalysts; (e) SEM images of MoO₂/NF; (f) diagrammatic drawing of the configuration of H-type cell; (g) chronoamperometry test results from 0 to –0.5 V; (h) NH₃ yield rate and Faradaic efficiency. Reproduced with permission from Ref. 216. Copyright 2021, Elsevier.

of NRR by PEC. Xu et al.[217] reported an innovative strategy for customizing the cathode Bi position with B doping and rolling curvature. The B doping in Bi matrix greatly reduced the energy barrier → *NNH in the determination step of N₂ potential energy, while the high curvature surface on the nanoroll contributed to the adsorption of N₂. When the TiO₂ nanorods array acted as a photoanode to collect light and provide photogenerated electrons, the integration of B doping and rolling curvature in a single cathode catalyst improved PEC NRR performance. Such a PEC system provided a NH₃ yield of 29.2 mg$_{NH3}$ g$_{cat}$⁻¹ h⁻¹ and the Faraday efficiency of 8.3% at a bias of 0.48 V (vs. RHE) in nitrogen fixation.

8.3.6 Urea Oxidation Reaction (UOR)

Urea is a suitable hydrogen carrier with abundant hydrogen content (6.67 wt.%). Electrolytic urea hydrogen production technology can not only degrade urea effectively by applying voltage to urea containing wastewater, but also achieve the purpose of environmental governance.[28,218] At the same time, clean hydrogen can be produced efficiently at the cathode to replace fossil energy, which is very consistent with the concept of sustainable development. However, UOR at the anode is a kinetic slow six electron transfer process, which limits the overall process of the reaction. Therefore, the development of highly active UOR catalyst materials is very important to improve the efficiency of UOR. In the early studies, although noble metal catalysts represented by Pt, Ir and Ru were found to have excellent catalytic activity, their large-scale commercial application was limited due to their high price and scarce reserves on Earth. The development of low-cost UOR catalysts is a pivotal issue to break through the development of urea energy technology. Self-supported array materials not only possess various particular advantages brought by non-binder, but also own the characteristics of easy recovery and flexible application, and have a wide application prospect.

The doping engineering strategy was used to obtain ternary NiMoV LDH[219] nanosheet arrays supported on a 3D NF substrate (Fig. 8.9a). The synergistic effect of the unique 2D/3D hierarchy exposed more active sites and accelerates charge and mass transfer. In addition, experimental and theoretical results confirmed that

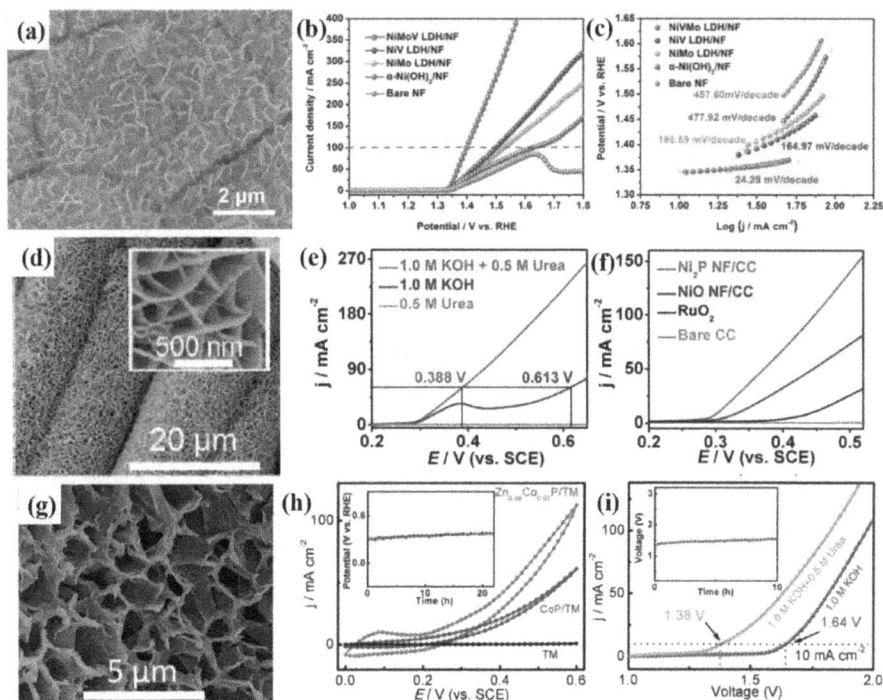

Figure 8.9. (a) SEM images of NiMoV LDH grown on NF; (b) LSV curves and (c) tafel plots of NiMoV LDH/NF, NiV LDH/NF, NiMo LDH/NF, α-Ni(OH)$_2$/NF and bare Ni foam in 1.0 M KOH electrolyte with 0.33 M urea. Reproduced with permission from Ref. 219. Copyright 2021, Elsevier. (d) SEM images of Ni$_2$P NF/CC; (e) Polarization curves for Ni$_2$P NF/CC||Ni$_2$P NF/CC couple in 1.0 M KOH with and without 0.5 M urea at scan rate of 5 mV s^{-1}. (f) Polarization curves for NiO NF/CC||NiO NF/CC, Ni$_2$P NF/CC||Ni$_2$P NF/CC, and Pt/C||RuO$_2$ in 1.0 M KOH with 0.5 M urea at scan rate of 5 mV s^{-1}. Reproduced with permission from Ref. 220. Copyright 2017, Royal Society of Chemistry. (g) SEM images of Zn$_{0.08}$Co$_{0.92}$P/TM; (h) Cyclic voltammograms of Zn$_{0.08}$Co$_{0.92}$P/TM, CoP/TM, and TM in 1.0 m KOH with 0.5 M urea at a scan rate of 10 mV s^{-1}; (i) Polarization curves for Zn$_{0.08}$Co$_{0.92}$P/TM||Zn$_{0.08}$Co$_{0.92}$P/TM in the presence and absence of 0.5 M urea in 1.0 M KOH at scan rate of 5 mV s^{-1}. Reproduced with permission from Ref. 221. Copyright 2017, Wiley.

Mo and V dopants can change the local electronic structure of the Ni site to optimize the adsorption energy of urea molecules. Therefore, the prepared NiMoV LDH/NF electrode achieved excellent catalytic efficiency, superb intrinsic activity (Fig. 8.9b, c) and strong durability of UOR due to its fast kinetics. A nickel phosphating nanosheet array on carbon cloth (Ni_2P NF/CC,[220] Fig. 8.9d) was reported as a highly active and durable 3D catalyst electrode for urea oxidation reaction (UOR) with a required potential of 0.447 V to achieve a geometric catalytic current density of 100 mA cm^{-2} in 1.0 M KOH and 0.5 M urea (Fig. 8.9e, f). It was noteworthy that the steady 3D array structure and promising electrocatalytic activities made it a bifunctional catalyst for UOR and HER, which could realize energy-saving electrochemical hydrogen production. The dual-electrode alkaline electrolytic cell only needed 1.35V battery voltage to reach 50 mA cm^{-2}, which was 0.58 V less than the voltage required for pure water to decompose to the same current density, with excellent long-term electrochemical durability and Faraday hydrogen production efficiency of nearly 100%. The $Zn_{0.08}Co_{0.92}P$ nanowall array ($Zn_{0.08}Co_{0.92}P$/TM[221], Fig. 8.9g) on the titanium mesh needed only 39 and 67 mV overpotential to drive the geometric catalytic current of 10 mA cm^{-2} in 0.5 M H_2SO_4 and 1.0 M KOH, respectively. The activity of $Zn_{0.08}Co_{0.92}P$/TM in urea oxidation reaction (UOR) was also better than that of CoP/TM, driving 115 mA cm^{-2} at 0.6 V in 0.5 M urea and 1.0 M KOH (Fig. 8.9h,i). This dual-function electrode presented wonderful HER and UOR activity, enabled a dual-electrode electrolytic cell based on $Zn_{0.08}Co_{0.92}P$/TM to save hydrogen generation, provided a low pressure of 10 mA cm^{-2} at a low pressure of 1.38 V, and served with strong electrochemical stability for long service life.

8.3.7 *Energy storage and conversion devices*

The application of array materials is not only limited to simple electrocatalytic semi-reaction, but also can be widely used in various energy storage and conversion devices. For example, rechargeable metal air batteries involve ORR and OER, while water splitting involves OER and HER. In order to simplify these energy devices, it is urgent to study and explore bifunctional electrocatalysts. In particular, in order to achieve reversible energy conversion of rechargeable metal air batteries, electrocatalysts are required to be active for both ORR and OER. To this end, the traditional method of preparing technology is to mix two powdered catalysts, such as reference Pt/C and IrO_2 or RuO_2, with conductive carbon in proportion and deposit them on a current collector. The operations are complex and it is difficult to make full use of them because one catalyst component does not work while the other does. In water electrolysis plants, although different OER and HER electrocatalysts can be coupled to achieve comprehensive water decomposition, the use of catalysts with bifunction OER and HER electrocatalysis can simplify the conversion equipment. In conclusion, the development of bifunctional electrocatalysts can effectively figure out these problems. As a result, the concept of bifunctional electrocatalysis has been widely developed, and many self-contained earth-rich nanoarrays with bifunctional electrocatalysis have been applied to both of these well-studied energy devices.

Metal air battery is composed of air electrode, electrolyte and negative electrode.[223–225] Oxygen in the air is the active substance of the positive electrode.

From this point of view, the capacity of the positive electrode is infinite. Metal, as a negative reactant, has a high theoretical energy density, and the theoretical capacity of metal air battery depends largely on the consumption of the negative metal. Li-air battery (LAB) and Zn-air battery (ZAB) demonstrate the most commercial prospects. LAB possess the highest energy density (3458 Wh kg^{-1}), but there are some safety risks and environmental problems in the process of use, and the reserves of metal Li are relatively low and the price is high. The most non-negligible is that LABs cannot work continuously and steadily in the air up till now. The ZAB with slightly lower energy density (1086 Wh kg^{-1}) is equipped with the superiority of stability, safety, low cost and low electrode overpotential. Disposable ZAB has welcomed commercial development in the field of low-current electrical equipment. Of course, these metal air battery devices all need bifunction catalysts to drive the ORR and OER at the solid-liquid-gas three-phase interface, so as to achieve the purpose of continuous and stable charge and discharge. Array catalysts as air electrode materials are not only provided with the merits of providing reaction sites and material transfer channels, but also have the advantages of flexibility, which also exhibits a good application prospect in flexible electronic devices.

The independent large macroporous air electrode with enhanced interfacial contact, fast mass transfer capability and tailored deposition space for large volumes of Li$_2$O$_2$ is essential to improve the rate performance of LABs. Ordered mesoporous carbon films with continuous macroporous channels were prepared by inverse topological transformation of ZnO nanorods arrays (Fig. 8.10a).[222] As freestanding air cathode of LABs, the layered porous carbon film showed excellent charge and discharge performance as well as rate performance (Fig. 8.10b,c). However, an increase in the cross-sectional area of continuous large holes on the cathode

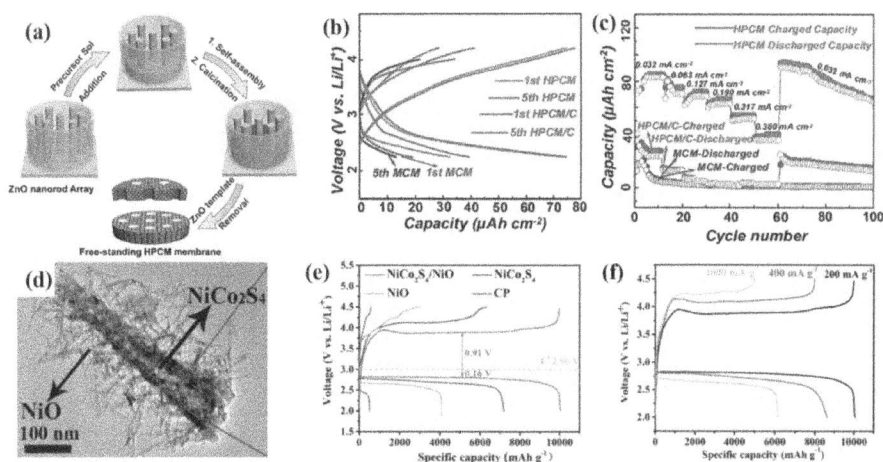

Figure 8.10. (a) Fabrication of free-standing hierarchically porous carbon membranes by inverse transformation from ZnO nanorod array; (b) The 1st and 5th discharge/charge curves of HPCM, HPCM/C and MCM at 0.032 mA cm^{-2}; (c) Rate performance. Reproduced with permission from Ref. 222. Copyright 2018, Wiley. (d) TEM images of NiCo2S4@NiO; (e) The initial deep discharge–charge curves of NiCo$_2$S$_4$@NiO, NiCo$_2$S$_4$, and NiO-based electrodes at a current density of 200 mA g^{-1}. (f) The discharge–charge curve of NiCo$_2$S$_4$@NiO-based electrodes at different current densities. Reproduced with permission from Ref. 86. Copyright 2019, Wiley.

surface resulted in dynamic overpotential with large voltage lags and linear voltage changes for Butler-volmer behavior. A carefully designed 3D layered heterostructure consisting of a $NiCo_2S_4$@NiO[86] core-shell array on conductive carbon paper was reported as an independent cathode for LABs (Fig. 8.10d–f). The unique layered array structure guaranteed the establishment of multidimensional channels for oxygen diffusion and electrolyte impregnation. The built-in interfacial potential NiO between $NiCo_2S_4$ could greatly enhance interfacial charge transfer dynamics. According to DFT calculations, the intrinsic LiO_2 affinity of $NiCo_2S_4$ and NiO played an important synergistic role in promoting the formation of large pea-like Li_2O_2, which was conducive to the construction of low impedance Li_2O_2/cathode contact interface.

In addition, array materials are widely used in ZABs, especially flexible ZABs, involving carbon material array, transition metal compound array, transition metal/ carbon array, etc. (Table 8.4) The preparation of single atom by growth on carbon nanofibers component N doping carbon flake array (M SA@NCF/CNF)[227] support (Fig. 8.11a) exhibited excellent catalytic activity, reversible oxygen overvoltage (0.75 V) and the double characteristics of high stability (Fig. 8.11b), which was owing

Table 8.4. Performance of recent nanoarray electrode electrocatalysts based on liquid and flexible ZABs.

Catalyst	Substrate	Electrolyte	Power density (mW cm^{-2})	Cycle performance	Year	References
Co(OH)$_2$@NC.	CC	PVA-KOH-ZnAc	36.9	over 400 min at 5 mA cm^{-3}	2021	128
D-ZIF	NF	6 M KOH-0.2 M ZnAc	/	380 h at 2 mA cm^{-2}	2020	129
CoFe@NCNT/ CFC	CFC	PVA-KOH	37.7	900 min at 1 mA cm^{-2}	2020	132
CuCoNC-500	Cu foam	6 M KOH-ZnAc	140	360 h at 10 mA cm^{-2}	2019	135
IOSHs-NSC-Co$_9$S$_8$	CF	6 M KOH-ZnAc	133	80 h at 10 mA cm^{-2}	2019	226
Fe-Co$_4$N@N-C	CC	6 M KOH-ZnAc	105	> 30 h at 5 mA cm^{-2}	2019	138
CoSA@NCF/CNF	CNF	PVA-KOH-ZnAc	/	90 cycles	2019	227
(Ni,Co)S$_2$	CC	6 M KOH-ZnAc	152.7	100 h at 5 mA cm^{-2}	2018	228
Co-N$_x$-YSC-600/ CC	CC	PVA-KOH-ZnAc	55.3	200 min under bending	2021	229
Ni$_{0.5}$Mo$_{0.5}$OSe	CFC	PVA-KOH-ZnAc	166.7	300 h at 10 mA cm^{-2}	2020	230
NiFe@N-CFs	CF	6 M KOH-ZnAc	102	350 cycles at 10 mA cm^{-2}	2020	231
N-GQDs/NiCo$_2$S$_4$	CC	PVA-KOH-ZnAc	26.2	6 h under bending	2019	232
Pt/NBF-ReS$_2$ / Mo$_2$CT$_x$	Mo$_2$CT$_x$	6 M KOH-ZnAc	180.2	100 h at 5 mA cm^{-2}	2021	233

to the greatly improved active sites' accessibility and optimized single-sites/pore-structures correlations. In addition, the wearable ZAB based on Co Sa@NCF/CNF air electrode presented excellent stability during deformation, satisfactory energy storage capacity and superb practicability for use as an integrated battery system. The introduction of joint defects can effectively trigger ORR activity of imidazolate skeleton (ZIF) of phyllozeolite by increasing the inherent activity and electrical conductivity of metal sites (Fig. 8.11c).[129] Experimental results showed that after low temperature heat treatment, part of the imidazole molecule is successfully removed from ZIF without damaging its structural integrity, resulting in the formation of unsaturated metal sites and faster electron transport rates. The homemade ZAB with D-ZIF as air cathode exhibited high open-circuit voltage and sensational cycling stability (Fig. 8.11d,e). The developed chain deficiency regulation strategy could provide a brand new prospect for MOF-based electrocatalysts to achieve high catalytic activity.

Electrode overpotential is generated in the cathode (hydrogen evolution) reaction and anode (oxygen evolution) reaction in the process of hydrolysis, which needs to be reduced by optimizing the electrode structure and improving the catalyst activity, and which is the most fundamental technical means to reduce the current and future hydrogen production power consumption.[234–236] The incorporation of metal single atoms into nanostructures is a novel approach that stimulates the number and type of active centers to improve the catalytic activity of HER and OER reactions in water decomposition. Here, Hoa et al. reported a continuous molybdenum sulfide-carbide heterostructure based nanosheet doped with Ruthenium atoms (2.02 at%) and coated on a highly conductive 1D Titanium nitride nanorods array (Ru-MoS$_2$-Mo$_2$C/TiN)[166] to form a 3D layered porous material by an efficient synthesis strategy (Fig. 8.12a,b).

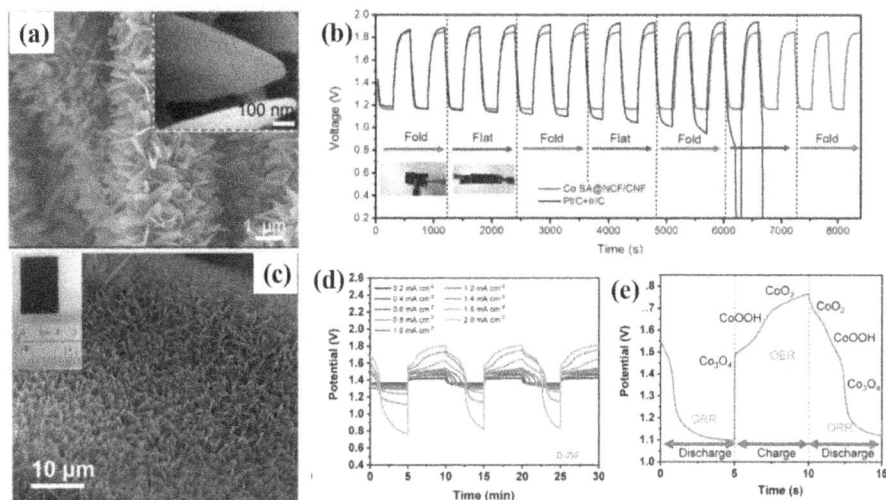

Figure 8.11. (a) FESEM of Co SA@NCF/CNF; (b) the charge-discharge curves of the wearable ZAB under alternately folding and releasing conditions. Reproduced with permission from Ref. 227. Copyright 2019, Wiley-VCH. (c) SEM of D-ZIF; (d) galvanostatic charge-discharge profiles of the D-ZIF-ZABs at various current densities. (e) Voltage curves of the discharge-charge tests. Reproduced with permission from Ref. 129. Copyright 2021, Wiley.

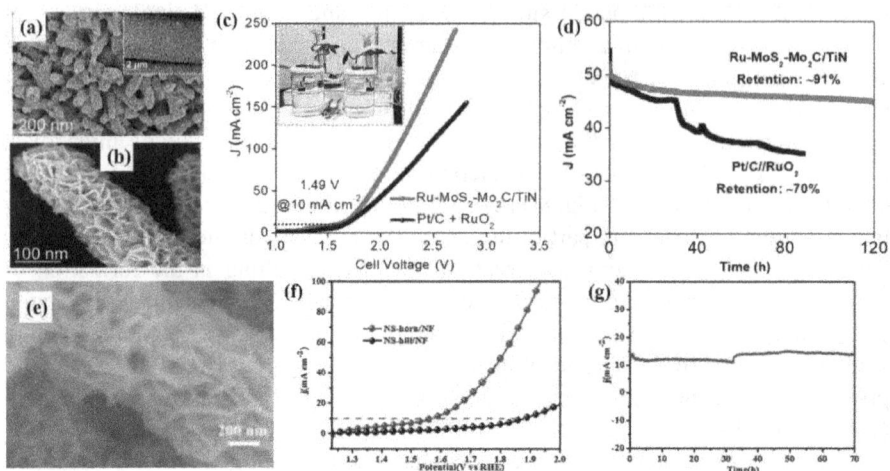

Figure 8.12. (a, b) FESEM images of Ru-MoS$_2$-Mo$_2$C/TiN on CC; (c) LSV measurements of the Ru-MoS$_2$_Mo$_2$C/TiN and Pt-C//RuO$_2$-based devices for overall water splitting in 1.0 M KOH medium; (d) stability of the Ru-MoS$_2$-Mo$_2$C/TiN and Pt-C//RuO$_2$-based devices measured at an initial current response of 50 mA cm^{-2}. Reproduced with permission from Ref. 166. Copyright 2021, Elsevier. (e) SEM of NS-Horn/NF; (d) polarization curve of NS-Horn/NF and NS-Hill/NF for overall water splitting in 1.0 M KOH; (e) chronopotentiometry curve of NS-Horn/NF for water splitting at 15 mA cm^{-2} in 1 M KOH solution. Reproduced with permission from Ref. 200. Copyright 2021, Wiley-VCH.

Materials with fine-tuning electronic structure and multi-integrated active sites showed an unimpressive overpotential of 25 and 280 mV at 10 mA cm^{-2} for HER and OER in 1.0 M KOH medium, respectively. The water electrolyzer provided by Ru-MoS$_2$-Mo$_2$C/TiN required only 1.49 V at the operating voltage of 10 mA cm^{-2}, which exceeded the system based on commercial catalysts and had been previously reported (Fig. 8.12c,d). Furthermore, a novel layered Ni$_3$S$_2$/VS$_4$ nano-angular array grown on NF (NS-Horn/NF) was prepared by self-driven synthesis strategy (Fig. 8.12e).[200] They demonstrated that the *in situ* generation of VS$_4$ in NS horn/NF not only triggered the formation of this unique hierarchy, but also promoted the grafting of the active bridge S$_2$$^{2-}$ enriched at the strongly coupled interface between Ni$_3$S$_2$ and VS$_4$, thus enhancing HER dynamics. More importantly, a large amount of active nickel oxide for OER effectively interfaced from the surface reconstruction NS-Horn/NF benefit from the partial leaching of vanadium (IV) of VS$_4$, which promoted OH$^-$ adsorption and leads in alkaline media quickly determine the OER rate step. When using an assembled alkaline electrolyzer as anode and cathode, the NS-Horn/NF electrode requires only a small voltage of 1.57 V to generate 10 mA cm^{-2} and remains in this state for at least 70 h (Fig. 8.12f,g).

8.4 Conclusions and prospects

With the unceasing joint efforts of researchers, significant progress has been made in self-supported earth-abundant nanoarrays from their synthesis to their application as promising electrocatalysts for a new generation of energy technologies. Satisfactory electrocatalytic effects were obtained by *in situ* doping, interface regulation, surface modification and defect structure of the electrode materials. A crowd of efficient and stable electrocatalysts have been developed for the applications of ORR, OER, HER, CRR, NRR, UOR and some energy devices. In this context, self-supported array catalysts are provided with their own unique virtue and characteristics: abundant active sites, strong material diffusion capacity, flexible and controllable structure. Although a lot of specially designed nanoarrays have achieved satisfactory electrocatalytic performance in some energy conversion reactions, there are still some severe challenges in considering practical industrial applications of electrocatalysis in this area of research. First of all, most of the current studies are guided by morphology and performance analysis, and lack of basic understanding of related reaction mechanisms, especially the clarification of active sites and reaction pathways, which severely restricts the rational design of advanced catalysts. A clear understanding of these things will allow us to consciously adjust the composition, morphology and structure of the target array electrocatalysts. Secondly, the lack of raw material cost, preparation process complexity, pollution and other considerations should be given immensely attention. The application of electrocatalysis should not only stay at the laboratory level, but also go out of the laboratory to try large-scale and efficient catalytic applications. In view of the above problems, some development prospects of array materials as electrocatalysts in the future are proposed.

(1) From the perspective of material design, we should not only rationally design efficient and stable electrocatalyst arrays that meet the needs through model construction and theoretical calculation, but also make efforts in the direction of raw material reserves, cost, process, yield and environment, etc. Array electrocatalysts always own the advantages of flexibility and recycling, so the development vision should be put into large-scale production in the future. We can totally try to cooperate with some hydrogen energy companies or sewage treatment enterprises to put the feasibility of laboratory implementation into the industry line of related application fields, which has far-reaching and practical significance for the application of array catalysts in electrocatalysis.

(2) From the perspective of material synthesis, the development of more efficient synthesis methods, including component construction, morphology control and structural engineering, is a major challenge to overcome. *In situ* characterization techniques are indispensable for discovering the relationship between physicochemical properties and electrocatalytic properties in order to better understand the relationship between experimental results and theoretical calculations. Therefore, extensive exploration of *in situ* characterization, such as *in situ* XAESE, XPS, Raman spectroscopy, infrared spectroscopy and TEM, is expected to guide future electrocatalyst design.

(3) From the perspective of SACs, which have been the focus of recent research, SACs possess sky-high atomic utilization ratio due to their isolated atoms, showing high stability, high selectivity and high activity in various catalytic reactions in the field of electrocatalysis. The common synthesis strategies are the bottom-up synthesis strategies mainly composed of wet chemistry, ALD technology, space limitation and the top-down synthesis strategies mainly composed of high

temperature atomic migration and trapping and high temperature pyrolysis. Advanced characterization techniques such as HADDF-STEM and XAS provide great help for understanding the structure and catalytic properties of single-atom site catalysts. DFT and first principles play an important role in the design and selection of single-atom site catalysts and the study of catalytic mechanism. SACs are equipped with a promising development in the field of electrocatalysis, but there are still some subsistent problems, such as: (I) in order to avoid thermodynamic instability, most of the single atom loading is low; (II) for some electrocatalytic reactions, metal atoms tend to agglomerate during thermal treatment, which will reduce the catalytic activity and stability of the final electrocatalyst; (III) due to the harsh preparation conditions, the mass production is limited; (IV) the catalytic mechanism of SAC's active site is not clear. The future development of SACs is both promising and challenging.

In summary, categories, synthesis methods and applications of self-supported nanoarrays in electrochemical reactions and energy conversion devices are introduced in this section. A variety of nanoarrays based on earth-rich compounds have gained attention as efficient and robust alternatives to precious metal electrocatalysts in different energy conversion reactions (i.e., ORR, OER, HER, CRR, NRR and UOR) and energy devices (rechargeable metal air batteries and electrolytic water devices). Although there are still some challenges, the ongoing development of related fields, especially in theoretical computing and materials science, will create potential new opportunities for the exploration of more advanced electrocatalysts in the future. We believe that sustainable development based on the next generation of clean energy technologies can ultimately be achieved through further efforts in the exciting field of self-support nanoarray electrocatalysts.

References

[1] Gao, P., S. Li, X. Bu, S. Dang, Z. Liu, H. Wang, L. Zhong, M. Qiu, C. Yang, J. Cai, W. Wei and Y. Sun. 2017. Direct conversion of CO_2 into liquid fuels with high selectivity over a bifunctional catalyst, *Nat. Chem.*, 9: 1019–1024.

[2] Verma, P., T. Kumari and A. S. Raghubanshi. 2021. Energy emissions, consumption and impact of urban households: A review. *Renew. and Sustain. Energy Rev.*, 147: 111210.

[3] Ok, Y. S., K. N. Palansooriya, X. Yuan and J. Rinklebe. 2021. Special issue on biochar technologies, production, and environmental applications in critical reviews in environmental science & technology during 2017–2021. *Environ. Sci. and Technol.*, DOI: 10.1080/10643389.2021.1990446, 1-9.

[4] Liu, J., D. Zhu, Y. Zheng, A. Vasileff and S.-Z. Qiao. 2018. Self-supported earth-abundant nanoarrays as efficient and robust electrocatalysts for energy-related reactions. *ACS Catal.*, 8: 6707–6732.

[5] Li, R., K. Xiang, Z. Peng, Y. Zou and S. Wang. 2021. Recent advances on electrolysis for simultaneous generation of valuable chemicals at both anode and cathode. *Adv. Energy Mater.*, 11: 2102292.

[6] He, Q., S. Qiao, Y. Zhou, R. Vajtai, D. Li, P. M. Ajayan, L. Ci and L. Song. 2021. Carbon nanotubes-based electrocatalysts: Structural regulation, support effect, and synchrotron-based characterization. *Adv. Funct. Mater.*, 2106684.

[7] Li, Z., N. H. Attanayake, J. L. Blackburn and E. M. Miller. 2021. Carbon dioxide and nitrogen reduction reactions using 2D transition metal dichalcogenide (TMDC) and carbide/nitride (MXene) catalysts. *Energy & Environ. Sci.*, 14: 6242–6286.

[8] Tang, L., Q. Xu, Y. Zhang, W. Chen and M. Wu. 2021. MOF/PCP-based electrocatalysts for the oxygen reduction reaction. *Electrochem. Energy Rev.*, DOI: 10.1007/s41918-021-00113-7.

[9] Hu, C., R. Paul, Q. Dai and L. Dai. 2021. Carbon-based metal-free electrocatalysts: From oxygen reduction to multifunctional electrocatalysis. *Chem. Soc. Rev.*, 50: 11785–11843.

[10] Chandrasekaran, S., C. Zhang, Y. Shu, H. Wang, S. Chen, T. Nesakumar Jebakumar Immanuel Edison, Y. Liu, N. Karthik, R. D. K. Misra, L. Deng, P. Yin, Y. Ge, O. A. Al-Hartomy, A. Al-Ghamdi, S. Wageh, P. Zhang, C. Bowen and Z. Han. 2021. Advanced opportunities and insights on the influence of nitrogen incorporation on the physico-/electro-chemical properties of robust electrocatalysts for electrocatalytic energy conversion. *Coordin. Chem. Rev.*, 449: 214209.

[11] Zhang, Y.-C., C. Han, J. Gao, L. Pan, J. Wu, X.-D. Zhu and J.-J. Zou. 2021. NiCo-based electrocatalysts for the alkaline oxygen evolution reaction: A Review. *ACS Catal.*, 11: 12485–12509.

[12] Ding, H., H. Liu, W. Chu, C. Wu and Y. Xie. 2021. Structural transformation of heterogeneous materials for electrocatalytic oxygen evolution reaction. *Chem. Rev.*, 121: 13174–13212.

[13] Gao, L., X. Cui, C. D. Sewell, J. Li and Z. Lin. 2021. Recent advances in activating surface reconstruction for the high-efficiency oxygen evolution reaction. *Chem. Soc. Rev.*, 50: 8428–8469.

[14] Gao, J., H. Tao and B. Liu. 2021. Progress of nonprecious-metal-based electrocatalysts for oxygen evolution in acidic media. *Adv. Mater.*, 33: 2003786.

[15] Jadhav, H. S., H. A. Bandal, S. Ramakrishna and H. Kim. 2021. Critical Review, Recent updates on Zeolitic Imidazolate Framework-67 (ZIF-67) and its derivatives for electrochemical water splitting. *Adv. Mater.*, DOI: 10.1002/adma.202107072, 2107072.

[16] Qiao, J., L. Kong, S. Xu, K. Lin, W. He, M. Ni, Q. Ruan, P. Zhang, Y. Liu, W. Zhang, L. Pan and Z. Sun. 2021. Research progress of MXene-based catalysts for electrochemical water-splitting and metal-air batteries. *Energy Storage Mater.*, 43: 509–530.

[17] Xu, Q., J. Zhang, H. Zhang, L. Zhang, L. Chen, Y. Hu, H. Jiang and C. Li. 2021. Atomic heterointerface engineering overcomes the activity limitation of electrocatalysts and promises highly-efficient alkaline water splitting. *Energy & Environ. Sci.*, 14: 5228–5259.

[18] Zhang, L.-N., R. Li, H.-Y. Zang, H.-Q. Tan, Z.-H. Kang, Y.-H. Wang and Y.-G. Li. 2021. Advanced hydrogen evolution electrocatalysts promising sustainable hydrogen and chlor-alkali co-production. *Energy & Environ. Sci.*, DOI: 10.1039/D1EE02798K.

[19] Liang, F., K. Zhang, L. Zhang, Y. Zhang, Y. Lei and X. Sun. 2021. Recent development of electrocatalytic CO_2 reduction application to energy conversion. *Small*, 17: 2100323.

[20] Zhi, X., A. Vasileff, Y. Zheng, Y. Jiao and S.-Z. Qiao. 2021. Role of oxygen-bound reaction intermediates in selective electrochemical CO_2 reduction. *Energy & Environ. Sci.*, 14: 3912–3930.

[21] Wang, Q., Y. Lei, D. Wang and Y. Li. 2019. Defect engineering in earth-abundant electrocatalysts for CO_2 and N_2 reduction. *Energy & Environ. Sci.*, 12: 1730–1750.

[22] Kong, X., H.-Q. Peng, S. Bu, Q. Gao, T. Jiao, J. Cheng, B. Liu, G. Hong, C.-S. Lee and W. Zhang. 2020. Defect engineering of nanostructured electrocatalysts for enhancing nitrogen reduction. *J. of Mater. Chem. A*, 8: 7457–7473.

[23] Liu, D., M. Chen, X. Du, H. Ai, K. H. Lo, S. Wang, S. Chen, G. Xing, X. Wang and H. Pan. 2021. Development of electrocatalysts for efficient nitrogen reduction reaction under ambient condition. *Advan. Funct. Mater.*, 31: 2008983.

[24] Wan, Y., J. Xu and R. Lv. 2019. Heterogeneous electrocatalysts design for nitrogen reduction reaction under ambient conditions. *Mater. Today*, 27: 69–90.

[25] Rehman, F., M. Delowar Hossain, A. Tyagi, D. Lu, B. Yuan and Z. Luo. 2021. Engineering electrocatalyst for low-temperature N_2 reduction to ammonia. *Mater. Today*, 44: 136–167.

[26] Pu, Z., T. Liu, I. S. Amiinu, R. Cheng, P. Wang, C. Zhang, P. Ji, W. Hu, J. Liu and S. Mu. 2020. Transition-metal phosphides: Activity origin, energy-related electrocatalysis applications, and synthetic strategies. *Advan. Funct. Mater.*, 30: 2004009.

[27] Zhu, D., M. Qiao, J. Liu, T. Tao and C. Guo. 2020. Engineering pristine 2D metal-organic framework nanosheets for electrocatalysis. *J. of Mater. Chem. A*, 8: 8143–8170.

[28] Zhu, B., Z. Liang and R. Zou. 2020. Designing advanced catalysts for energy conversion based on urea oxidation reaction. *Small*, 16: 1906133.

[29]   Kim, H., T. Y. Yoo, M. S. Bootharaju, J. H. Kim, D. Y. Chung and T. Hyeon. 2021. Noble metal-based multimetallic nanoparticles for electrocatalytic applications. *Adv. Sci.*, DOI: 10.1002/advs.202104054, 2104054.

[30]   Lyu, F., Q. Wang, S. M. Choi and Y. Yin. 2019. Noble-metal-free electrocatalysts for oxygen evolution. *Small*, 15: 1804201.

[31]   Su, Z. and T. Chen. 2021. Porous noble metal electrocatalysts: synthesis, performance, and development. *Small*, 17: 2005354.

[32]   Cai, B., S. Henning, J. Herranz, T. J. Schmidt and A. Eychmüller. 2017. Nanostructuring noble metals as unsupported electrocatalysts for polymer electrolyte fuel cells. *Adv. Energy Mater.*, 7: 1700548.

[33]   Kwon, T., T. Kim, Y. Son and K. Lee. 2021. Dopants in the design of noble metal nanoparticle electrocatalysts and their effect on surface energy and coordination chemistry at the nanocrystal surface. *Adv. Energy Mater.*, 11: 2100265.

[34]   Lei, Y., Q. Wang, S. Peng, S. Ramakrishna, D. Zhang and K. Zhou. 2020. Electrospun inorganic nanofibers for oxygen electrocatalysis: design, fabrication, and progress. *Adv. Energy Mater.*, 10: 1902115.

[35]   Shi, F., X. Zhu and W. Yang. 2020. Micro-nanostructural designs of bifunctional electrocatalysts for metal-air batteries. *Chinese J. Catal.*, 41: 390–403.

[36]   Mohamed, A. G. A., Y. Huang, J. Xie, R. A. Borse, G. Parameswaram and Y. Wang. 2020. Metal-free sites with multidimensional structure modifications for selective electrochemical CO_2 reduction. *Nano Today*, 33: 100891.

[37]   Zhu, C., H. Wang and C. Guan. 2020. Recent progress on hollow array architectures and their applications in electrochemical energy storage. *Nanoscale Horiz.*, 5: 1188–1199.

[38]   Li, L., W. Liu, H. Dong, Q. Gui, Z. Hu, Y. Li and J. Liu. 2021. Surface and interface engineering of nanoarrays toward advanced electrodes and electrochemical energy storage devices. *Adv. Mater.*, 33: e2004959.

[39]   Hou, J., Y. Wu, B. Zhang, S. Cao, Z. Li and L. Sun. 2019. Rational design of nanoarray architectures for electrocatalytic water splitting. *Adv. Funct. Mater.*, 29: 201808367.

[40]   Lou, W., A. Ali and P. K. Shen. 2022. Recent development of Au arched Pt nanomaterials as promising electrocatalysts for methanol oxidation reaction. *Nano Res.*, 15: 18–37.

[41]   Wang, Y., Q. Cao, C. Guan and C. Cheng. 2020. Recent advances on self-supported arrayed Bifbnctional oxygen electrocatalysts for flexible solid-state Zn-air batteries. *Small*, 16: 2002902.

[42]   Wang, J., N. Zang, C. Xuan, B. Jia, W. Jin and T. Ma. 2021. Self-supporting electrodes for gas-involved key energy reactions. *Adv. Funct. Mater.*, 31: 2104620.

[43]   Ma, T. Y., S. Dai and S. Z. Qiao. 2016. Self-supported electrocatalysts for advanced energy conversion processes. *Mater. Today*, 19: 265–273.

[44]   Wen, Q., Y. Zhao, Y. Liu, H. Li and T. Zhai. 2021. Ultrahigh-current-density and long-term-durability electrocatalysts for water splitting. *Small*, DOI: 10.1002/smll.202104513, 2104513.

[45]   Hou, J., Y. Wu, B. Zhang, S. Cao, Z. Li and L. Sun. 2019. Rational design of nanoarray architectures for electrocatalytic water splitting. *Adv. Funct. Mater.*, 29: 1808367.

[46]   Wang, C., B. Yan, Z. Chen, B. You, T. Liao, Q. Zhang, Y. Lu, S. Jiang and S. He. 2021. Recent advances in carbon substrate supported nonprecious nanoarrays for electrocatalytic oxygen evolution. *J. Mater. Chem. A*, 9: 25773–25795.

[47]   Yang, H., M. Driess and P. W. Menezes. 2021. Self-supported electrocatalysts for practical water electrolysis. *Adv. Energy Mater.*, 11: 2102074.

[48]   Singh, B., A. Singh, A. Yadav and A. Indra. 2021. Modulating electronic structure of metal-organic framework derived catalysts for electrochemical water oxidation. *Coordin. Chem. Rev.*, 447: 214144.

[49]   Wang, J., H. Kong, J. Zhang, Y. Hao, Z. Shao and F. Ciucci. 2021. Carbon-based electrocatalysts for sustainable energy applications. *Prog. Mater. Sci.*, 116: 100717.

[50]   Zhang, X.-Y., W.-L. Yu, J. Zhao, B. Dong, C.-G. Liu and Y.-M. Chai. 2021. Recent development on self-supported transition metal-based catalysts for water electrolysis at large current density. *Appl. Mater. Today*, 22: 100913.

[51] Wu, M., G. Zhang, M. Wu, J. Prakash and S. Sun. 2019. Rational design of multifunctional air electrodes for rechargeable Zn-Air batteries: Recent progress and future perspectives. *Energy Storage Mater.*, 21: 253–286.

[52] Sun, H., Z. Yan, F. Liu, W. Xu, F. Cheng and J. Chen. 2020. Self-supported transition-metal-based electrocatalysts for hydrogen and oxygen evolution. *Adv. Mater.*, 32: 1806326.

[53] Zhou, Y., J. Li, X. Gao, W. Chu, G. Gao and L.-W. Wang. 2021. Recent advances in single-atom electrocatalysts supported on two-dimensional materials for the oxygen evolution reaction. *J. Mater. Chem. A*, 9: 9979–9999.

[54] Yang, Y., Y. Yang, Z. Pei, K.-H. Wu, C. Tan, H. Wang, L. Wei, A. Mahmood, C. Yan, J. Dong, S. Zhao and Y. Chen. 2020. Recent progress of carbon-supported single-atom catalysts for energy conversion and storage. *Matter.*, 3: 1442–1476.

[55] Tran, D. T., D. C. Nguyen, H. T. Le, T. Kshetri, V. H. Hoa, T. L. L. Doan, N. H. Kim and J. H. Lee. 2021. Recent progress on single atom/sub-nano electrocatalysts for energy applications. *Prog. in Mater. Sci.*, 115: 100711.

[56] Wang, H., M. Zhou, P. Choudhury and H. Luo. 2019. Perovskite oxides as bifunctional oxygen electrocatalysts for oxygen evolution/reduction reactions-A mini review. *Appl. Mater. Today*, 16: 56–71.

[57] Wei, C., S. Sun, D. Mandler, X. Wang, S. Z. Qiao and Z. J. Xu. 2019. Approaches for measuring the surface areas of metal oxide electrocatalysts for determining their intrinsic electrocatalytic activity. *Chem. Soc. Rev.*, 48: 2518–2534.

[58] Chandrasekaran, S., D. Ma, Y. Ge, L. Deng, C. Bowen, J. Roscow, Y. Zhang, Z. Lin, R. D. K. Misra, J. Li, P. Zhang and H. Zhang. 2020. Electronic structure engineering on two-dimensional (2D) electrocatalytic materials for oxygen reduction, oxygen evolution, and hydrogen evolution reactions. *Nano Energy*, 77: 105080.

[59] Chen, Z., W. Wei and B.-J. Ni. 2021. Transition metal chalcogenides as emerging electrocatalysts for urea electrolysis. *Curr. Opin. in Electrochem.*, DOI: 10.1016/j.coelec.2021.100888, 100888.

[60] Chen, Z., D. Higgins, A. Yu, L. Zhang and J. Zhang. 2011. A review on non-precious metal electrocatalysts for PEM fuel cells. *Energy & Environ. Sci.*, 4: 3167–3192.

[61] Sun, Y., T. Zhang, C. Li, K. Xu and Y. Li. 2020. Compositional engineering of sulfides, phosphides, carbides, nitrides, oxides, and hydroxides for water splitting. *J. Mater. Chem. A*, 8: 13415–13436.

[62] Rossini, P. d. O., A. Laza, N. F. B. Azeredo, J. M. Gonçalves, F. S. Felix, K. Araki and L. Angnes. 2020. Ni-based double hydroxides as electrocatalysts in chemical sensors: A review. *TrAC Trends in Analytical Chemistry*, 126: 115859.

[63] Mohammed-Ibrahim, J. 2020. A review on NiFe-based electrocatalysts for efficient alkaline oxygen evolution reaction. *J. Power Sources*, 448: 227375.

[64] Cai, Z., X. Bu, P. Wang, J. C. Ho, J. Yang and X. Wang. 2019. Recent advances in layered double hydroxide electrocatalysts for the oxygen evolution reaction. *J. Mater. Chem. A*, 7: 5069–5089.

[65] Mahmood, A., W. Guo, H. Tabassum and R. Zou. 2016. Metal-organic framework-based nanomaterials for electrocatalysis. *Adv. Energy Mater.*, 6: 1600423.

[66] Chen, S., M. Cui, Z. Yin, J. Xiong, L. Mi and Y. Li. 2021. Single-atom and dual-atom electrocatalysts derived from metal organic frameworks: Current progress and perspectives. *Chem. Sus. Chem.*, 14: 73–93.

[67] Yang, L., X. Zeng, W. Wang and D. Cao. 2018. Recent progress in MOF-derived, heteroatom-doped porous carbons as highly efficient electrocatalysts for oxygen reduction Reaction in fuel cells. *Adv. Funct. Mater.*, 28: 1704537.

[68] Shao, M., Q. Chang, J.-P. Dodelet and R. Chenitz. 2016. Recent advances in electrocatalysts for oxygen reduction reaction. *Chem. Rev.*, 116: 3594–3657.

[69] Zhou, W., J. Jia, J. Lu, L. Yang, D. Hou, G. Li and S. Chen. 2016. Recent developments of carbon-based electrocatalysts for hydrogen evolution reaction. *Nano Energy*, 28: 29–43.

[70] Amiinu, I. S., X. Liu, Z. Pu, W. Li, Q. Li, J. Zhang, H. Tang, H. Zhang and S. Mu. 2018. From 3D ZIF nanocrystals to Co-N$_x$/C nanorod array electrocatalysts for ORR, OER, and Zn-air Batteries. *Adv. Funct. Mater.*, 28: 201704638.

[71] He, B.-C., C. Zhang, P.-P. Luo, Y. Li and T.-B. Lu. 2020. Integrating Z-scheme heterojunction of Co-C$_3$N$_4$@α-Fe$_2$O$_3$ for efficient visible-light-driven photocatalytic CO$_2$ reduction. *Green Chem.*, 22: 7552–7559.

[72]   Wang, G., W. Chen, G. Chen, J. Huang, C. Song, D. Chen, Y. Du, C. Li and K. K. Ostrikov. 2020. Trimetallic Mo-Ni-Co selenides nanorod electrocatalysts for highly-efficient and ultra-stable hydrogen evolution. *Nano Energy*, 71: 104637.

[73]   Zhou, J., Y. Li, L. Yu, Z. Li, D. Xie, Y. Zhao and Y. Yu. 2020. Facile *in situ* fabrication of $Cu_2O@Cu$ metal-semiconductor heterostructured nanorods for efficient visible-light driven CO_2 reduction. *Chem. Eng. J.*, 385: 123940.

[74]   Liu, T. and P. Diao. 2020. Nickel foam supported Cr-doped $NiCo_2O_4/FeOOH$ nanoneedle arrays as a high-performance bifunctional electrocatalyst for overall water splitting. *Nano Res.*, 13: 3299–3309.

[75]   Han, J., J. Zhang, T. Wang, Q. Xiong, W. Wang, L. Cao and B. Dong. 2019. Zn doped FeCo layered double hydroxide nanoneedle arrays with partial amorphous phase for efficient oxygen evolution reaction. *ACS Sustain. Chem. & Eng.*, 7: 13105–13114.

[76]   Fu, Q., T. Wu, G. Fu, T. Gao, J. Han, T. Yao, Y. Zhang, W. Zhong, X. Wang and B. Song. 2018. Skutterudite-type ternary $Co_{1-x}Ni_xP_3$ nanoneedle array electrocatalysts for enhanced hydrogen and oxygen evolution. *ACS Energy Letters*, 3: 1744–1752.

[77]   Gong, Y., Z. Xu, H. Pan, Y. Lin, Z. Yang and X. Du. 2018. Hierarchical Ni_3S_2 nanosheets coated on Co_3O_4 nanoneedle arrays on 3D nickel foam as an efficient electrocatalyst for the oxygen evolution reaction. *J. Mater. Chem. A*, 6: 5098–5106.

[78]   Wang, X., J. He, B. Yu, B. Sun, D. Yang, X. Zhang, Q. Zhang, W. Zhang, L. Gu and Y. Chen. 2019. $CoSe_2$ nanoparticles embedded MOF-derived Co-N-C nanoflake arrays as efficient and stable electrocatalyst for hydrogen evolution reaction. *Appl. Catals. B: Environ.*, 258: 117996.

[79]   Jiang, Y., Y. P. Deng, R. Liang, J. Fu, D. Luo, G. Liu, J. Li, Z. Zhang, Y. Hu and Z. Chen. 2019. Multidimensional ordered bifunctional air electrode enables flash reactants shuttling for high-energy flexible Zn-air batteries. *Adv. Energy Mater.*, 9: 201900911.

[80]   Gao, T., C. Zhou, X. Chen, Z. Huang, H. Yuan and D. Xiao. 2020. Surface *in situ* self-reconstructing hierarchical structures derived from ferrous carbonate as efficient bifunctional iron-based catalysts for oxygen and hydrogen evolution reactions. *J. Mater. Chem. A*, 8: 18367–18375.

[81]   Xu, Y., T. Ren, K. Ren, S. Yu, M. Liu, Z. Wang, X. Li, L. Wang and H. Wang. 2021. Metal-organic frameworks-derived Ru-doped Co_2P/N-doped carbon composite nanosheet arrays as bifunctional electrocatalysts for hydrogen evolution and urea oxidation. *Chem. Engin. J.*, 408: 127308.

[82]   Sun, H., C. Tian, Y. Li, J. Wu, Q. Wang, Z. Yan, C. P. Li, F. Cheng and M. Du. 2020. Coupling NiCo alloy and CeO_2 to enhance electrocatalytic Hhydrogen evolution in alkaline solution. *Adv. Sustain. Systems*, 4: 202000122.

[83]   Zhou, P., Y. Zhang, B. Ye, S. Qin, R. Zhang, T. Chen, H. Xu, L. Zheng and Q. Yang. 2019. MoP/Co_2P hybrid nanostructure anchored on carbon fiber paper as an effective electrocatalyst for hydrogen evolution. *Chem. Cat. Chem.*, 11: 6086–6091.

[84]   Zhu, C., Y. Ma, W. Zang, C. Guan, X. Liu, S. J. Pennycook, J. Wang and W. Huang. 2019. Conformal dispersed cobalt nanoparticles in hollow carbon nanotube arrays for flexible Zn-air and Al-air batteries. *Chem. Engin. J.*, 369: 988–995.

[85]   Guo, X., R.-M. Kong, X. Zhang, H. Du and F. Qu. 2017. $Ni(OH)_2$ nanoparticles embedded in conductive microrod array: an efficient and durable electrocatalyst for alkaline oxygen evolution reaction. *ACS Catal.*, 8: 651–655.

[86]   Wang, P., C. Li, S. Dong, X. Ge, P. Zhang, X. Miao, R. Wang, Z. Zhang and L. Yin. 2019. Hierarchical $NiCo_2S_4@NiO$ core-shell heterostructures as catalytic cathode for long-life Li-O_2 batteries. *Adv. Energy Mater.*, 9: 201900788.

[87]   Guo, X., X. Hu, D. Wu, C. Jing, W. Liu, Z. Ren, Q. Zhao, X. Jiang, C. Xu, Y. Zhang and N. Hu. 2019. Tuning the bifunctional oxygen electrocatalytic properties of core-shell $Co_3O_4@$ NiFe LDH catalysts for Zn-air batteries: Effects of interfacial cation valences. *ACS Appl. Mater. Interfaces*, 11: 21506–21514.

[88]   Doan, T. L. L., D. C. Nguyen, S. Prabhakaran, D. H. Kim, D. T. Tran, N. H. Kim and J. H. Lee. 2021. Single-atom Co-decorated MoS_2 nanosheets assembled on metal nitride nanorod arrays as an efficient bifunctional electrocatalyst for pH-universal water splitting. *Adv. Funct. Mater.*, 31: 202100233.

[89]   Liu, C., T. Liu, Y. Li, Z. Zhao, D. Zhou, W. Li, Y. Zhao, H. Yang, L. Sun, F. Li and Z. Li. 2020. A dendritic Sb_2Se_3/In_2S_3 heterojunction nanorod array photocathode decorated with a MoS_x catalyst for efficient solar hydrogen evolution. *J. Mater. Chem. A*, 8: 23385–23394.

[90] Xu, H., Z. X. Shi, Y. X. Tong and G. R. Li. 2018. Porous microrod arrays constructed by carbon-confined NiCo@NiCoO$_2$ core@shell nanoparticles as efficient electrocatalysts for oxygen evolution. *Adv. Mater.*, 30: e1705442.

[91] Xue, J., J. Liu, Y. Liu, H. Li, Y. Wang, D. Sun, W. Wang, L. Huang and J. Tang. 2019. Recent advances in synthetic methods and applications of Ag$_2$S-based heterostructure photocatalysts. *J. Mater. Chem. C*, 7: 3988–4003.

[92] Albero, J., J. N. Clifford and E. Palomares. 2014. Quantum dot based molecular solar cells. *Coordin. Chem. Rev.*, 263-264: 53–64.

[93] Consonni, V. and A. M. Lord. 2021. Polarity in ZnO nanowires: A critical issue for piezotronic and piezoelectric devices. *Nano Energy*, 83: 105789.

[94] Liu, M.-L., B.-B. Chen, C.-M. Li and C.-Z. Huang. 2019. Carbon dots prepared for fluorescence and chemiluminescence sensing. *Sci. China Chem.*, 62: 968–981.

[95] Ge, M., Q. Li, C. Cao, J. Huang, S. Li, S. Zhang, Z. Chen, K. Zhang, S. S. Al-Deyab and Y. Lai. 2017. One-dimensional TiO$_2$ nanotube photocatalysts for solar water splitting. *Adv. Sci.*, 4: 1600152.

[96] Song, B., M. Chen, G. Zeng, J. Gong, M. Shen, W. Xiong, C. Zhou, X. Tang, Y. Yang and W. Wang. 2020. Using graphdiyne (GDY) as a catalyst support for enhanced performance in organic pollutant degradation and hydrogen production: A review. *J. Hazard. Mater.*, 398: 122957.

[97] Zhang, X., K. Wan, P. Subramanian, M. Xu, J. Luo and J. Fransaer. 2020. Electrochemical deposition of metal-organic framework films and their applications. *J. Mater. Chem. A*, 8: 7569–7587.

[98] Lu, Z., J. Zheng, J. Shi, B.-F. Zeng, Y. Yang, W. Hong and Z.-Q. Tian. 2021. Application of micro/nanofabrication techniques to on-chip molecular electronics. *Small Methods*, 5: 2001034.

[99] Rao, C. R. K. and D. C. Trivedi. 2005. Chemical and electrochemical depositions of platinum group metals and their applications. *Coordin. Chem. Rev.*, 249: 613–631.

[100] Chuan, X.-y. and S.-h. Zhou. 2011. Preparation of mesoporous carbons by a template method. *Carbon*, 49: 3708–3716.

[101] Niu, H., H. Zhang, W. Yue, S. Gao, H. Kan, C. Zhang, C. Zhang, J. Pang, Z. Lou, L. Wang, Y. Li, H. Liu and G. Shen. 2021. Micro-nano processing of active layers in flexible tactile sensors via template methods: A Review. *Small*, 17: 2100804.

[102] Hou, P., C. Liu, C. Shi and H. Cheng. 2012. Carbon nanotubes prepared by anodic aluminum oxide template method. *Chinese Sci. Bullet.*, 57: 187–204.

[103] Wang, H.-F., L. Chen, H. Pang, S. Kaskel and Q. Xu. 2020. MOF-derived electrocatalysts for oxygen reduction, oxygen evolution and hydrogen evolution reactions. *Chem. Society Rev.*, 49: 1414–1448.

[104] Zhong, M., L. Kong, N. Li, Y.-Y. Liu, J. Zhu and X.-H. Bu. 2019. Synthesis of MOF-derived nanostructures and their applications as anodes in lithium and sodium ion batteries. *Coordin. Chem. Rev.*, 388: 172–201.

[105] Wen, X. and J. Guan. 2019. Recent progress on MOF-derived electrocatalysts for hydrogen evolution reaction. *Appl. Mater. Today*, 16: 146–168.

[106] Zhang, T. and L. Fu. 2018. Controllable chemical vapor deposition growth of two-dimensional heterostructures. *Chem.*, 4: 671–689.

[107] Li, M., D. Liu, D. Wei, X. Song, D. Wei and A. T. S. Wee. 2016. Controllable synthesis of graphene by plasma-enhanced chemical vapor deposition and its related applications. *Adv. Sci.*, 3: 1600003.

[108] Hou, P.-X., F. Zhang, L. Zhang, C. Liu and H.-M. Cheng. 2021. Synthesis of carbon nanotubes by floating catalyst chemical vapor deposition and their applications. *Adv. Funct. Mater.*, DOI: 10.1002/adfm.202108541, 2108541.

[109] Dou, T., Y. Qin, F. Zhang and X. Lei. 2021. CuS nanosheet arrays for electrochemical CO$_2$ reduction with surface reconstruction and the effect on selective formation of formate. *ACS Appl. Energy Mater.*, 4: 4376–4384.

[110] Li, Z., S. Feng, S. Liu, X. Li, L. Wang and W. Lu. 2015. A three-dimensional interconnected hierarchical FeOOH/TiO$_{(2)}$/ZnO nanostructural photoanode for enhancing the performance of photoelectrochemical water oxidation. *Nanoscale*, 7: 19178–19183.

[111] Zhang, X., X. Li, R. Li, Y. Lu, S. Song and Y. Wang. 2019. Highly active core-shell carbon/NiCo$_2$O$_4$ double microtubes for efficient oxygen evolution reaction: Ultralow overpotential and superior cycling stability. *Small*, 15: e1903297.

[112] Yang, D., L. Cao, J. Huang, K. Kajiyoshi, L. Feng, L. Kou, Q. Liu and L. Feng. 2020. Generation of Ni_3S_2 nanorod arrays with high-density bridging S2(2-) by introducing a small amount of $Na_3VO_4.12H_2O$ for superior hydrogen evolution reaction. *Nanoscale*, 12: 2063–2070.

[113] Jin, B. Ren, J. Chen, H. Cui and C. Wang. 2019. A facile method to conduct 3D self-supporting Co-FeCo/N-doped graphene-like carbon bifunctional electrocatalysts for flexible solid-state Zinc-air battery. *Appl. Catal. B: Environ.*, 256: 117887.

[114] Lee, C., K. Shin, C. Jung, P.-P. Choi, G. Henkelman and H. M. Lee. 2019. Atomically embedded Ag via electrodiffusion boosts oxygen evolution of CoOOH nanosheet arrays. *ACS Catal.*, 10: 562–569.

[115] Li, D., G. Hao, W. Guo, G. Liu, J. Li and Q. Zhao. 2020. Highly efficient Ni nanotube arrays and Ni nanotube arrays coupled with NiFe layered-double-hydroxide electrocatalysts for overall water splitting. *J. Power Sources*, 448: 227434.

[116] Li, R., X. Li, D. Yu, L. Li, G. Yang, K. Zhang, S. Ramakrishna, L. Xie and S. Peng. 2019. $Ni_3ZnC_{0.7}$ nanodots decorating nitrogen-doped carbon nanotube arrays as a self-standing bifunctional electrocatalyst for water splitting. *Carbon*, 148: 496–503.

[117] Gao, W., W. Gou, R. Wei, X. Bu, Y. Ma and J. C. Ho. 2020. *In situ* electrochemical conversion of cobalt oxide@MOF-74 core-shell structure as an efficient and robust electrocatalyst for water oxidation, *Appl. Mater. Today*, 21: 100820.

[118] Jiang, H., M. Sun, S. Wu, B. Huang, C. S. Lee and W. Zhang. 2021. Oxygen-incorporated NiMoP nanotube arrays as efficient bifunctional electrocatalysts for urea-assisted energy-saving hydrogen production in alkaline electrolyte. *Adv. Funct. Mater.*, 31: 202104951.

[119] Luo, J.-T., G.-L. Zang and C. Hu. 2020. An efficient 3D ordered mesoporous Cu sphere array electrocatalyst for carbon dioxide electrochemical reduction. *J. Mater. Sci. & Technol.*, 55: 95–106.

[120] Ding, J., Z. Liu, X. Liu, J. Liu, Y. Deng, X. Han, C. Zhong and W. Hu. 2019. Mesoporous decoration of freestanding palladium nanotube arrays boosts the electrocatalysis capabilities toward formic acid and formate oxidation. *Adv. Energy Mater.*, 9: 201900955.

[121] Geng, B., F. Yan, X. Zhang, Y. He, C. Zhu, S. L. Chou, X. Zhang and Y. Chen. 2021. Conductive CuCo-based bimetal organic framework for efficient hydrogen evolution. *Adv. Mater.*, DOI: 10.1002/adma.202106781, e2106781.

[122] Xu, Y., M. Liu, M. Wang, T. Ren, K. Ren, Z. Wang, X. Li, L. Wang and H. Wang. 2022. Methanol electroreforming coupled to green hydrogen production over bifunctional NiIr-based metal-organic framework nanosheet arrays. *Appl. Catal. B: Environ.*, 300: 120753.

[123] Zhong, Y., Z. Pan, X. Wang, J. Yang, Y. Qiu, S. Xu, Y. Lu, Q. Huang and W. Li. 2019. Hierarchical Co_3O_4 nano-micro arrays featuring superior activity as cathode in a flexible and rechargeable Zinc-air battery. *Adv. Sci. (Weinh)*, 6: 1802243.

[124] Huang, N., R. Peng, Y. Ding, S. Yan, G. Li, P. Sun, X. Sun, X. Liu and H. Yu. 2019. Facile chemical-vapour-deposition synthesis of vertically aligned co-doped MoS_2 nanosheets as an efficient catalyst for triiodide reduction and hydrogen evolution reaction. *J. Catal.*, 373: 250–259.

[125] Hu, J., W. Liu, C. Xin, J. Guo, X. Cheng, J. Wei, C. Hao, G. Zhang and Y. Shi. 2021. Carbon-based single atom catalysts for tailoring the ORR pathway: A concise review. *J. Mater. Chem. A*, 9: 24803–24829.

[126] Wang, Y.-J., W. Long, L. Wang, R. Yuan, A. Ignaszak, B. Fang and D. P. Wilkinson. 2018. Unlocking the door to highly active ORR catalysts for PEMFC applications: Polyhedron-engineered Pt-based nanocrystals. *Energy & Environ. Sci.*, 11: 258–275.

[127] Zhang, J., J. Zhang, F. He, Y. Chen, J. Zhu, D. Wang, S. Mu and H. Y. Yang. 2021. Defect and doping Co-engineered non-metal nanocarbon ORR electrocatalyst. *Nano-Micro Letters*, 13: 65–73.

[128] Wang, Y., A. Li and C. Cheng. 2021. Ultrathin $Co(OH)_2$ nanosheets@nitrogen-doped carbon nanoflake arrays as efficient air cathodes for rechargeable Zn-air batteries. *Small*, 17: e2101720.

[129] Yang, F., J. Xie, X. Liu, G. Wang and X. Lu. 2021. Linker defects triggering boosted oxygen reduction activity of Co/Zn-ZIF nanosheet arrays for rechargeable Zn-air batteries *Small*, 17: e2007085.

[130] Shao, L., Z.-X. Liang, H. Chen, Z.-X. Song, X.-H. Deng, G. Huo, X.-M. Kang, L. Wang, X.-Z. Fu and J.-L. Luo. 2020. $CuCo_2S_4$ hollow nanoneedle arrays supported on Ni foam as efficient trifunctional electrocatalysts for overall water splitting and Al-Air batteries. *J. Alloys Compd.*, 845: 155392.

[131] Wang, Y., F. Yan, X. Ma, C. Zhu, X. Zhang and Y. Chen. 2021. Hierarchically 3D bifunctional catalysts assembled with 1D MoC core/branched N-doped CNT arrays for Zinc-air batteries. *Electrochim. Acta*, 367: 137522.

[132] Liu, L., X. Zhang, F. Yan, B. Geng, C. Zhu and Y. Chen. 2020. Self-supported N-doped CNT arrays for flexible Zn-air batteries. *J. Mater. Chem. A*, 8: 18162–18172.

[133] Zhang, Y., B. Ouyang, G. Long, H. Tan, Z. Wang, Z. Zhang, W. Gao, R. S. Rawat and H. J. Fan. 2020. Enhancing bifunctionality of CoN nanowires by Mn doping for long-lasting Zn-air batteries. *Sci. China Chem.*, 63: 890–896.

[134] Li, S., W. Xie, Y. Song and M. Shao. 2019. Layered double hydroxide@polydopamine core-shell nanosheet arrays-derived bifunctional electrocatalyst for efficient, flexible, all-solid-state Zinc-air battery. *ACS Sustain. Chem. & Engin.*, 8: 452–459.

[135] Sun, H., Q. Li, Y. Lian, C. Zhang, P. Qi, Q. Mu, H. Jin, B. Zhang, M. Chen, Z. Deng and Y. Peng. 2020. Highly efficient water splitting driven by Zinc-air batteries with a single catalyst incorporating rich active species. *Appl. Catal. B: Environ.*, 263: 118139.

[136] Xie, W., Y. Song, S. Li, J. Li, Y. Yang, W. Liu, M. Shao and M. Wei. 2019. Single-atomic-Co electrocatalysts with self-supported architecture toward oxygen-involved reaction. *Adv. Funct. Mater.*, 29: 201906477.

[137] Wu, M., G. Zhang, N. Chen, W. Chen, J. Qiao and S. Sun. 2020. A self-supported electrode as a high-performance binder- and carbon-free cathode for rechargeable hybrid zinc batteries. *Energy Storage Mater.*, 24: 272–280.

[138] Xu, Q., H. Jiang, Y. Li, D. Liang, Y. Hu and C. Li. 2019. *In-situ* enriching active sites on co-doped Fe-Co$_4$N@N-C nanosheet array as air cathode for flexible rechargeable Zn-air batteries. *Appl. Catal. B: Environ.*, 256: 117893.

[139] Ji, D., L. Fan, L. Li, N. Mao, X. Qin, S. Peng and S. Ramakrishna. 2019. Hierarchical catalytic electrodes of cobalt-embedded carbon nanotube/carbon flakes arrays for flexible solid-state Zn-air batteries. *Carbon*, 142: 379–387.

[140] Zhang, H., J. Xu, Y. Jin, Y. Tong, Q. Lu and F. Gao. 2018. Quantum effects allow the construction of two-dimensional Co$_3$O$_4$-embedded nitrogen-doped porous carbon nanosheet arrays from bimetallic MOFs as bifunctional oxygen electrocatalysts. *Chem.*, 24: 14522–14530.

[141] Niu, W., S. Pakhira, K. Marcus, Z. Li, J. L. Mendoza-Cortes and Y. Yang. 2018. Apically dominant mechanism for improving catalytic activities of N-doped carbon nanotube arrays in rechargeable Zinc-air battery. *Adv. Energy Mater.*, 8: 201800480.

[142] Zang, W., A. Sumboja, Y. Ma, H. Zhang, Y. Wu, S. Wu, H. Wu, Z. Liu, C. Guan, J. Wang and S. J. Pennycook. 2018. Single Co atoms anchored in porous N-doped carbon for efficient Zinc-air battery cathodes. *ACS Catal.*, 8: 8961–8969.

[143] Li, S., J. Liu, G. Zhu and H. Han. 2019. Pd@Pt core-shell nanodots arrays for efficient electrocatalytic oxygen reduction. *ACS Appl. Nano Mater.*, 2: 3695–3700.

[144] Yu, M., Z. Wang, C. Hou, Z. Wang, C. Liang, C. Zhao, Y. Tong, X. Lu and S. Yang. 2017. Nitrogen-doped Co$_3$O$_4$ mesoporous nanowire arrays as an additive-free air-cathode for flexible solid-state Zinc-air batteries. *Adv. Mater.*, 29: 201602868.

[145] Yan, S., C. Luo, H. Zhang, L. Yang, N. Huang, M. Zhang, H. Yu, P. Sun, L. Wang, X. Lv and X. Sun. 2021. *In-Situ* derived Co$_{1-x}$S@nitrogen-doped carbon nanoneedle array as a bifunctional electrocatalyst for flexible Zinc-air battery. *J. Electroanal. Chem.*, 900: 115711.

[146] Anantharaj, S., S. Kundu and S. Noda. 2021. "The Fe Effect": A review unveiling the critical roles of Fe in enhancing OER activity of Ni and Co based catalysts. *Nano Energy*, 80: 105514.

[147] Tahir, M., L. Pan, F. Idrees, X. Zhang, L. Wang, J.-J. Zou and Z. L. Wang. 2017. Electrocatalytic oxygen evolution reaction for energy conversion and storage: A comprehensive review. *Nano Energy*, 37: 136–157.

[148] Zhang, T., F. Song, Y. Wang, J. Yuan, L. Niu, A.-j. Wang and K. Fang. 2021. Bifunctional WS$_2$@ Co$_3$S$_4$ core-shell nanowire arrays for efficient water splitting, *Electrochim. Acta*, DOI: 10.1016/j. electacta.2021.139648.

[149] Cai, X., T. Jiang and M. Wu. 2021. Confined growth of NiFe LDH with hierarchical structures on copper nanowires for long-term stable rechargeable Zn-air batteries. *Appl. Surf. Sci.*, DOI: 10.1016/j.apsusc.2021.151911.

[150]   Zhang, Y., G. Zhang, M. Zhang, X. Zhu, P. Shi, S. Wang and A.-L. Wang. 2021. Synergistic electronic and morphological modulation by trace Ir introduction boosting oxygen evolution performance over a wide pH range. *Chem. Engin. J.*, DOI: 10.1016/j.cej.2021.133577.

[151]   Li, X., C. Liu, Z. Fang, L. Xu, C. Lu and W. Hou. 2021. Ultrafast room-temperature dynthesis of self-supported NiFe-layered double hydroxide as large-current-density oxygen evolution electrocatalyst. *Small*, DOI: 10.1002/smll.202104354.

[152]   Li, S., Z. Liu, F. Wang, F. Yuan and Y. Ni. 2021. Ce-doped FeNi-layered double hydroxide nanosheets grown on an open-framework nickel phosphate nanorod array for oxygen evolution reaction. *ACS Appl. Energy Mater.*, 4: 12836–12847.

[153]   Xie, W., J. Huang, L. Huang, S. Geng, S. Song, P. Tsiakaras and Y. Wang. 2022. Novel fluorine-doped cobalt molybdate nanosheets with enriched oxygen-vacancies for improved oxygen evolution reaction activity. *Appl. Catal. B: Environ.*, 303: 120871.

[154]   Dong, G., F. Xie, F. Kou, T. Chen, F. Wang, Y. Zhou, K. Wu, S. Du, M. Fang and J. C. Ho. 2021. NiFe-layered double hydroxide arrays for oxygen evolution reaction in fresh water and seawater. *Mater. Today Energy*, 22: 100883.

[155]   Yang, J., C. Chai, C. Jiang, L. Liu and J. Xi. 2021. MoS_2-CoS_2 heteronanosheet arrays coated on porous carbon microtube textile for overall water splitting. *J. Power Sources*, 514: 230580.

[156]   Lin, S., Y. Yu, D. Sun, F. Meng, W. Chu, L. Huang, J. Ren, Q. Su, S. Ma and B. Xu. 2021. $FeNi_2P$ three-dimensional oriented nanosheet array bifunctional catalysts with better full water splitting performance than the full noble metal catalysts. *J. Colloid Interface Sci.*, DOI: 10.1016/j.jcis.2021.09.166.

[157]   Yan, Y., H. Liu, C. Liu, Y. Zhao, S. Liu, D. Wang, M. Fritz, A. Ispas, A. Bund, P. Schaaf and X. Wang. 2021. Efficient preparation of Ni-M (M = Fe, Co, Mo) bimetallic oxides layer on Ni nanorod arrays for electrocatalytic oxygen evolution. *Appl. Mater. Today*, 25: 101185.

[158]   Yao, J., M. Zhang, X. Ma, L. Xu, F. Gao, J. Xiao and H. Gao. 2022. Interfacial electronic modulation of CoP-CoO p-p type heterojunction for enhancing oxygen evolution reaction. *J. Colloid Interface Sci.*, 607: 1343–1352.

[159]   Nazar, N., S. Manzoor, Y. u. Rehman, I. Bibi, D. Tyagi, A. H. Chughtai, R. S. Gohar, M. Najam-Ul-Haq, M. Imran and M. N. Ashiq. 2022. Metal-organic framework derived CeO_2/C nanorod arrays directly grown on nickel foam as a highly efficient electrocatalyst for OER. *Fuel*, 307: 121823.

[160]   Wu Y., Y. Li, M. Yuan, H. Hao, X. San, Z. Lv, L. Xu and B. Wei. 2022. Operando capturing of surface self-reconstruction of Ni_3S_2/$FeNi_2S_4$ hybrid nanosheet array for overall water splitting. *Chem. Engin. J.*, 427: 131944.

[161]   Yan, X.-T., M. Yang, M.-X. Li, Z.-Y. Lin, Y.-N. Zhen, R.-Y. Fan, J.-F. Yu, Y.-M. Chai and B. Dong. 2021. Uniform W-NiCoP microneedles by molten salt decomposition as bifunctional electrocatalyst for alkaline water splitting. *Appl. Mater. Today*, 24: 101154.

[162]   Zhang, Y., H. Guo, J. Ren, X. Li, W. Ren and R. Song. 2021. MoO_3 crystal facets modulation by doping heteroatom Fe from polyoxometalate for quasi-industrial oxygen evolution reaction. *Appl. Catal. B: Environ.*, 298: 120582.

[163]   Hu, L., R. Xiao, X. Wang, X. Wang, C. Wang, J. Wen, W. Gu and C. Zhu. 2021. MXene-induced electronic optimization of metal-organic framework-derived CoFe LDH nanosheet arrays for efficient oxygen evolution. *Appl. Catal. B: Environ.*, 298: 120599.

[164]   Shi, Z.-X., J.-W. Zhao, C.-F. Li, H. Xu and G.-R. Li. 2021. Fully exposed edge/corner active sites in Fe substituted-Ni(OH)$_2$ tube-in-tube arrays for efficient electrocatalytic oxygen evolution. *Appl. Catal. B: Environ.*, 298: 120558.

[165]   Chen, M. T., J. J. Duan, J. J. Feng, L. P. Mei, Y. Jiao, L. Zhang and A. J. Wang. 2022. Iron, rhodium-codoped Ni_2P nanosheets arrays supported on nickel foam as an efficient bifunctional electrocatalyst for overall water splitting. *J. Colloid Interface Sci.*, 605: 888–896.

[166]   Hoa, V. H., D. T. Tran, S. Prabhakaran, D. H. Kim, N. Hameed, H. Wang, N. H. Kim and J. H. Lee. 2021. Ruthenium single atoms implanted continuous MoS_2-Mo_2C heterostructure for high-performance and stable water splitting. *Nano Energy*, 88: 106277.

[167]   Roh, H., H. Jung, H. Choi, J. W. Han, T. Park, S. Kim and K. Yong. 2021. Various metal (Fe, Mo, V, Co)-doped Ni_2P nanowire arrays as overall water splitting electrocatalysts and their applications in unassisted solar hydrogen production with STH 14 %, *Appl. Catal. B: Environ.*, 297: 120434.

[168] Chen, B., D. Kim, Z. Zhang, M. Lee and K. Yong. 2021. MOF-derived NiCoZnP nanoclusters anchored on hierarchical N-doped carbon nanosheets array as bifunctional electrocatalysts for overall water splitting. *Chem. Engin. J.*, 422: 130533.

[169] Yang, Y., Y. Xie, Z. Yu, S. Guo, M. Yuan, H. Yao, Z. Liang, Y. R. Lu, T.-S. Chan, C. Li, H. Dong and S. Ma. 2021. Self-supported NiFe-LDH@CoS$_x$ nanosheet arrays grown on nickel foam as efficient bifunctional electrocatalysts for overall water splitting. *Chem. Engin. J.*, 419: 129512.

[170] Tran, D. T., H. T. Le, V. H. Hoa, N. H. Kim and J. H. Lee. 2021. Dual-coupling ultrasmall iron-Ni$_2$P into P-doped porous carbon sheets assembled Cu$_x$S nanobrush arrays for overall water splitting. *Nano Energy*, 84: 105861.

[171] Liu, P., B. Chen, C. Liang, W. Yao, Y. Cui, S. Hu, P. Zou, H. Zhang, H. J. Fan and C. Yang. 2021. Tip-enhanced electric field: a new mechanism promoting mass transfer in oxygen evolution reactions. *Adv. Mater.*, 33: e2007377.

[172] Wang, Y., L. Yan, K. Dastafkan, C. Zhao, X. Zhao, Y. Xue, J. Huo, S. Li and Q. Zhai. 2021. Lattice matching growth of conductive hierarchical porous MOF/LDH heteronanotube arrays for highly efficient water oxidation. *Adv. Mater.*, 33: e2006351.

[173] Lei, Y., T. Xu, S. Ye, L. Zheng, P. Liao, W. Xiong, J. Hu, Y. Wang, J. Wang, X. Ren, C. He, Q. Zhang, J. Liu and X. Sun. 2021. Engineering defect-rich Fe-doped NiO coupled Ni cluster nanotube arrays with excellent oxygen evolution activity. *Appl. Catal. B: Environ.*, 285: 119809.

[174] Fei, B., Z. Chen, J. Liu, H. Xu, X. Yan, H. Qing, M. Chen and R. Wu. 2020. Ultrathinning nickel sulfide with modulated electron density for efficient water splitting. *Adv. Energy Mater.*, 10: 202001963.

[175] Li, X., Z. Kou, S. Xi, W. Zang, T. Yang, L. Zhang and J. Wang. 2020. Porous NiCo$_2$S$_4$/FeOOH nanowire arrays with rich sulfide/hydroxide interfaces enable high OER activity, *Nano Energy*, 78: 105230.

[176] Yang, Y., H. Yao, Z. Yu, S. M. Islam, H. He, M. Yuan, Y. Yue, K. Xu, W. Hao, G. Sun, H. Li, S. Ma, P. Zapol and M. G. Kanatzidis. 2019. Hierarchical nanoassembly of MoS$_2$/Co$_9$S$_8$/Ni$_3$S$_2$/Ni as a highly efficient electrocatalyst for overall water splitting in a wide pH range. *J. Am. Chem. Soc.*, 141: 10417–10430.

[177] Xia, Z., H. Sun, X. He, Z. Sun, C. Lu, J. Li, Y. Peng, S. Dou, J. Sun and Z. Liu. 2019. *In situ* construction of CoSe$_2$@vertical-oriented graphene arrays as self-supporting electrodes for sodium-ion capacitors and electrocatalytic oxygen evolution. *Nano Energy*, 60: 385–393.

[178] Gong, L., H. Yang, A. I. Douka, Y. Yan and B. Y. Xia. 2021. Recent progress on NiFe-based electrocatalysts for alkaline oxygen evolution, *Adv. Sustain. Syst.*, 5: 2000136.

[179] Lv, L., Z. Yang, K. Chen, C. Wang and Y. Xiong. 2019. Electrocatalysts: 2D layered double hydroxides for oxygen evolution reaction: from fundamental design to application. *Adv. Energy Mater.*, 9: 1970057.

[180] Wen, Q., K. Yang, D. Huang, G. Cheng, X. Ai, Y. Liu, J. Fang, H. Li, L. Yu and T. Zhai. 2021. Schottky heterojunction nanosheet array achieving high-current-density oxygen evolution for industrial water splitting electrolyzers. *Adv. Energy Mater.*, 11: 2102353.

[181] Wang, B., X. Han, C. Guo, J. Jing, C. Yang, Y. Li, A. Han, D. Wang and J. Liu. 2021. Structure inheritance strategy from MOF to edge-enriched NiFe-LDH array for enhanced oxygen evolution reaction. *Appl. Catal. B: Environ.*, 298: 120580.

[182] Cao, Y. 2021. Roadmap and direction toward high-performance MoS$_2$ hydrogen evolution catalysts. *ACS Nano*, 15: 11014–11039.

[183] Li, Z., M. Hu, P. Wang, J. Liu, J. Yao and C. Li. 2021. Heterojunction catalyst in electrocatalytic water splitting. *Coordin. Chem. Rev.*, 439: 213953.

[184] Yan, Y., B. Y. Xia, B. Zhao and X. Wang. 2016. A review on noble-metal-free bifunctional heterogeneous catalysts for overall electrochemical water splitting. *J. Mater. Chem. A*, 4: 17587–17603.

[185] Pei, H., L. Zhang, G. Zhi, D. Kong, Y. Wang, S. Huang, J. Zang, T. Xu, H. Wang and X. Li. 2021. Rational construction of hierarchical porous FeP nanorod arrays encapsulated in polypyrrole for efficient and durable hydrogen evolution reaction. *Chem. Engin. J.*, DOI: 10.1016/j.cej.2021.133643.

[186] Gao, D., J. Guo, H. He, P. Xiao and Y. Zhang. 2022. Geometric and electronic modulation of fcc NiCo alloy by Group-VI B metal doping to accelerate hydrogen evolution reaction in acidic and alkaline media. *Chem. Engin. J.*, 430: 133110.

[187] Peng, H., K. Zhou, Y. Jin, Q. Zhang, J. Liu and H. Wang. 2022. Hierarchical nanostructure with ultrafine MoO_3 particles-decorated $Co(OH)_2$ nanosheet array on Ag nanowires for promoted hydrogen evolution reaction. *Chem. Engin. J.*, 429: 132477.

[188] Wang, C., Y. Li, C. Gu, L. Zhang, X. Wang and J. Tu. 2022. Active Co@CoO core/shell nanowire arrays as efficient electrocatalysts for hydrogen evolution reaction. *Chem. Engin. J.*, 429: 132226.

[189] Zhang, L., J. Zhang, J. Fang, X. Y. Wang, L. Yin, W. Zhu and Z. Zhuang. 2021. Cr-doped CoP nanorod arrays as high-performance hydrogen evolution reaction catalysts at high current density. *Small*, 17: e2100832.

[190] Li, L. R., B. H. Hu, T. W. Yu, Z. P. Shao, Y. Wang and S. Q. Song. 2021. New TiO_2-based oxide for catalyzing alkaline hydrogen evolution reaction with noble metal-like performance, DOI: 10.1002/smtd.202100246.

[191] Nairan, A., C. Liang, S.-W. Chiang, Y. Wu, P. Zou, U. Khan, W. Liu, F. Kang, S. Guo, J. Wu and C. Yang. 2021. Proton selective adsorption on Pt-Ni nano-thorn array electrodes for superior hydrogen evolution activity. *Energy & Environ. Sci.*, 14: 1594–1601.

[192] Li, R.-Q., Q. Liu, Y. Zhou, M. Lu, J. Hou, K. Qu, Y. Zhu and O. Fontaine. 2021. 3D self-supported porous vanadium-doped nickel nitride nanosheet arrays as efficient bifunctional electrocatalysts for urea electrolysis. *J. Mater. Chem. A*, 9: 4159–4166.

[193] Li, Y., B. Zhang, W. Wang, X. Shi, J. Zhang, R. Wang, B. He, Q. Wang, J. Jiang, Y. Gong and H. Wang. 2021. Selective-etching of MOF toward hierarchical porous Mo-doped CoP/N-doped carbon nanosheet arrays for efficient hydrogen evolution at all pH values. *Chem. Engin. J.*, 405: 126981.

[194] Huang, H., A. Cho, S. Kim, H. Jun, A. Lee, J. W. Han and J. Lee. 2020. Structural design of amorphous $CoMoP_x$ with abundant active sites and synergistic catalysis effect for effective water splitting. *Adv. Funct. Mater.*, 30: 202003889.

[195] Yu, L., L. Wu, S. Song, B. McElhenny, F. Zhang, S. Chen and Z. Ren. 2020. Hydrogen generation from seawater electrolysis over a sandwich-like $NiCoN|Ni_xP|NiCoN$ microsheet array catalyst. *ACS Energy Letters*, 5: 2681–2689.

[196] Wang, C. and L. Qi. 2020. Heterostructured inter-doped Ruthenium-cobalt oxide hollow nanosheet arrays for highly efficient overall water splitting. *Angew. Chem. Int. Ed. Engl.*, 59: 17219–17224.

[197] Feng, Q., M. Li, T. Wang, Y. Chen, X. Wang, X. Zhang, X. Li, Z. Yang, L. Feng, J. Zheng, H. Xu, T. Zhai and Y. Jiang. 2020. Low-temperature growth of three dimensional ReS_2/ReO_2 metal-semiconductor heterojunctions on graphene/polyimide film for enhanced hydrogen evolution reaction. *Appl. Catal. B: Environ.*, 271: 118924.

[198] Li, Y. K., G. Zhang, W. T. Lu and F. F. Cao. 2020. Amorphous Ni-Fe-Mo suboxides coupled with Ni network as porous nanoplate array on nickel foam: a highly efficient and durable bifunctional electrode for overall water splitting. *Adv. Sci. (Weinh)*, 7: 1902034.

[199] Chen, Z., B. Fei, M. Hou, X. Yan, M. Chen, H. Qing and R. Wu. 2020. Ultrathin prussian blue analogue nanosheet arrays with open bimetal centers for efficient overall water splitting. *Nano Energy*, 68: 104371.

[200] Yang, D., L. Cao, L. Feng, J. Huang, K. Kajiyoshi, Y. Feng, Q. Liu, W. Li, L. Feng and G. Hai. 2019. Formation of hierarchical Ni_3S_2 nanohorn arrays driven by in-situ generation of VS_4 nanocrystals for boosting alkaline water splitting. *Appl. Catal. B: Environ.*, 257: 117911.

[201] Yang, Y., H. Yao, Z. Yu, S. M. Islam, H. He, M. Yuan, Y. Yue, K. Xu, W. Hao, G. Sun, H. Li, S. Ma, Peter Zapol, and Mercouri G. Kanatzidis. 2019. Hierarchical nanoassembly of $MoS_2/Co_9S_8/Ni_3S_2/$ Ni as a highly efficient wide-pH range electrocatalyst for overall water splitting. *J. Amer. Chem. Soc.*, 10.1021/jacs.9b04492.

[202] Liu, Z., L. Zhao, Y. Liu, Z. Gao, S. Yuan, X. Li, N. Li and S. Miao. 2019. Vertical nanosheet array of 1T phase MoS_2 for efficient and stable hydrogen evolution. *Appl. Catal. B: Environ.*, 246: 296–302.

[203] Zhang, X., Y.-Y. Zhang, Y. Zhang, W.-J. Jiang, Q.-H. Zhang, Y.-G. Yang, L. Gu, J.-S. Hu and L.-J. Wan. 2019. Phase-controlled synthesis of 1T-MoSe$_2$/NiSe heterostructure nanowire arrays

via electronic injection for synergistically enhanced hydrogen evolution. *Small Methods*, 3: 201800317.

[204] Zhang, Y.-Y., X. Zhang, Z.-Y. Wu, B.-B. Zhang, Y. Zhang, W.-J. Jiang, Y.-G. Yang, Q.-H. Kong and J.-S. Hu. 2019. Fe/P dual doping boosts the activity and durability of CoS₂ polycrystalline nanowires for hydrogen evolution. *J. Mater. Chem. A*, 7: 5195–5200.

[205] Zhang, X., F. Zhou, W. Pan, Y. Liang and R. Wang. 2018. General construction of molybdenum-based nanowire arrays for pH-universal hydrogen Evolution electrocatalysis. *Adv. Funct. Mater.*, 28: 201804600.

[206] Yang, R., Y. Zhou, Y. Xing, D. Li, D. Jiang, M. Chen, W. Shi and S. Yuan. 2019. Synergistic coupling of CoFe-LDH arrays with NiFe-LDH nanosheet for highly efficient overall water splitting in alkaline media. *Appl. Catal. B: Environ.*, 253: 131–139.

[207] Liu, M., Z. Sun, S. Li, X. Nie, Y. Liu, E. Wang and Z. Zhao. 2021. Hierarchical superhydrophilic/superaerophobic CoMnP/Ni₂P nanosheet-based microplate arrays for enhanced overall water splitting. *J. Mater. Chem. A*, 9: 22129–22139.

[208] Liu, J., Y.-Y. Ren, J. Wu, W. Xia, B.-Y. Deng and F. Wang. 2021. Hybrid artificial photosynthetic systems constructed using quantum dots and molecular catalysts for solar fuel production: development and advances. *J. Mater. Chem. A*, 9: 19346–19368.

[209] Zhang, T., X. Han, H. Yang, A. Han, E. Hu, Y. Li, X. Q. Yang, L. Wang, J. Liu and B. Liu. 2020. Atomically dispersed nickel(I) on an alloy-encapsulated nitrogen-doped carbon nanotube array for high-performance electrochemical CO₂ reduction reaction. *Angew. Chem. Int. Ed. Engl.*, 59: 12055–12061.

[210] Fan, J., X. Zhao, X. Mao, J. Xu, N. Han, H. Yang, B. Pan, Y. Li, L. Wang and Y. Li. 2021. Large-area vertically aligned bismuthene nanosheet arrays from galvanic replacement reaction for efficient electrochemical CO₂ conversion. *Adv. Mater.*, 33: e2100910.

[211] Qi, F., K. Liu, D.-K. Ma, F. Cai, M. Liu, Q. Xu, W. Chen, C. Qi, D. Yang and S. Huang. 2021. Dual active sites fabricated through atomic layer deposition of TiO₂ on MoS₂ nanosheet arrays for highly efficient electroreduction of CO₂ to ethanol. *J. Mater. Chem. A*, 9: 6790–6796.

[212] Zhang, L.-H., F. Yu and N. R. Shiju. 2021. Carbon-based catalysts for selective electrochemical nitrogen-to-ammonia conversion. *ACS Sustain. Chem. & Engin.*, 9: 7687–7703.

[213] Shi, L., Y. Yin, S. Wang and H. Sun. 2020. Rational catalyst design for N₂ reduction under ambient conditions: Strategies toward enhanced conversion efficiency. *ACS Catal.*, 10: 6870–6899.

[214] Yan, X., D. Liu, H. Cao, F. Hou, J. Liang and S. X. Dou. 2019. Nitrogen reduction to ammonia on atomic-scale active sites under mild conditions. *Small Methods*, 3: 1800501.

[215] Sun, Y., T. Jiang, J. Duan, L. Jiang, X. Hu, H. Zhao, J. Zhu, S. Chen and X. Wang. 2020. Two-dimensional nanomesh arrays as bifunctional catalysts for N₂ electrolysis. *ACS Catal.*, 10: 11371–11379.

[216] Wang, H.-Y., J.-T. Ren, C.-C. Weng, X.-W. Lv and Z.-Y. Yuan. 2021. Insight into the valence state of sisal-like MoO₂ nanosheet arrays for N₂ electrolysis. *Chem. Engin. J.*, 426: 130761.

[217] Xu, F., F. Wu, K. Zhu, Z. Fang, D. Jia, Y. Wang, G. Jia, J. Low, W. Ye, Z. Sun, P. Gao and Y. Xiong. 2021. Boron doping and high curvature in Bi nanorolls for promoting photoelectrochemical nitrogen fixation. *Appl. Catal. B: Environ.*, 284: 119689.

[218] Singh, R. K., K. Rajavelu, M. Montag and A. Schechter. 2021. Advances in catalytic electrooxidation of urea: A review. *Energy Technol.*, 9: 2100017.

[219] Wang, Z., W. Liu, J. Bao, Y. Song, X. She, Y. Hua, G. Lv, J. Yuan, H. Li and H. Xu. 2021. Modulating electronic structure of ternary NiMoV LDH nanosheet array induced by doping engineering to promote urea oxidation reaction. *Chem. Engin. J.*, DOI: 10.1016/j.cej.2021.133100.

[220] Liu, D., T. Liu, L. Zhang, F. Qu, G. Du, A. M. Asiri and X. Sun. 2017. High-performance urea electrolysis towards less energy-intensive electrochemical hydrogen production using a bifunctional catalyst electrode. *J. Mater. Chem. A*, 5: 3208–3213.

[221] Liu, T., D. Liu, F. Qu, D. Wang, L. Zhang, R. Ge, S. Hao, Y. Ma, G. Du, A. M. Asiri, L. Chen and X. Sun. 2017. Enhanced electrocatalysis for energy-efficient hydrogen production over CoP catalyst with nonelectroactive Zn as a promoter. *Adv. Energy Mater.*, 7: 201700020.

[222] Xu, S. M., X. Liang, Z. C. Ren, K. X. Wang and J. S. Chen. 2018. Free-standing air cathodes based on 3D hierarchically porous carbon membranes: kinetic overpotential of continuous macropores in Li-O₂ Batteries. *Angew. Chem. Int. Ed. Engl.*, 57: 6825–6829.

[223]   Wang, H.-F. and Q. Xu. 2019. Materials design for rechargeable metal-air batteries. *Matter.*, 1: 565–595.

[224]   Liu, Q., Z. Pan, E. Wang, L. An and G. Sun. 2020. Aqueous metal-air batteries: Fundamentals and applications. *Energy Storage Mater.*, 27: 478–505.

[225]   Yu, J., B.-Q. Li, C.-X. Zhao and Q. Zhang. 2020. Seawater electrolyte-based metal-air batteries: From strategies to applications. *Energy & Environ. Sci.*, 13: 3253–3268.

[226]   Tang, K., C. Yuan, Y. Xiong, H. Hu and M. Wu. 2020. Inverse-opal-structured hybrids of N, S-codoped-carbon-confined Co_9S_8 nanoparticles as bifunctional oxygen electrocatalyst for on-chip all-solid-state rechargeable Zn-air batteries. *Appl. Catal. B: Environ.*, 260: 118209.

[227]   Ji, D., L. Fan, L. Li, S. Peng, D. Yu, J. Song, S. Ramakrishna and S. Guo. 2019. Atomically transition metals on self-supported porous carbon flake arrays as binder-free air cathode for wearable Zinc-air batteries. *Adv. Mater.*, 31: e1808267.

[228]   Zhang, J., X. Bai, T. Wang, W. Xiao, P. Xi, J. Wang, D. Gao and J. Wang. 2019. Bimetallic nickel cobalt sulfide as efficient electrocatalyst for Zn-air battery and water splitting. *Nanomicro. Lett.*, 11: 2–12.

[229]   Li, Z., J. Yang, X. Ge, Y.-P. Deng, G. Jiang, H. Li, G. Sun, W. Liu, Y. Zheng, H. Dou, H. Jiao, J. Zhu, N. Li, Y. Hu, M. Feng and Z. Chen. 2021. Self-assembly of colloidal MOFs derived yolk-shelled microcages as flexible air cathode for rechargeable Zn-air batteries. *Nano Energy*, 89: 106314.

[230]   Balamurugan, J., T. T. Nguyen, D. H. Kim, N. H. Kim and J. H. Lee. 2021. 3D nickel molybdenum oxyselenide ($Ni_{1-x}Mo_xOSe$) nanoarchitectures as advanced multifunctional catalyst for Zn-air batteries and water splitting. *Appl. Catal. B: Environ.*, 286: 119909.

[231]   Niu, Y., X. Teng, S. Gong and Z. Chen. 2020. A bimetallic alloy anchored on biomass-derived porous N-doped carbon fibers as a self-supporting bifunctional oxygen electrocatalyst for flexible Zn-air batteries. *J. Mater. Chem. A*, 8: 13725–13734.

[232]   Liu, W., B. Ren, W. Zhang, M. Zhang, G. Li, M. Xiao, J. Zhu, A. Yu, L. Ricardez-Sandoval and Z. Chen. 2019. Defect-enriched nitrogen doped-graphene quantum dots engineered $NiCo_2S_4$ nanoarray as high-efficiency bifunctional catalyst for flexible Zn-air battery. *Small*, 15: e1903610.

[233]   Yi, M., N. Li, B. Lu, L. Li, Z. Zhu and J. Zhang. 2021. Single-atom Pt decorated in heteroatom (N, B, and F)-doped ReS_2 Grown on Mo_2CT_x for efficient pH-universal hydrogen evolution reaction and flexible Zn-air batteries. *Energy Storage Mater.*, 42: 418–429.

[234]   Zhang, Z., X. Wu, Z. Kou, N. Song, G. Nie, C. Wang, F. Verpoort and S. Mu. 2022. Rational design of electrospun nanofiber-typed electrocatalysts for water splitting: A review. *Chem. Engin. J.*, 428: 131133.

[235]   Li, X., L. Zhao, J. Yu, X. Liu, X. Zhang, H. Liu and W. Zhou. 2020. Water splitting: From electrode to green energy system. *Nano-Micro Lett.*, 12: 131–143.

[236]   Li, D., J. Shi and C. Li. 2018. Transition-metal-based electrocatalysts as cocatalysts for photoelectrochemical water splitting: A mini review. *Small*, 14: 1704179.

9

Single-Metal-Atom Electrocatalysts for Clean Energy Conversion

Xun Cui,[1,5,6,*] *Likun Gao,*[2,*] *Zhengnan Wei,*[3] *Jingying Li,*[3] *Rui Ma,*[4] *Yingkui Yang*[4] *and Huaming Yang*[1,5,6,*]

9.1 Introduction

Electrochemical clean energy conversion plays a vital role in tackling the ever-growing worldwide challenges in climate change, environmental pollution and energy crisis.[1-3] Accordingly, great efforts have been made for advanced electrochemical clean energy conversion and generation technologies, such as water electrolysis, fuel cells, metal-air batteries, and carbon dioxide conversion, that are critical for clean and sustainable energy generation, conversion, storage and utilization.[4-8] In fact, the overall performances of aforementioned technologies strongly depend on the physicochemical characteristics of the materials employed in each component.[9,10] Particularly, the electrocatalysts that are commonly employed at the electrode/electrolyte interface to improve the energy conversion efficiency have always been the focus of the researchers in the last few decades as a result of the complicated reaction mechanisms and generally sluggish reaction kinetics of the associated electrochemical processes.[11-13]

[1] Engineering Research Centre of Nano-Geomaterials of Ministry of Education, China University of Geosciences, Wuhan 430074, China.

[2] Key Laboratory of Bio-based Material Science and Technology of Ministry of Education, Northeast Forestry University, Harbin 150040, China.

[3] New Energy Development Centre of Shengli Petroleum Administration Co., Ltd., SINOPEC, Dongying 257000, China.

[4] Key Laboratory of Catalysis and Energy Materials Chemistry of Ministry of Education, Hubei Key Laboratory of Catalysis and Materials Science, South-Central University for Nationalities, Wuhan 430074, China.

[5] Faculty of Materials Science and Chemistry, China University of Geosciences, Wuhan 430074, China.

[6] Key Laboratory of Functional Geomaterials in China Nonmetallic Minerals Industry, China University of Geosciences, Wuhan 430074, China.

* Corresponding authors: cuixun@cug.edu.cn; gaolk@nefu.edu.cn; hm.yang@cug.edu.cn

Recent years have witnessed continuous endeavor and massive accomplishments in the development of cost-effective and high-efficiency electrocatalysts. Numerous electrocatalysts based on noble metals, non-noble metals, and their alloys, have been widely investigated to apply for various electrocatalytic processes.[14–19] Unfortunately, the noble metal-based electrocatalysts suffer natural scarcity and prohibitive cost and the non-noble metal-based electrocatalysts face inadequate electrocatalytic activities and stability, respectively, therefore greatly hampering their potential large-scale commercial applications.[20,21] Accordingly, the development of novel and robust electrocatalysts with unique characteristics for maximizing the metal utilization, minimizing the cost, upgrading the intrinsic activity, and improving the long-term durability is extremely required.

With the rapid development of nanoscience and nanotechnology, single-metal-atom catalysts (SMACs) recently have been widely acknowledged as a new family of electrocatalysts in which the atomically isolated single-metal-atom sites anchored on specific supporting materials serve directly as the active sites.[22,23] In fact, the concept of "single-metal-atom catalysts" historically can be traced back to the pioneering research work reported by Zhang and co-workers in 2011.[24,25] Because of the single-atom features of active sites, the SMACs offer great advantages in maximizing the utilization of metal with a nearly 100% efficiency (Fig. 9.1a), consequently leading to significantly lowered cost and remarkably improved electrocatalytic activity.[26] Meanwhile, since the performance of an electrocatalyst is essentially dependent on the electronic structures of active sites that are largely influenced by the size and morphology, downsizing metal nanoclusters (NCs) or metal nanoparticles (NPs) to single-metal-atom sites with unsaturated coordination environment and quantized orbital of active sites is acknowledged to be highly effective in promoting the intrinsic activity of the active sites (Fig. 9.1b).[23,27]

In the past couple of years, numerous SMACs stabilized on different metal compounds- or carbon-based supporting materials have been therefore successfully developed for applications in a variety of prospective areas, such as organic electrosynthesis, thermocatalytic reaction, energy conversion and storage technologies, photocatalysis, electrocatalysis and biosensors.[28–32] The effects of their

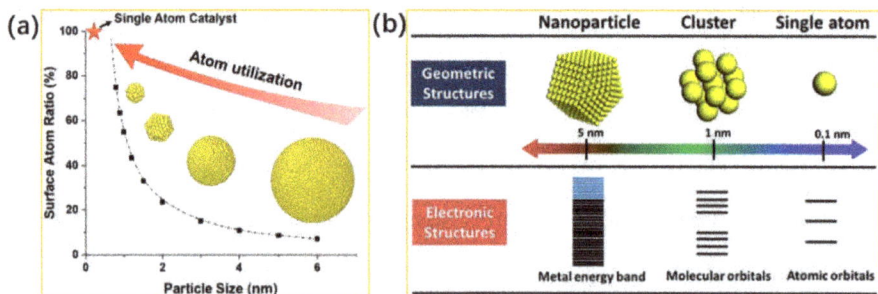

Figure 9.1. (a) The ratio of surface atoms as different particle sizes. Reproduced with permission from Ref. 26. Copyright 2020, American Chemical Society. (b) The geometric and electronic structures of single atom, cluster, and nanoparticle. Reproduced with permission from Ref. 23. Copyright 2018, American Chemical Society.

unique electronic properties, physical structures, geometry, and functions toward specific performances have been excitingly explored as well.[33,34] Particularly, the SMACs impressively exhibited great potentials in the progress of electrocatalytic clean energy conversion, including water electrolysis, fuel cells, metal-air batteries, and carbon dioxide conversion. However, the controllable and large-scale synthesis of SMACs still remains challenging due to the strong tendency of migration and aggregation of active single atoms during synthesis and the subsequent activation and application processes. To this end, the rational utilization of interactions between metal sites and supporting materials plays a considerable role to immobilize the single-metal-atom sites on the supporting materials. Moreover, the development of effective synthetic approaches is indispensable to achieve the atomic dispersion and chemical bonding of metal active sites on supporting materials for the further large-scale practical applications of SMACs.

In this chapter, we first give an overview of SMACs, covering the properties and origins of electrocatalytic activity. Then, the advanced synthetic strategies and characterization techniques for the preparation of SMACs are introduced. The most recent advances regarding SMACs and their potential applications for reactions involved in clean energy conversion technologies, including oxygen reduction reaction (ORR), oxygen evolution reaction (OER), hydrogen evolution reaction (HER), and carbon dioxide reduction reaction (CO_2RR) are subsequently briefly performed. Particular attentions are paid to the structure-performance relationships and intrinsic origins of electrocatalytic activity. At last, future perspectives with regard to the challenges and opportunities awaiting this particular research field are briefly offered and discussed.

9.2 Unique features of SMACs

It has been well-acknowledged that improving the density of exposed active sites and optimizing their intrinsic activity are two general and practical approaches to significantly increase the catalytic efficiency of an electrocatalyst.[35] According to the definition, the SMACs only consist of atomically dispersed metal sites that are anchored on supporting materials (i.e., chemically bonded to surrounding atoms in the supporting material) without any appreciable interaction between them. In comparison to the traditional metal NCs or metal NPs-based electrocatalysts with extremely complex structures and compositions and poorly defined active sites, the SMACs offer the advantages of both homogeneous and heterogeneous catalysts and bridge the gap between them.

In general, the single-metal-atom sites have high surface energy and tend to agglomerate; therefore, they usually need to be stabilized onto supporting materials via strong coordination bindings to maintain atomic dispersion during the synthesis and subsequent electrocatalytic processes. Many types of heteroatoms (e.g., pyridine-N, pyrrole-N and thiophene-S) and defects (e.g., single and double vacancy carbon defects) on supporting materials have been verified as effective coordination sites to anchor various single-metal-atom sites. Intrinsically, the center metal atom together with its surrounding atoms or functional groups located on the supporting materials can be considered as the actual active site.[24,33,36,37] The electrocatalytic activities of SMACs largely originate from the adsorption/desorption of reaction intermediates onto the metal centers with unsaturated d orbitals. In contrast with metal NCs and metal NPs-based electrocatalysts, the homogeneously dispersed active metal centers are apparently favorable for gaining deep insights into the structure-property relationships and electrocatalytic mechanism of specific process. Besides, the strong chemical hybridization and electronic interaction between the center metal atom and the surrounding atoms located on the supporting materials results in unique electronic properties of single-metal-atom sites that can be further modulated to optimize the adsorption properties.[38–40] Therefore, the local atomic environment of SMACs plays a crucial role in determining the activity, selectivity, as well as long-term stability in electrocatalytic processes.[41–43]

It is worth noting that the structural and electronic characteristics of the supporting materials are also modulated by the immobilized single metal atoms.[44,45] Up to now, the commonly employed supporting materials to synthesize SMACs are mainly composed of two categories, i.e., carbon-based materials such as graphene, carbon nanotubes and porous carbon matrixes, and metal compounds-based materials containing metal oxides, metal dichalcogenides, metal carbides, and metal nitrides.[23,38,39,44,45,46–48] Distinctly, the abundance of supporting materials renders great possibilities and vast opportunities for engineering the local atomic environment of single-metal-atom sites to achieve desired electronic characteristics toward various electrocatalytic processes. Thus, optimizing the local atomic environment by tuning the supporting materials in terms of the atomic and electronic structures on the surface is of great significance for the development of highly-efficient SMACs.[49,50]

9.3 Synthesis and characterization of SMACs

Appropriate synthetic approaches play extremely important role in the controllable synthesis of high-quality SMACs because of the strong aggregation and migration tendency of metal sites during the synthetic procedure, which originates from the high surface energy of the isolated single-metal-atom sites on supporting materials.[51,52] In this regard, over the past several years, various strategies have been proposed and certified highly-effective for constructing SMACs with high density of exposed active sites and well-defined local atomic environment on various substrates, including coordinative pyrolysis, defects anchoring, atomic layer deposition, chemical vapor deposition, wet-chemistry method, electrochemical deposition, chemical reduction, photochemical reduction, and so forth. In this section, the aforementioned synthetic strategies will be briefly introduced.[26,27,53–55] In addition, considering the significance of the characterization in comprehending the structure-performance correlations and gaining insights in mechanisms of electrocatalytic processes, a brief overview of the most advanced microscopic and spectroscopic techniques that are capable of depicting the local atomic environments of SMACs will also be offered in this section.

9.3.1 Synthetic strategies of SMACs

9.3.1.1 Coordinative pyrolysis strategy

Coordinative pyrolysis strategy is the most widely adopted approach to prepare SMACs, which can be realized by decomposing various precursors that contain metal sources at high-temperature environments under certain gas atmosphere conditions (e.g., N_2, Ar and NH_3).[56-58] Generally, the SMACs can be acquired from varieties of precursors, including metal-ligand complexes, metal-polymers mixtures, metal-organic frameworks (MOFs), and so forth.[59-63] During the pyrolysis process, the organic moieties are completely converted into highly conductive nitrogen (N)-doped carbon matrix, with the single-metal-atom centers immobilized by the surrounding N sites. It is worth noting that the coordination in the precursor mixtures may play a vital role in deciding the local atomic environment of the SMACs obtained by pyrolysis. Obviously, this strategy possesses some distinct advantages, such as large-scale preparation, convenient manipulation and employment of commonly inexpensive raw materials. Nevertheless, the inevitable formation of metal NCs and metal NPs, especially at relatively high metal concentrations, is still a great challenge. Noting that, most of the reported carbon-based SMACs are synthesized by this coordinative pyrolysis method to date.

For instance, the commonly employed MOFs, which contain abundant anchoring sites, allow for the introduction of uniformly dispersed single-metal-atom sites via either coordination and adsorption or metal ion exchange. In a recently reported work, Chen and coworkers successfully prepared Fe SMACs via the pyrolysis treatment of ZIF-8 impregnated with iron acetylacetonate ($Fe(acac)_3$) precursor (Fig. 9.2a).[64] Typically, a solvothermal treatment at 120°C for 4 h was first employed to generate the $Fe(acac)_3$@ZIF-8 hybrid (i.e., ZIF-8 encapsulating $Fe(acac)_3$). The produced $Fe(acac)_3$@ZIF-8 subsequently subjected to a high-temperature pyrolysis treatment under Ar atmosphere condition directly resulted in the Fe-SMACs. During the necessary pyrolysis, the ZIF converted into highly conductive N-doped carbon matrix, the Zn species vaporized and the Fe atoms bonded to neighboring N sites in the obtained porous N-doped carbon matrix. Besides, Zheng and co-author recently proposed a simple yet scalable approach for the preparation of single-atom Ni catalysts through directly employing the commercial carbon black as the supporting material.[65] As exhibited in Fig. 9.2b, the adsorbed Ni^{2+} ions on the surface of carbon black by interactions between Ni^{2+} ions and the oxygen-containing functional groups or defect sites could be further immobilized onto the carbon black during the subsequent high-temperature pyrolysis treatment with the added urea as the supplementary nitrogen source.

9.3.1.2 Defects anchoring strategy

In addition to the coordinative pyrolysis, the defects that are commonly present in metal compounds such as oxides and hydroxides, and even carbon-based materials, have been widely explored to immobilize the single-metal-atom sites.[24,33,66-69] It has

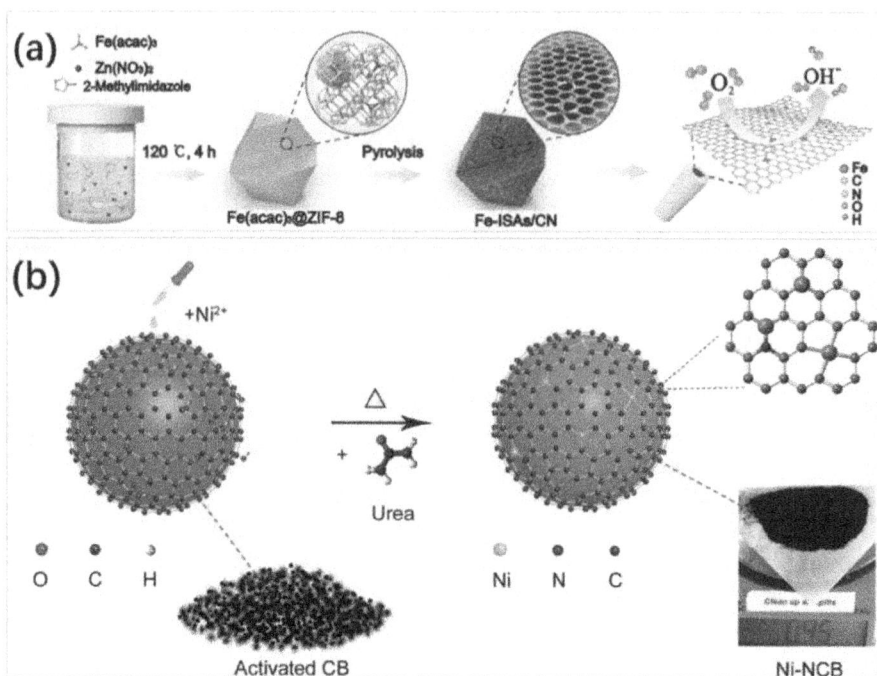

Figure 9.2. (a) Schematic synthesis of Fe-SMACs through the pyrolysis of ZIF-8 impregnated with metal precursors. Reproduced with permission from Ref. 64. Copyright 2017, Wiley-VCH. (b) Schematic synthesis of Ni-SMACs with commercial carbon black as the supporting material. Reproduced with permission from Ref. 65. Copyright 2018, Elsevier.

already been acknowledged that the single-metal-atom sites can be tightly stabilized on the surface of the metal compounds by capitalizing on the strong metal-oxygen interactions. Particularly, the single-metal-atom sites also have been certified to be stabilized more easily at coordinatively unsaturated steps and vacancy sites. Besides, the carbon-based materials, such as the most common graphene, can also be engineered to contain rich defect sites to enable the immobilization of single-metal-atom sites.[70–72] Accordingly, many research efforts have been dedicated to constructing defects on graphene to realize the graphene-supported SMACs. For example, Yao et al. proposed a melamine-assisted high-temperature annealing approach to create defects on graphene, and the subsequent successive wetness impregnation of Ni^{2+} cations, annealing reduction, and acid leaching treatment resulted in graphene defects-trapped Ni single-atom sites (Fig. 9.3a–c).[70] Note that, a thorough acid leaching treatment was employed to eliminate the inevitably aggregated Ni clusters in this work.

9.3.1.3 Atomic layer deposition strategy

Atomic layer deposition (ALD), which also has been widely known as atomic layer epitaxy, is an emerging gas-phase deposition technique for the synthesis of SMACs.[73–80] In a typical process of ALD, the gas-phase metal precursor would flow to the reaction chamber and be exposed to another gas reactant to generate an

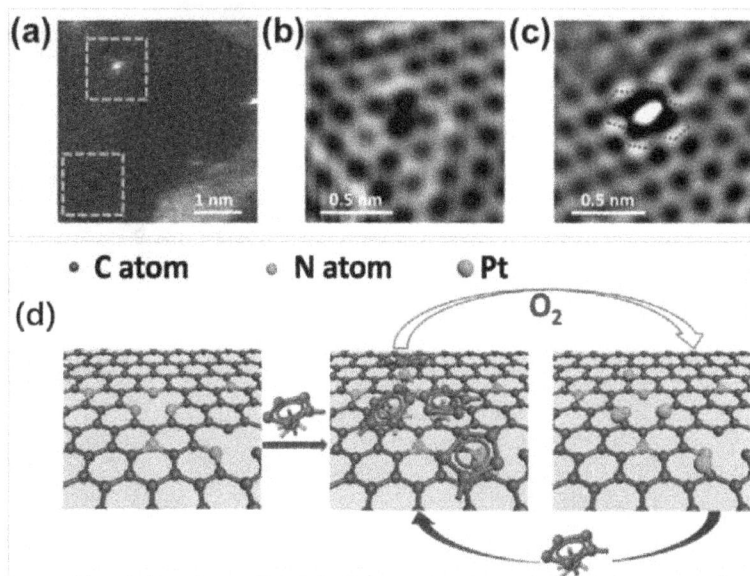

Figure 9.3. (a–c) HADDF-STEM images of single-atom Ni sites anchored by defective graphene. Reproduced with permission from Ref. 70. Copyright 2018, Cell Press. (d) Schematic illustration of the mechanism of atomic layer deposition of single-atom Pt sites. Reproduced with permission from Ref. 73. Copyright 2016, Nature Publishing Group.

atomic layer onto the supporting materials. Therefore, the deposition process can be accurately controlled by readily tuning the parameters of ALD process. Besides, this deposition strategy offers the as-synthesized SMACs uniform geometrical structures. Nevertheless, the costly equipment and low yield of target SMACs largely hamper its practical applications and large-scale synthesis. A typical example by Sun and co-workers recently reported the preparation of graphene nanosheet (GNS) supported single-atom Pt catalyst based on the ALD (Fig. 9.3d).[73] During the synthetic process, the Pt precursor (i.e., (methylcyclopentadienyl)-trimethylplatinum (MeCpPtMe$_3$)) was firstly exposed to GNS and then reacted with the adsorbed oxygen (either on GNS or Pt), to produce CO_2, H_2O, and hydrocarbon fragments. A subsequent oxygen exposure leads to the formation of single-atom Pt sites coating with a new adsorbed oxygen layer.

9.3.1.4 Other synthetic strategies

In addition to the strategies introduced above, other methods such as chemical vapor deposition, electrochemical deposition, chemical reduction, photochemical reduction, and, etc., have also been designed for the preparation of SMACs.[81–87] To date, a wide variety of SMACs have been realized based on these advanced synthetic strategies. For instance, Kauppinen and co-workers proposed a one-step catalytic chemical vapor deposition approach to prepare a highly graphitized graphene nanoflake (GF)-carbon nanotube (CNT) hybrid material doped simultaneously with single atoms of N, Co, and Mo (N-Co-Mo-GF/CNT), as depicted in Fig. 9.4a.[88] Toshima et al. synthesized Pd nanoclusters supported single-atom Au catalysts through a successive chemical

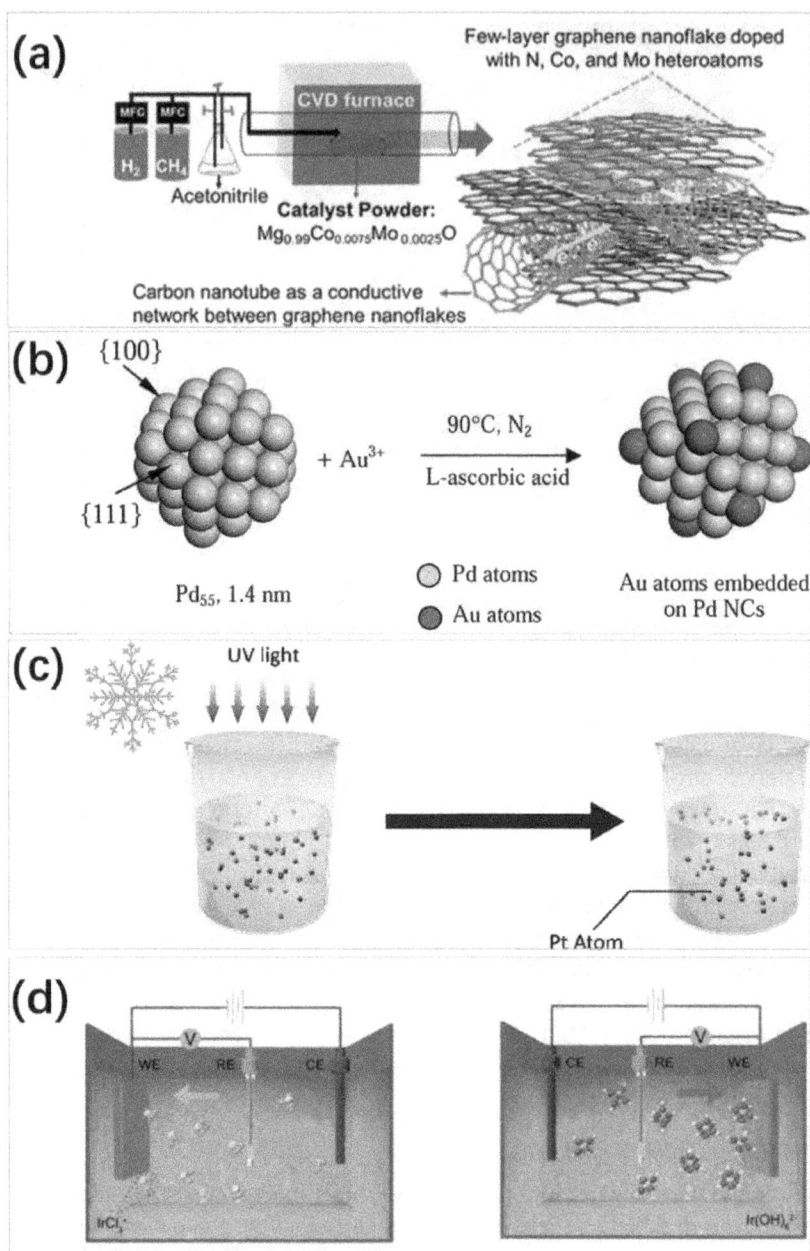

Figure 9.4. (a) Scheme of the chemical vapor deposition of N-Co-Mo-GC/CNT. Reproduced with permission from Ref. 88. Copyright 2020, American Chemical Society. (b) Scheme of the chemical reduction of single-atom Au sites on Pd mother clusters. Reproduced with permission from Ref. 85. Copyright 2014, Royal Society of Chemistry. (c) Scheme of the ultralow-temperature photochemical synthesis of atomically dispersed Pt. Reproduced with permission from Ref. 87. Copyright 2019, Royal Society of Chemistry. (d) Schematic of cathodic (left) and anodic (right) electrochemical deposition of single-atom Ir catalysts. Reproduced with permission from Ref. 89. Copyright 2020, Nature Publishing Group.

reduction strategy using L-ascorbic acid as the reducing reagent (Fig. 9.4b).[85] Wu et al. employed a ultralow-temperature ($-60°C$) photochemical synthesis route to prepare the atomically dispersed Pt catalysts (Fig. 9.4c).[87] Zeng and co-author reported both cathodic and anodic electrochemical deposition (Fig. 9.4d, taking the deposition of noble Ir metal as an example) as a general strategy to produce more than 30 different SMACs including common noble and non-noble metals.[89]

Nevertheless, it remains challenging to precisely control the coordination structure of the atomically dispersed metal sites in the preparation of SMACs. Most of these derived SMACs usually suffer from low loading of single-atom metal sites, poorly defined coordination structure with a diversity of coordination number with the supporting material, and low yields, which impede their further large-scale practical applications to some extent. In this regard, new synthetic methods, novel substrates and effective ligands for developing stable SMACs with well-defined coordination environment and high metal loading are still important and urgently needed.

9.3.2 Characterization of SMACs

Modern atomic-resolution characterization techniques play an extremely considerable role in the design and synthesis of SMACs and the mechanical understanding of the electrocatalytic processes. The methods to characterize SMACs are briefly introduced in this section and for more detailed information, the reader is referred to some recently published review articles.[45,90,91] The aberration-corrected high-angle annular dark field scanning transmission electron microscopy (HAADF-STEM) is a powerful tool for identifying the morphological information of SMACs.[92,93] More detailed information concerning the local coordination structure, chemical binding, and oxidation states of SMACs can also be clarified based on extended X-ray absorption fine structure (EXAFS) and X-ray absorption near-edge spectroscopy (XANES) analysis.[94,95] The curve fitting of the Fourier transformation (FT)-EXAFS analysis results is the most straightforward approach for recognizing the coordination environment of SMACs. Nevertheless, the information offered by EXAFS and XANES is only based on the average values. Recently, the electron energy loss spectroscopy (EELS) and scanning tunneling microscopy (STM) with atomic resolution have also been employed to directly elucidate SMACs.[96–98] Based on these characterization techniques, the presence of individual atoms in the SMACs can be confirmed. In addition to these aforementioned experimental characterization, the density functional theory (DFT)-based computational calculation is also a highly effective approach to not only study the pathway of specific electrocatalytic processes but to elucidate the coordination information of SMACs, thereby revealing and predicting the most favorable and stable configurations of SMACs.[99,100]

In addition, the significance of *in situ* and *operando* characterization techniques is ever-increasing to gain deep insight into the nature of SMACs during the electrocatalytic processes. Recently, the *in situ* electrochemical X-ray absorption fine structure (XAFS) measurements have been used to observe the structural evolutions of SMACs during the ORR process.[101] Besides, the Fourier transform infrared (FTIR) spectroscopy, surface-enhanced Raman spectroscopy (SERS) and Mössbauer spectroscopy have also been well established for *in situ* observations of

SMACs. Nevertheless, the greatest disadvantage of these techniques is that only the average spectra can be acquired.

9.4 Applications of SMACs for electrochemical clean energy conversion

Owing to their unique coordination structures and electronic characteristics of single-metal-atom active sites, SMACs as electrocatalysts have been widely employed in the studies of electrochemical clean energy conversion technologies such as water electrolysis, fuel cells, metal-air batteries and carbon dioxide conversion. In this section, the most recent developments of SMACs and their potential applications for reactions involved in these clean energy conversion technologies, including ORR, OER, HER, and CO_2RR are briefly outlined through enumerating typical examples in the following four parts. Particular attentions are paid to the structure-performance correlations and intrinsic origins of electrocatalytic activity.

9.4.1 Oxygen reduction reaction

The ORR is a crucial cathodic reaction involved in many promising clean energy conversion technologies, including fuel cells and metal-air batteries. Generally, the overall ORR process is quite complex and involves the adsorption and desorption of multiple oxygen intermediate species as well as the multielectron transfer processes, thereby possessing sluggish reaction kinetics.[102] Typically, the ORR process follows a four-electron (4e⁻) pathway for the reduction of molecular oxygen into H_2O (acidic condition) or OH⁻ (alkaline condition) with an equilibrium potential of 1.23 V versus reversible hydrogen electrode (RHE). Nevertheless, hydrogen peroxide (H_2O_2, acidic condition) or peroxide species (OOH⁻, alkaline condition) can also be produced via a two-electron (2e⁻) pathway. The generation of H_2O_2 or OOH⁻ on the one side lowers the overall energy conversion efficiency, on the other side leads to the corrosion of membrane and supporting materials in the fuel cells.[103,104] Accordingly, the 4e⁻ pathway is highly preferred and a superior electrocatalyst toward ORR should selectively eliminate the 2e⁻ pathway. In fact, the selectivity of either 4e⁻ or 2e⁻ pathway for a specific electrocatalyst largely relies on the adsorption strength between the reactants/intermediates and the active sites as well as the reaction barriers. Up to now, noble metal-based materials such as Pt and its alloys are still the best ORR electrocatalysts as a result of their excellent activity and 4e⁻ pathway selectivity. However, the resource scarcity, prohibitive cost, and susceptibility to CO poisoning of these noble metal-based materials greatly impede their large-scale practical applications.

Recently, it has been widely verified that the noble metal-based SMACs as electrocatalysts could significantly accelerate the electrocatalytic ORR activity, while remarkably maximizing the atomic utilization of noble metal and lowering the cost. Theoretical computational calculations have identified that the moderate adsorption strength of oxygen intermediate species on single-metal-atom sites are responsible for the improved electrocatalytic ORR performance, which furthermore can be manipulated to impact the reaction pathway (4e- or 2e- pathways).[105] Up to now, many noble metal-based SMACs have been proposed and explored as the catalysts for the electrocatalytic ORR process.[106–108] Notably, some of the reported SMACs reveal impressive electrocatalytic abilities toward ORR. As a typical example, Xu and co-workers recently proposed a low-cost carbon-defect-stabilized Pt SMACs as the catalysts toward high-efficiency electrocatalytic ORR process (Fig. 9.5a).[106] The ORR performance of the optimal carbon-defect-anchored Pt SMACs ($Pt_{1.1}/BP_{defect}$) is much higher than that of the control sample $Pt_{1.1}/BP$ (BP refers to the carbon black (Black Pearls 2000)) as well as the commercial Pt/C in terms of the much higher onset potential (E_{onset}) and half-wave potential ($E_{1/2}$) (Fig. 9.5b). Moreover, the H_2/O_2 single cell measurement further verified the significant ORR activity of the $Pt_{1.1}/BP_{defect}$ as a maximum power density of 520 mW cm^{-2} at 80°C can be reached, which corresponds to an ultrahigh Pt utilization of 0.09 g_{Pt} kW^{-1}. To deeply reveal the effect of carbon defects on electrocatalytic ORR activity of single-metal-atom Pt sites, computational DFT-based calculations were also conducted to optimize the structures and study the adsorption properties of oxygenated intermediates on these different substrates. Based on both experimental and theoretical studies, the high electrocatalytic ORR performance of carbon-defect-anchored Pt SMACs can be ascribed to the unique configuration of active sites (i.e., $Pt-C_4$ sites that are stably anchored in the carbon divacancies) (Fig. 9.5c,d).

In addition, inspired by the design and preparation of noble metal-based SMACs, low-cost non-noble metal-based SMACs have also been investigated to realize high ORR performance. Among these non-noble metal-based SMACs, Fe- and Co-SMACs which possess the apparent merits of unique electronic structures

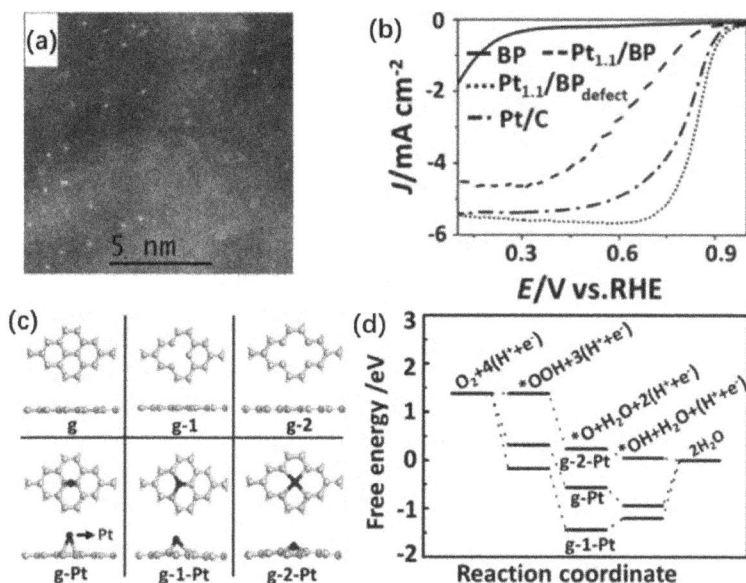

Figure 9.5. (a) HAADF-STEM image of carbon-defect-anchored Pt sites ($Pt_{1.1}/BP_{defect}$). (b) Linear sweep voltammetry (LSV) curves of $Pt_{1.1}/BP_{defect}$, reference samples ($Pt_{1.1}/BP$ and BP, BP refers to the carbon black (Black Pearls 2000)), and commercial Pt/C in 0.1 M HClO$_4$. (c) Optimized structures of different substrates. (d) Free energy diagram for ORR on these substrates. Reproduced with permission from Ref. 106. Copyright 2019, Wiley-VCH.

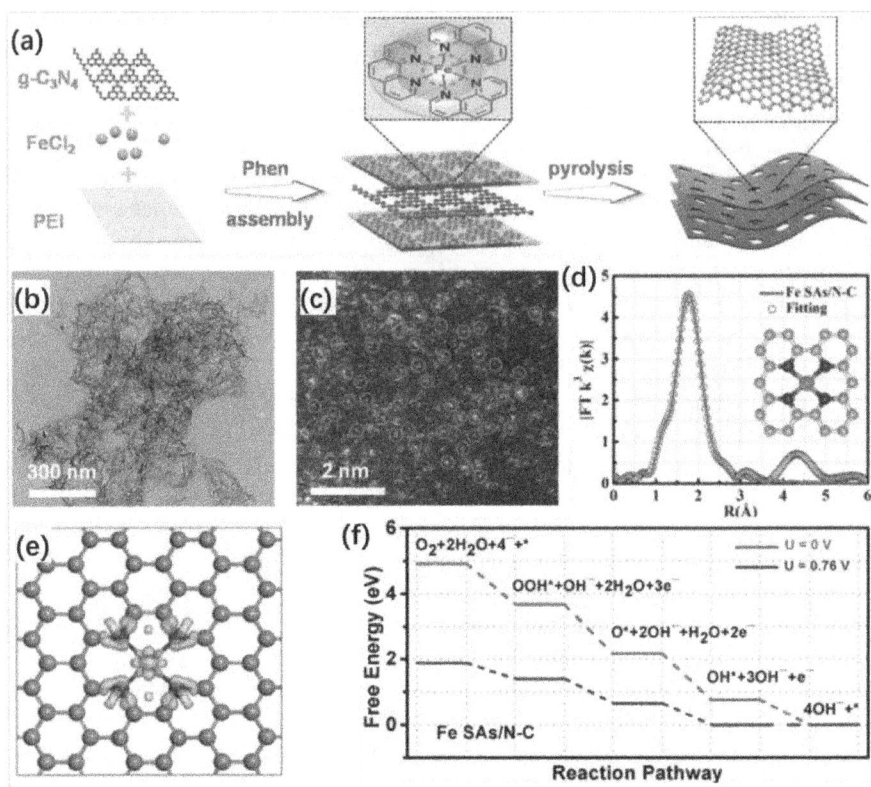

Figure 9.6. (a) Scheme of the synthesis of Fe SAs/N-C. (b) TEM image of Fe SAs/N-C. (c) HAADF-STEM image of Fe SAs/N-C. (d) FT-EXAFS fitting curve of Fe SAs/N-C, inset in (d) shows the proposed model of Fe-N$_4$ sites. (e) Calculated charge density difference of Fe SAs/N-C. (f) Free energy diagrams of Fe SAs/N-C toward ORR. Reproduced with permission from Ref. 111. Copyright 2019, American Chemical Society.

and optimal molecular oxygen adsorption strength have garnered the most attention and interest in the past few years for the development of high-efficiency catalysts toward the electrocatalytic ORR.[64,109,110] For instance, Wu and co-workers reported a scalable molecules-confined pyrolysis strategy to synthesize the Fe SMACs on porous N-doped carbon nanosheets (Fe SAs/N-C), as exhibited in Fig. 9.6a–c.[111] The XANES analysis and FT-EXAFS fitting results (Fig. 9.6d) confirmed the Fe-N$_4$ configuration of the atomically isolated Fe sites. As a result, the obtained Fe SAs/N-C exhibited a remarkable $E_{1/2}$ of 0.91 V vs RHE in 0.1 M KOH solution. Markedly, the Fe SAs/N-C also offered good electrocatalytic ORR performance comparable to that of commercial Pt/C catalyst even in a more challenging acidic solution. Besides, the Fe SAs/N-C displayed remarkably high stability in both alkaline and acidic conditions. The excellent ORR performance can be ascribed to the high 4e$^-$ pathway selectivity of the single-metal-atom Fe sites with the configuration of Fe-N$_4$ based on DFT-based computational calculations (Fig. 9.6e,f). In another work reported by Wu and co-workers, a nitrogen-coordinated single-metal-atom Co catalyst (20Co-NC-1100) derived from Co-doped MOFs was synthesized via a one-step thermal activation.[57] The well-dispersed atomically Co-N$_4$ active sites incorporated in the 3D porous MOF-derived carbon structures endowed both considerable activity and stability for the electrocatalytic ORR process in challenging acidic condition (e.g., E$_{1/2}$ of 0.80 V). It is also worth noting that besides metallic Fe and Co, carbon supported Cu, Zn, Mn, et al. SMACs have been reported

Table 9.1. ORR performance of some representative advanced SMACs. Note: a, b, and c refer to electron transfer number, half-wave potential, and onset potential, respectively.

Catalyst	Electrolyte	n^a	E$_{1/2}$^{b} (V vs. RHE)	E$_{onset}$^{c} (V vs. RHE)	References
Pt$_1$-N/BP	0.1 M HClO$_4$	4	0.76	~ 0.9	107
Pt$_1$-N/BP	0.1 M KOH	4	0.87	> 0.9	107
Fe-ISAs/CN	0.1 M KOH	4	0.9	0.986	64
1.5Fe-ZIF	0.5 M H$_2$SO$_4$	4	0.88	0.98	112
Fe-N$_4$ SAs/NPC	0.1 M KOH	4	0.885	0.972	113
FeN$_x$-PNC	0.1 M KOH	4	0.86	0.997	114
Co-N-C-10	0.1 M HClO$_4$	4	0.79	0.92	57
Co-N-C@F127	0.5 M H$_2$SO$_4$	4	0.84	0.93	115
Co SAs/N-C(800)	0.1 M KOH	4	0.881	0.982	60
Mn@NG	0.1 M KOH	4	0.82	0.95	116
Mn/C-NO	0.1 M KOH	4	0.86	0.94	117
Cu-N-C	0.1 M KOH	4	0.869	> 0.9	118
Zn-N-C-1	0.1 M KOH	4	0.873	--	119
Zn-N-C-1	0.1 M HClO$_4$	4	0.746	--	119

to show significant electrocatalytic ORR performance. The ORR performance of some representative advanced SMACs are selectively listed in Table 9.1.

9.4.2 Oxygen evolution reaction

The OER process is the anodic reaction for electrochemical water splitting and essentially the reverse version to that of the ORR process. Compared with the rather fast rate of cathodic HER process, the proposed overall reaction pathways for the anodic OER involves multiproton-coupled electron-transfer steps, thereby possessing intrinsically sluggish kinetics and essentially determining the performance of overall water splitting.[120,121] Besides the electrochemical water splitting, the OER process also plays a key role in many rechargeable metal-air batteries, such as zinc-air and lithium-air batteries.[122] At present, both the noble metal-based IrO$_2$ and RuO$_2$ are still the best electrocatalysts toward OER; however, their large-scale practical applications are seriously restricted by their high cost and resource scarcity. Over the past few decades, numerous OER electrocatalysts have been proposed, such as metal oxides, (oxy)hydroxides, sulfides, senides, perovskites, and spinels.[1] Very recently, the reasonably designed SMACs with unique electronic properties and well-defined structures have also been proposed as advanced OER electrocatalysts.

Downsizing the noble metal-based clusters or NPs into atomically isolated atoms is a feasible approach for lowering the loading amount and maximizing the atomic utilization of noble metal in OER electrocatalysts. A typical instance by Yang and co-workers recently reported the synthesis and electrocatalytic OER applications of NiV layered double hydroxide (LDH) supported noble Ir and Ru SMACs (i.e., NiVIr-LDH and NiVRu-LDH, respectively) via a one-pot hydrothermal strategy.[123] Based on comparative studies, the authors confirmed that the doping of noble Ru or Ir with optimal doping concentrations could synergistically modulate the atomic and electronic structure of NiV-LDH, therefore boosting the OER activity in alkaline condition. As exhibited in Fig. 9.7a, the electrocatalytic OER performance of both NiVRu-LDH and NiVIr-LDH was markedly enhanced in comparison with that of the pristine NiV-LDH. Notably, the NiVIr-LDH (Ir content: 0.62 at%) delivered the highest electrocatalytic OER performance, only requiring low overpotentials of 180 mV and 272 mV to attain current densities of 10 mA cm^{-2} and 100 mA cm^{-2}, respectively (Fig. 9.7b). Note that the isomorphic substitution of V by Ir or Ru could cause possible lattice deformation and V vacancies, leading to a seriously distorted octahedral V environment. As the Ir doping improves both the M-OH and M-O processes, while Ru doping is conducive to the M-OOH process, the OER activity of both NiVIr-LDH and NiVRu-LDH was therefore significantly improved compared

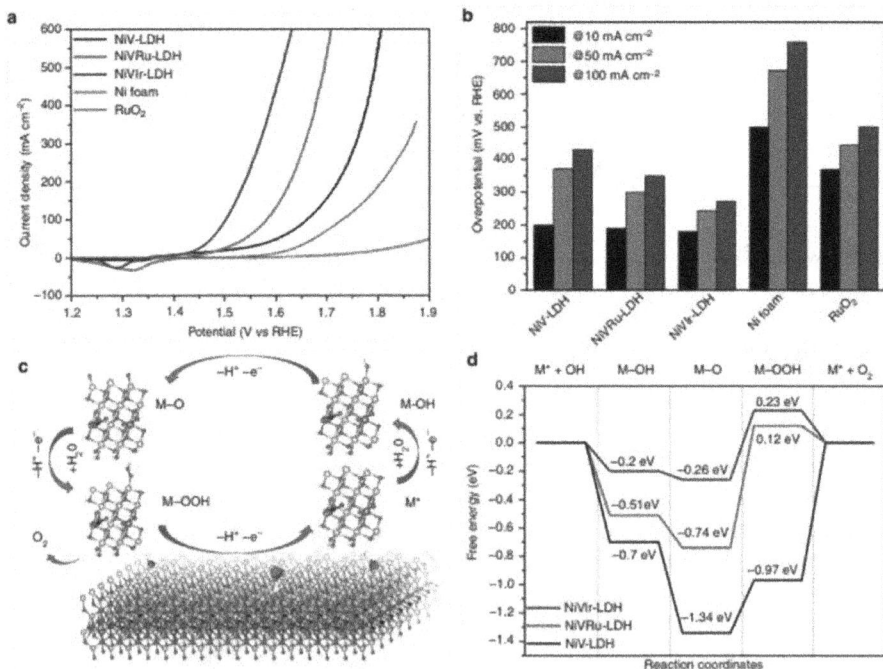

Figure 9.7. (a) OER polarization curves, (b) the overpotentials at current densities of 10, 50, and 100 mA cm^{-2} of NiV-LDH, NiVRu-LDH, NiVIr-LDH, Ni foam, and RuO$_2$. (c) The atomic model of NiVIr-LDH and the proposed OER pathway. (d) The free energy diagram of OER on the NiV-LDH, NiVRu-LDH, and NiVIr-LDH electrocatalysts. Reproduced with permission from Ref. 123. Copyright 2019, Nature Publishing Group.

with that of the pristine NiV-LDH. Theoretical calculations further identified that both Ir and Ru doping could markedly improve the overall OER kinetics (Fig. 9.7c,d). In addition to the high-efficiency Ir and Ru SMACs, single-atom Pt and Au sites have also been proposed and explored to remarkably improve the OER performance of common transition-metal-based OER electrocatalysts.[124,125]

In addition, besides noble metal-based SMACs, non-noble metal-based SMACs with transition metal single-atom sites (e.g., Fe, Co, Ni, etc.) anchored on carbon-based supporting materials have also been verified to be highly effective toward electrocatalytic OER. Generally, the defect engineering and heteroatom doping of the supporting materials have been widely accepted as practical strategies to regulate the coordination environment and adjust the adsorption properties of the active single-metal sites, thus improving their electrocatalytic OER performance. As recently reported by Zhang and coworkers, the divacancies anchored square-planar single-atom Ni sites on defective graphene with a low density of state near the Fermi level is beneficial for the adsorption of various reaction intermediates, therefore leading to a high electrocatalytic OER performance.[70] Feng et al. reported the atomically anchored single-atom Ni sites coordinated with both S and N heteroatoms in porous carbon (PC) nanosheets ($S|NiN_x$-PC) as electrocatalysts toward OER.[126] Benefitting from the abundant atomically dispersed $S|NiN_x$ active sites, the as-achieved $S|NiN_x$-PC electrocatalysts promoted the OER performance by delivering low overpotential (1.51 V to attain $10\,mA\,cm^{-2}$), small Tafel slope ($45\,mV\,dec^{-1}$), and excellent durability in alkaline condition, which represent the best among all reported carbon-based electrocatalysts and is even superior to benchmark Ir/C catalyst.

Besides, according to the DFT-based simulations, the electrocatalytic OER process is mainly affected by the electronic properties of the d orbital, and the intermediate species were believed to be preferentially adsorbed on both the single-metal sites and carbon sites, resulting in a dual-site mechanisms.[127] Very recently, Hu and coworkers reported the synthesis of uniformly dispersed binary Co-Ni sites incorporated in N-doped hollow carbon matrix (i.e., CoNi-SAs/NC) through the pyrolysis treatment of dopamine-coated MOFs (Fig. 9.8a–c).[128] As a result of the abundant atomically anchored single-metal sites with high intrinsic activity and the synergistic effects from dual-metal Co-Ni sites dispersed in porous N-doped hollow carbon matrix, the energetic barrier of electrocatalytic OER is significantly reduced, substantially speeding up the OER kinetics and contributing to the improved electrocatalytic OER activity. Accordingly, the CoNi-SAs/NC required a lower overpotential of 340 mV to achieve an anodic OER current density of 10 mA cm^{-2} (Fig. 9.8d) compared with those of the control sample (CoNi-NPs/NC, 440 mV) and IrO$_2$ (400 mV), demonstrating its superior performance for the OER process in alkaline condition. Except for the carbon-based supporting materials, some conventional OER electrocatalysts such as metal oxides (e.g., LiCoO$_2$) and hydroxides (e.g., Ni(OH)$_2$) have also been employed as the host materials to anchor the non-noble single-metal-atom sites and improve the performance.[129,130] The OER performance of some representative advanced SMACs are selectively summarized in Table 9.2.

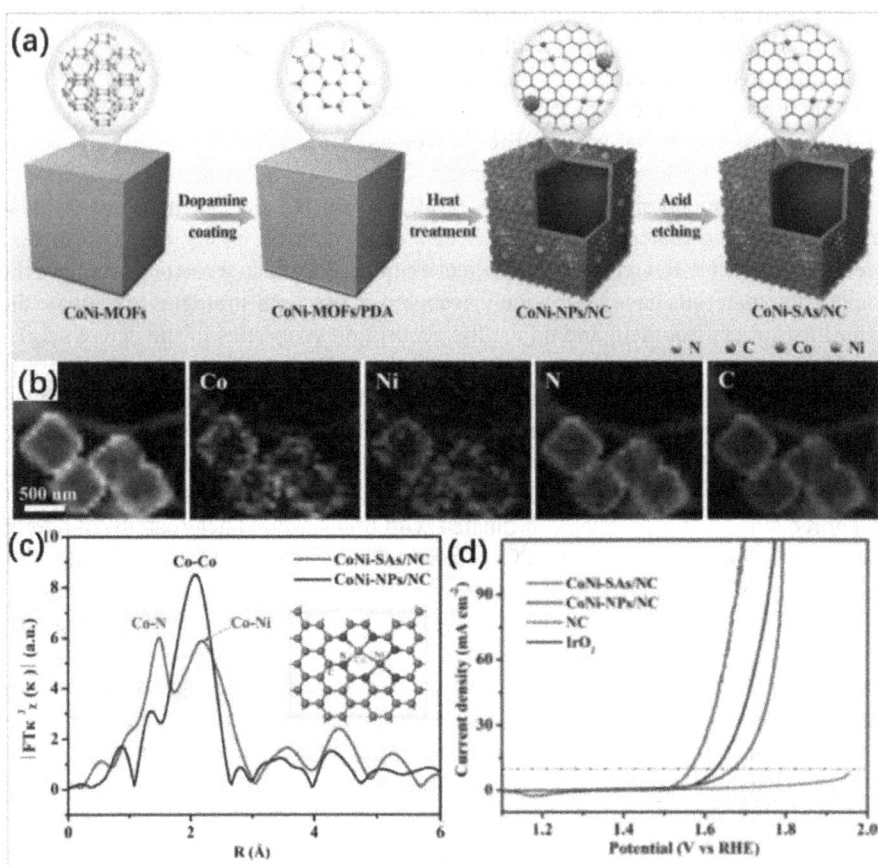

Figure 9.8. (a) Schematic illustration of the synthesis of CoNi-SAs/NC. (b) Elemental mappings of CoNi-SAs/NC. (c) FT-EXAFS spectra of CoNi-SAs/NC and CoNi-NPs/NC (control sample). Inset in (c) shows the proposed model of Ni-Co dual sites. (d) OER polarization curves of CoNi-SAs/NC, control samples (CoNi-NPs/NC and NC), and IrO_2. Reproduced with permission from Ref. 128. Copyright 2019, Wiley-VCH.

9.4.3 Hydrogen evolution reaction

The electrochemical water splitting by the HER process is an efficient strategy for the production of hydrogen gas (H_2) for promising clean and sustainable energy technologies.[132,133] Generally, the electrocatalysts play a vital role during the HER process and mainly determine the efficiency. In principle, the electrocatalytic HER process with only one intermediate (H*) mainly consists of three possible reaction steps containing the adsorption of molecular hydrogen (Volmer reaction) and subsequently the desorption of H* (Heyrovsky or Tafel reactions). Besides, the electrocatalytic HER process generally proceeds via either a Volmer-Heyrovsky mechanism or a Volmer-Tafel mechanism.[134,135] For a specific catalyst, the intrinsic electronic properties of the catalytic active sites that significantly affect the adsorption free energy of H* (ΔG_H) play a dominant role in the HER performance.[136,137] Presently, the most effective electrocatalysts toward HER are based on noble

Table 9.2. OER performance of some representative advanced SMACs. Note: a and b refer to the overpotential at current density of 10 mA cm^{-2} and turnover frequency at specific overpotential, respectively.

Catalyst	Electrolyte	Overpotential^a (mV)	Tafel slope (mV dec^{-1})	TOF^b (s^{-1})	References
NiVIr-LDH	1 M KOH	180	38	--	123
NiVRu-LDH	1 M KOH	190	83	--	123
CoSe$_{2-x}$-Pt	1 M KOH	255	31	--	124
^{s}Au/NiFe LDH	1 M KOH	237	36	0.11 at 280 mV	125
A-Ni@DG	1 M KOH	270	47	13.4 at 300 mV	70
S|NiN$_x$-PC/EG	1 M KOH	280	45	10.9 at 300 mV	126
CoNi-SAs/NC	1 M KOH	340	58.7	--	128
La-LiCoO$_2$	1 M KOH	330	48	--	129
W-Ni(OH)$_2$	1 M KOH	237	33	0.74 at 250 mV	130
Mn-NG	1 M KOH	337	55	--	131

Pt-based nanostructures with low overpotential, small Tafel slope and high exchange current density.[138,139]

In view of the high-cost and natural scarcity of noble Pt metal as aforementioned, downsizing the size of Pt NPs to atomically dispersed Pt sites is a highly desirable approach to minimize the cost and maximize the utilization efficiency. Very recently, Song and co-workers reported a highly curved onion-like carbon nanospheres (OLC) supported single-atom Pt sites (Pt$_1$/OLC) as electrocatalysts toward HER through an ALD approach (Fig. 9.9a).[140] During the synthesis, the OLC was firstly subjected to a thermal treatment at different conditions to accurately adjust the configuration and distribution density of oxygen species. Then, a single cycle ALD was employed to generate single-metal-atom Pt sites on the supporting material of OLC with a multishell fullerene structure (Fig. 9.9b,c). The subsequent electrochemical measurements verified the excellent HER performance of Pt$_1$/OLC with Pt loading of 0.27 wt% in acidic media in terms of the low overpotential of 38 mV at 10 mA cm^{-2} and high turnover frequency (40.78 s^{-1} at 100 mV), which is comparable to the commercial Pt/C (Pt loading of 20 wt%) catalyst and superior to the commercial Pt/C (Pt loading of 5 wt%) catalyst and the control sample (i.e., graphene-supported single-atom Pt sites with a similar Pt loading (Pt$_1$/graphene)) (Fig. 9.9d). Based on the DFT-based first-principle calculations, the significant facilitation effect of a tip-enhanced local electric field at the single-atom Pt site toward the reaction kinetics of HER was proposed (Fig. 9.9e). As shown in Fig. 9.9f, a remarkably high local proton concentration (> 1.99 mol l^{-1}) near the single-atom Pt site induced by the tip-enhancement effect may play a crucial role in optimizing the performance for electrocatalytic HER. Besides noble Pt metal, the SMACs involving other Pt group metals, such as Ru, Rh, Pd, and Ir, have also been explored for efficient electrocatalytic HER.[141,142]

To further reduce the cost, the SMACs based on non-noble transition metals such as Fe, Co, and Ni have been proposed as electrocatalysts toward HER. These SMACs generally involve non-noble transition metal single-atom sites anchored on

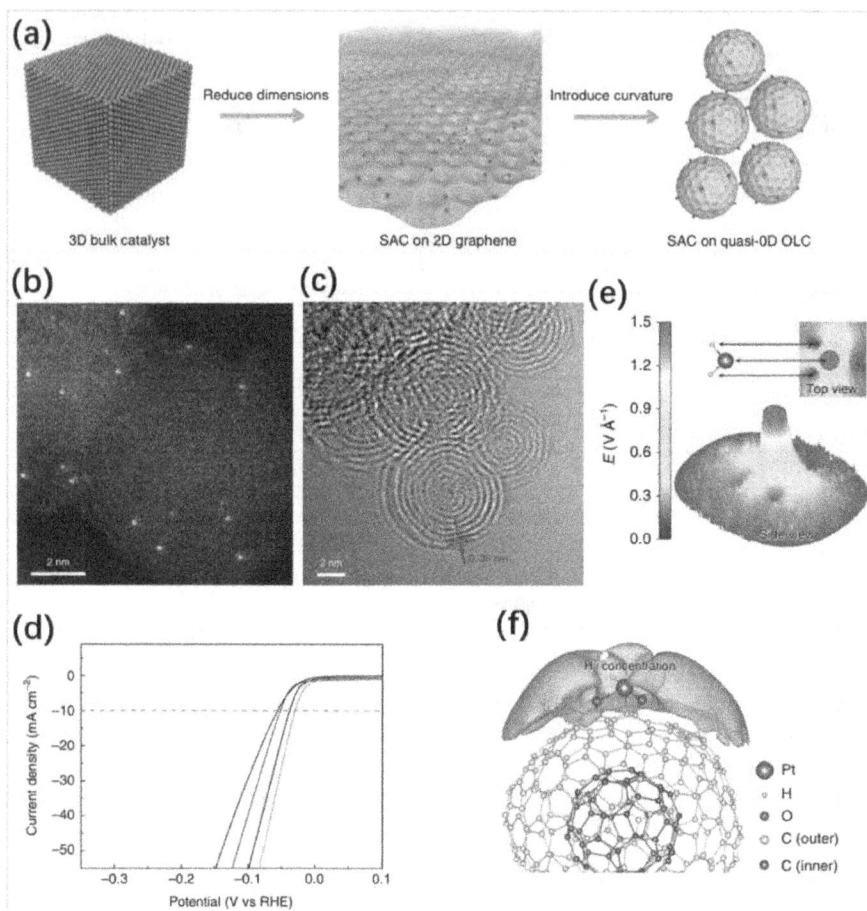

Figure 9.9. (a) Schematic illustration of the synthesis of Pt$_1$/OLC. (b) HAADF-STEM image of Pt$_1$/OLC. (c) TEM image of Pt$_1$/OLC. (d) Polarization curves of Pt$_1$/OLC (black), commercial Pt/C catalysts (5 wt% (second from left) and 20 wt% (far right), respectively) and Pt$_1$/graphene (far left) in a 0.5 M H$_2$SO$_4$ electrolyte. (e) The map of the strong local electric field around the single-atom Pt site at an equilibrium potential form. (f) The remarkable enrichment of protons (> 1.99 mol l^{-1}) around the Pt site induced by the local electric field at the single-atom Pt site. Reproduced with permission from Ref. 140. Copyright 2019, Nature Publishing Group.

supporting materials like N-doped carbon and metal chalcogenides. However, it is still challenging to ensure the stability of these SMACs as the non-noble transition metals are usually unstable during the electrocatalytic HER process, tending to be oxidized, dissolved, or leached depending upon the conditions. Recently, studies based on theoretical calculations have revealed that non-noble metal single-atom sites could exhibit metallic properties or reduced bandgap, thus boosting electron transfer during electrocatalytic reactions.[143] In addition, the single-atom M-N-C sites produces by N-heteroatom doping have also been confirmed to be capable of properly adjusting the electronic properties of the single-metal atoms and boosting HER performance as a result from the N-induced charge redistribution.[144] Noting

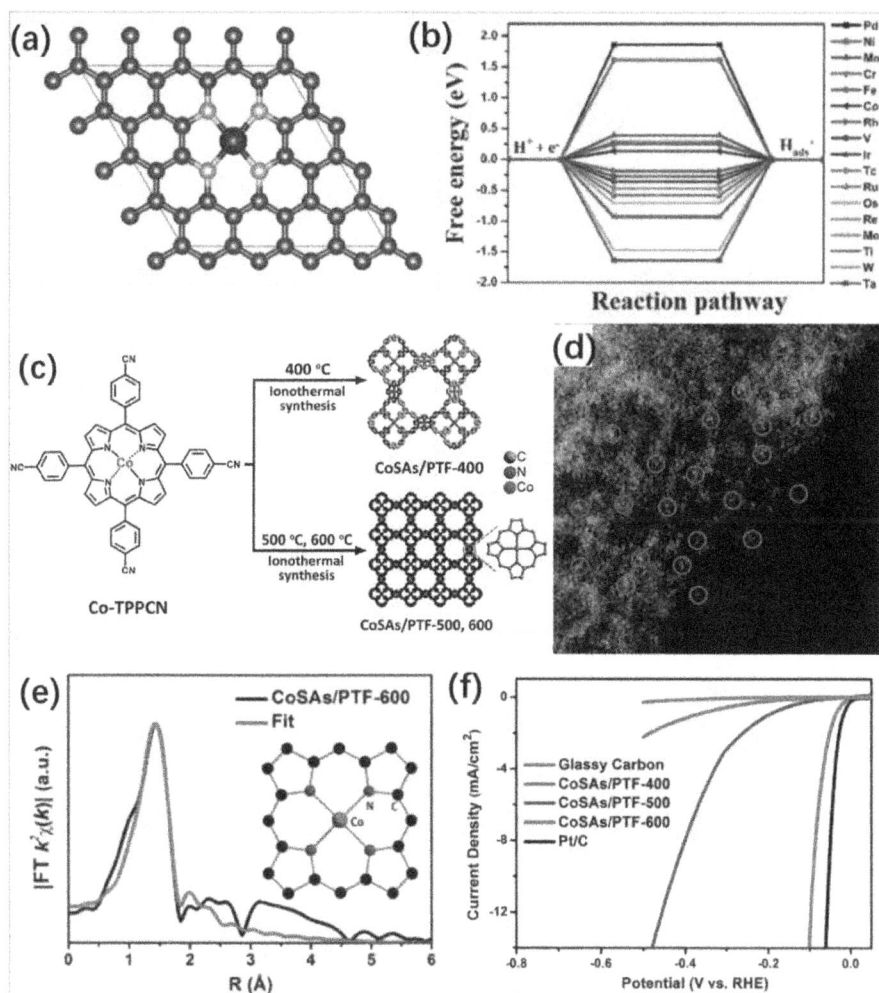

Figure 9.10. (a) Metal active site coordinated with four nitrogen atoms in graphene sheet. (b) ΔG_H^* diagram for hydrogen adsorption reaction (Volmer reaction) toward a series of transition metals used as single-atom catalysts. Reproduced with permission from Ref. 145. Copyright 2019, Wiley-VCH. (c) Schematic illustration of the fabrication of CoSAs/PTFs. (d) HAADF-STEM image of CoSAs/PTF-600. (e) FT-EXAFS fitting curves of CoSAs/PTF-600, inset in (e) shows the proposed model of Co-N$_4$ sites. (f) HER polarization curves of CoSAs/PTFs, Pt/C and glassy carbon in 0.5 M H$_2$SO$_4$ electrolyte. Reproduced with permission from Ref. 146. Copyright 2019, Royal Society of Chemistry.

that, among these non-noble metal SMACs, the Co-based SMACs have been theoretically predicted to be most efficient toward HER (Fig. 9.10a,b) due to its unique electronic structure that the anti-bonding state orbital of single-atom Co sites is neither entirely empty nor completely filled.[145] Recently, Cao and co-workers reported the preparation of Co SMACs toward electrocatalytic HER based on porous porphyrinic triazine-based frameworks (CoSAs/PTFs) by an ionothermal synthesis approach (Fig. 9.10c).[146] The XANES and HAADF-STEM characterizations clearly confirmed the highly dispersed Co atoms with the configuration of Co-N$_4$

(Fig. 9.10d,e). The optimal CoSAs/PTFs-600 obtained at ionothermal temperature of 600°C displayed the highest HER performance with the lowest overpotential of 94 mV to attain current density of 10 mA cm^{-2} in acid condition compared with the other two control samples (CoSAs/PTFs-400 and CoSAs/PTFs-500 obtained at ionothermal temperatures of 400°C and 500°C, respectively) (Fig. 9.10f). The superior HER activity was credited to the production of abundant atomic dispersion of coordinatively unsaturated Co-N$_4$ sites anchored on the hierarchically micro- and

Table 9.3. HER performance of some representative advanced SMACs. Note: a and b refer to the overpotential at current density of 10 mA cm^{-2} and turnover frequency at specific overpotential, respectively.

Catalyst	Electrolyte	Overpotential[a] (mV)	Tafel slope (mV dec^{-1})	TOF[b] (s^{-1})	References
Pt$_1$/OLC	0.5 M H$_2$SO$_4$	38	36	40.78 at 100 mV	140
Ru-NC-700	0.1 M KOH	47	14	--	141
IrP$_2$@NC	0.5 M H$_2$SO$_4$	8	28	--	142
IrP$_2$@NC	1 M KOH	28	50	--	142
CoSAs/PTFs	0.5 M H$_2$SO$_4$	94	50	--	146
Mo$_1$N$_1$C$_2$	0.1 M KOH	132	90	0.148 at 50 mV 0.465 at 100 mV 1.460 at 150 mV	147
W-SAC	0.1 M KOH	85	53	1.16 at 40 mV 3.40 at 80 mV 6.35 at 120 mV	148
Cu@MoS$_2$	0.5 M H$_2$SO$_4$	131	51	--	149
Fe/GD	0.5 M H$_2$SO$_4$	66	37.8	1.59 at 100 mV	150
Ni/GD	0.5 M H$_2$SO$_4$	88	45.8	4.15 at 100 mV	150

mesoporous carbon substrate. Except for Co metal, other non-noble metals including Mo, W, Cu, Fe, and Ni have also been reported for highly efficient HER.[147–150] The HER performance of some representative advanced SMACs are selectively summarized in Table 9.3.

9.4.4 CO₂ reduction reaction

As one of the typical greenhouse gases, the CO$_2$ released by both natural and artificial processes has caused a series of environmental and sustainable development issues.[151,152] Particularly, since the Industrial Revolution, the anthropological consumption of fossil fuels has resulted in an explosion in CO$_2$ emissions, seriously disrupting the global carbon cycle.[153] Electrochemical reduction of CO$_2$ to value-added fuels and chemical molecules is one of the most promising strategies to alleviate the ever-growing energy crisis and the warm-house effect and close the carbon cycle. Nevertheless, significant overpotentials are generally required to surmount the activation barriers of electrocatalytic CO$_2$RR as a result of its sluggish kinetics. In addition, a wide range of products (e.g., CO, CH$_3$OH, CH$_4$, CH$_3$CH$_2$OH, etc.)

can be produced as a result from the multiple possible reaction pathways during the electrocatalytic CO_2RR.[154] Notably, a competitive HER process is usually involved since the electrocatalytic CO_2RR commonly covers the potential region for HER.[155] Consequently, the progress of highly active, selective and stable electrocatalysts is of vital importance to efficiently catalyze the CO_2RR process toward a particular product of interest for practical applications. Over the past few decades, various metal-based materials have been explored as electrocatalysts toward CO_2RR such as metals, metal oxides, carbon-based materials, and molecular compounds by optimizing their size, morphology, structure and composition.[156,157]

Recently, SMACs as promising catalysts with distinctive catalytic performance have attracted growing interest in the research field of electrocatalytic CO_2RR. Compared with conventional electrocatalysts, SMACs with unique single-metal-atom sites are advantageous for CO_2 reduction in terms of high utilization of the metal sites and strong interactions with carbonaceous reaction intermediates, therefore restricting the configurations of adsorbates and leading to enhanced electrocatalytic activity and selectivity. Among various SMACs, Ni-based SMACs are the most widely studied SMACs toward CO_2RR in aqueous solutions with the CO as the dominant product. The carbon-based supporting materials are commonly employed to stabilize the single-atom metal sites, including graphene, carbon nanotubes, and porous carbon nanostructures.[158–160] For example, Wang and co-workers recently reported 2D graphene nanosheets supported single-atom Ni sites as highly active and selective catalyst for the electrocatalytic CO_2RR to CO (Fig. 9.11a,b).[158] On account of the abundant single-atom Ni sites either anchored in the graphene vacancies or coordinated with neighboring N atoms, the resulting Ni SMACs (Ni-NG) delivered

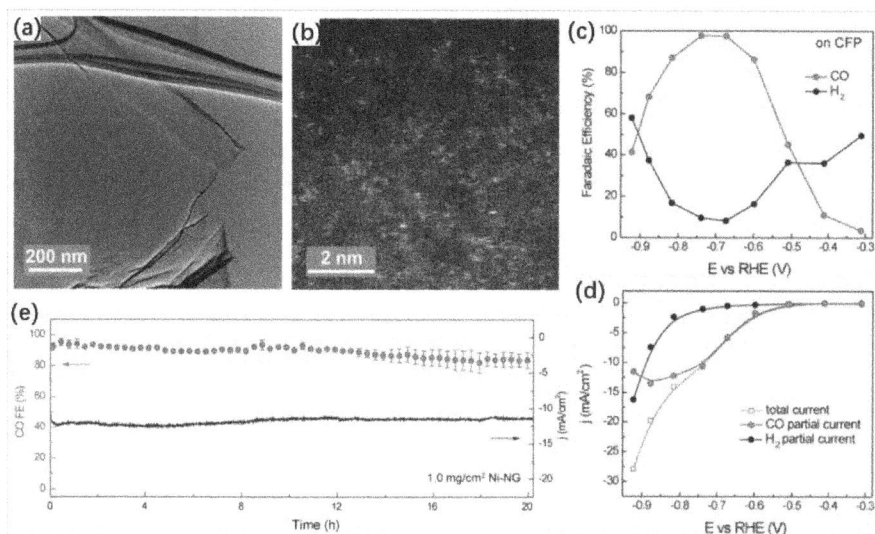

Figure 9.11. (a) TEM image of Ni-NG nanosheets. (b) HAADF-STEM image of Ni-NG nanosheets. (c) Faradaic efficiencies (FEs) of H_2 and CO and (d) the corresponding steady-state current densities of Ni-NG on CFP in CO_2-saturated 0.5 M $KHCO_3$. (e) The CO_2RR stability test of the current density and CO FE of Ni-NG on CFP under an overpotential of 0.64 V for more than 20 h continuous operation. Reproduced with permission from Ref. 158. Copyright 2018, Royal Society of Chemistry.

a high selectivity of 90% within a wide potential window in aqueous solution, a maximal CO Faradaic efficiency (FE) of > 95% under an overpotential of 550 mV and a superior long-term durability (over 20 h continuous electrolysis) (Fig. 9.11c–e). DFT-based computational calculations revealed that the weak CO binding and high HER barrier account for the outstanding CO_2-to-CO selectivity of the single-atom Ni sites. Noting that the high activity and selectivity of Ni-based SMACs have provided possibilities for scalable applications, therefore it is crucial to prepare high-efficiency Ni SMACs with low cost on a large scale. Recently, Zhang and co-workers proposed a ligand-mediated approach to synthesize a series of SMACs including Ni SMACs supported on commercial carbon black in large quantities (> 1 kg scale) with high metal loadings. Among these SMACs, the obtained Ni SMACs displayed outstanding performance for electrocatalytic CO_2RR to CO with a high FE of 98.9% at −1.2 V.[161]

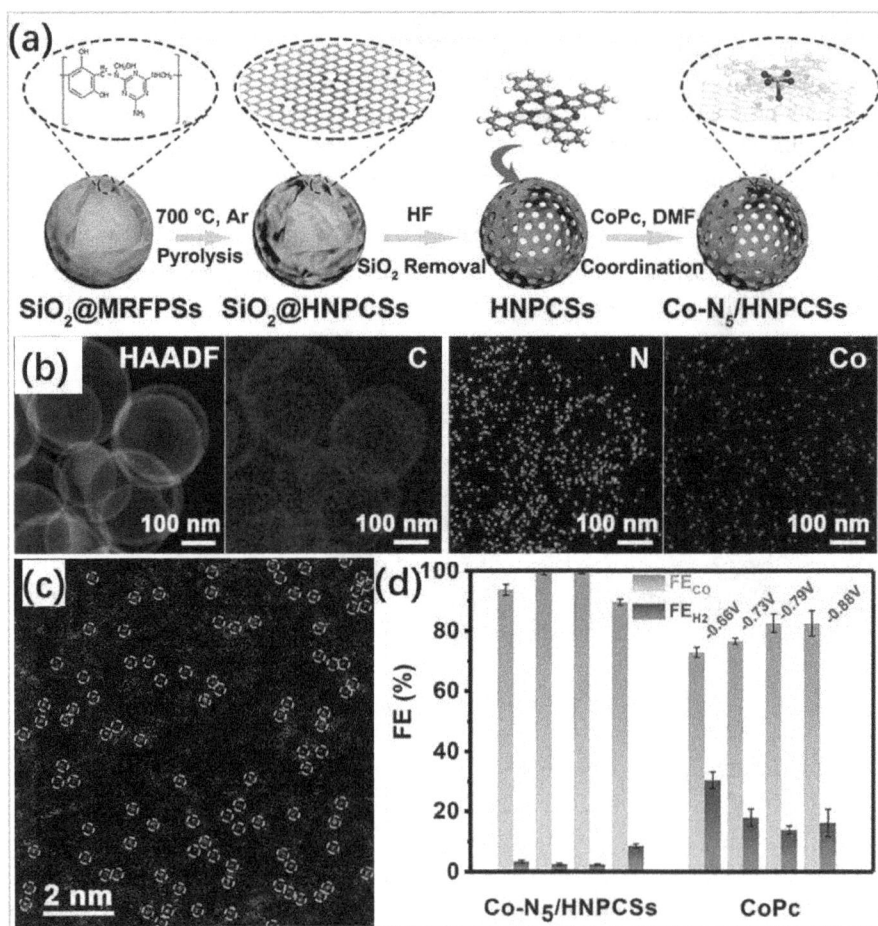

Figure 9.12. (a) Schematic illustration of Co-N$_5$/HNPCSs. (b) HAADF-STEM image and elemental mappings of Co-N$_5$/HNPCSs. (c) HAADF-STEM image of Co-N$_5$/HNPCSs. (d) FE$_{CO}$ and FE$_{H2}$ of M-N$_5$/HNPCSs and CoPc. Reproduced with permission from Ref. 162. Copyright 2018, American Chemical Society.

In addition to the Ni-based SMACs, other metals like Co, Fe, Cu, etc., have also been synthesized and employed for efficiently catalyzing CO_2RR.[162–166] For instance, the Co SMACs with abundant Co-N$_5$ sites stabilized on polymer-derived hollow N-doped porous carbon spheres (HNPCSs) was reported by Chen and co-workers via the pyrolysis treatment of the polymer precursors (Fig. 9.12a–c).[162] The obtained Co-N$_5$/HNPCSs with large surface area, abundant N sites and high electrical conductivity exhibited high CO_2-to-CO selectivity with FE above 90% over a wide potential range (–0.57 to –0.88 V) and even exceeded 99% at –0.73 and –0.79 V (Fig. 9.12d). Notably, the Co-N$_5$/HNPCSs also exhibited significant stability as the CO current density and FE remained almost unchanged after 10 h continuous operation. Based on both experiments and computational DFT calculations, the single-atom Co-N$_5$ sites were verified as the dominating active centers for electrocatalytic CO_2RR to CO. Besides, Fe SMACs with the configuration of Fe-N$_4$ supported by N-doped graphene (Fe/NG) was also prepared and reported for efficient electrocatalytic CO_2 reduction to CO. The nitrogen doping on graphene was proposed as a possible facilitator for the electrocatalytic CO_2RR to CO based on DFT simulations.[163] In the case of Cu SMACs, previous theoretical calculations have confirmed that the single-atom Cu sites can effectively facilitate the electrocatalytic CO_2RR process.[164] Feng and co-workers recently experimentally prepared and reported a high-efficiency Cu SMACs composed of coordinatively unsaturated single-atom Cu sites coordinated with neighboring N sites (Cu-N$_2$) supported by graphene matrix (Cu-N$_2$/GN) for

Table 9.4. CO_2RR performance of some representative advanced SMACs.

Catalyst	Electrolyte	FE (%)	Product	Potential (V)	References
Ni-NG	0.5 M KHCO$_3$	95	CO	~ –0.7	158
NC-CNTs (Ni)	0.1 M KHCO$_3$	90	CO	–0.8	159
SE-Ni SAs@PNC	0.5 M KHCO$_3$	> 90	CO	~ –0.7	160
Ni-SAC	0.1 M KHCO$_3$	98.9	CO	–1.2	161
Co-N$_5$/HNPCSs	0.2 M NaHCO$_3$	99	CO	–0.73 and –0.79	162
Fe/NG	0.1 M KHCO$_3$	80	CO	–0.6	163
Cu-N$_2$/GN	0.1 M KHCO$_3$	81	CO	–0.5	165

CO_2RR.[165] Owing to the unsaturated coordination environment of the single-atom Cu sites and the unique structure of ultrathin nanosheets, the Cu-N$_2$/GN exhibited a high CO_2RR activity and selectivity to produce CO with an onset potential of –0.33 V and the maximum FE of 81% at a low potential of -0.50 V. Table 9.4 summarizes some representative advanced SMACs for catalyzing the CO_2 reduction.

9.5 Summary and perspective

Single-atom sites stabilizing endows SMACs with unique physicochemical characteristics, making the SMACs promising candidates for electrocatalysts used in electrocatalytic clean energy conversion technologies, such as water electrolysis, fuel cells, metal-air batteries, and carbon dioxide conversion. In this chapter, an overview of SMACs covering the unique features and mechanistic origins of electrocatalytic activity is indicated firstly. The advanced synthetic strategies and characterization techniques for SMACs are then performed. With particular attention to the structure-property correlations and intrinsic origins of electrocatalytic activity, a focused review of the most recent advances of SMACs and their potential applications for reactions involved in various reactions, including ORR, OER, HER, and CO_2RR, are briefly presented through enumerating typical examples as well. Downsizing metal NPs to single-metal-atom sites with the unsaturated coordination and the quantized orbital of active sites has been widely acknowledged to be highly effective in promoting the intrinsic activity of the active sites and maximizing the metal atom-utilization. The reasonable selection of metal centers and precise control of coordination environments could effectively modify the electronic properties and construct high-efficiency electrocatalytic active sites, accordingly optimizing the adsorption and desorption properties for reactants, key intermediates, and products, boosting those electrocatalytic processes such as ORR, OER, HER, and CO_2RR.

Although tremendous advance has been made in the development of SMACs and understanding of the electrocatalytic mechanisms theoretically and experimentally for ORR, OER, HER, and CO_2RR, there are still some challenges in this field that need to be addressed. (1) Currently, SMACs with relatively low metal loadings are predominant in most of the reported studies, which largely restrict their practical applications as a result of the low density of exposed active sites. Therefore, the development of new general synthesis techniques to obtain SMACs with high metal loadings is highly desired to further improve the electrocatalytic activity to achieve the large-scale practical applications. (2) The design and development of SMACs can reliably facilitate the reaction kinetics and enhance the efficiency of various electrocatalytic processes. However, the deep insight into the structural-performance relationships and detailed electrocatalytic mechanisms of single-atom sites for boosting the catalytic reactions should be better understood and further figured out. Meanwhile, the key parameters that significantly affect the catalytic performance of SMACs toward specific electrocatalytic process should be further identified. Particular attention should also be paid to the possible synergistic enhancement and the specific mechanism. (3) The selectivity of SMACs on some reactions like ORR and CO_2RR for specific products is still worth further exploration. The high selectivity of SMACs is thought to be caused by the homogeneous structure of active single-atom sites; therefore, more efforts should be put into precisely regulating the coordination environment of metal centers. (4) The stability of SMACs still remains challenging due to the possible aggregation and leaching of atomically dispersed metal sites during the electrocatalytic reactions. The reasonable selection of supporting materials and further improving the interactions between single-atom sites and supporting materials are among the most promising strategies to enhance the long-term stability of SMACs. In summary, more theoretical and experimental studies are still required for the rational engineering of SMACs for the large-scale practical applications of electrocatalytic clean energy conversion technologies in the future.

Acknowledgements

We would like to acknowledge the financial support from the National Natural Science Foundation of China (Grant 52003300) and Fundamental Research Funds for the Central Universities, China University of Geosciences (Wuhan).

References

[1]   Gao, L., X. Cui, C. D. Sewell, J. Li and Z. Lin. 2021. Recent advances in activating surface reconstruction for the high-efficiency oxygen evolution reaction. *Chem. Soc. Rev.*, 50: 8428–8469.

[2]   Cui, X., P. Xiao, J. Wang, M. Zhou, W. Guo, Y. Yang, Y. He, Z. Wang, Y. Yang, Y. Zhang and Z. Lin. 2017. Highly branched metal alloy networks with superior activities for the methanol oxidation reaction. *Angew. Chem., Int. Ed.*, 56: 4488–4493.

[3]   Xue, W., Q. Zhou, X. Cui, S. Jia, J. Zhang and Z. Lin. 2021. Metal-organic frameworks-derived heteroatom-doped carbon electrocatalysts for oxygen reduction reaction. *Nano Energy*, 86: 106073.

[4]   Yan, Y., S. Liang, X. Wang, M. Zhang, S.-M. Hao, X. Cui, Z. Li and Z. Lin. 2021. Robust wrinkled MoS2/NC bifunctional electrocatalysts interfaced with single Fe atoms for wearable zinc-air batteries. *Proc. Natl. Acad. Sci. U. S. A.*, 118: e2110036118.

[5]   Sewell, C. D., Z. Wang, Y.-W. Harn, S. Liang, L. Gao, X. Cui and Z. Lin. 2021. Tailoring oxygen evolution reaction activity of metal-oxide spinel nanoparticles via judiciously regulating surface-capping polymers. *J. Mater. Chem. A*, 9: 20375–20384.

[6]   Cui, X., Y. Yang, Y. Li, F. Liu, H. Peng, Y. Zhang and P. Xiao. 2015. Electrochemical fabrication of porous $Ni_{0.5}Co_{0.5}$ alloy film and its enhanced electrocatalytic activity towards methanol oxidation. *J. Electrochem. Soc.*, 162: F1415–F1424.

[7]   Harn, Y.-W., S. Liang, S. Liu, Y. Yan, Z. Wang, J. Jiang, J. Zhang, Q. Li, Y. He, Z. Li, L. Zhu, H.-P. Cheng and Z. Lin. 2021. Tailoring electrocatalytic activity of in situ crafted perovskite oxide nanocrystals via size and dopant control. *Proc. Natl. Acad. Sci. U. S. A.*, 118: e2014086118.

[8]   Gao, L., X. Cui, Z. Wang, C. D. Sewell, Z. Li, S. Liang, M. Zhang, J. Li, Y. Hu and Z. Lin. 2021. Operando unraveling photothermal-promoted dynamic active-sites generation in $NiFe_2O_4$ for markedly enhanced oxygen evolution. *Proc. Natl. Acad. Sci. U. S. A.*, 118: e2023421118.

[9]   Wang, H., R. Liu, Y. Li, X. Lü, Q. Wang, S. Zhao, K. Yuan, Z. Cui, X. Li, S. Xin, R. Zhang, M. Lei and Z. Lin. 2018. Durable and efficient hollow porous oxide spinel microspheres for oxygen reduction. *Joule*, 2: 337–348.

[10]   Wang, T., C. Yang, Y. Liu, S. M. Yang, X. Li, M. Yang, Y. He, H. Li, H. Chen and Z. Lin. 2020. Dual-shelled multidoped hollow carbon nanocages with hierarchical porosity for high-performance oxygen reduction reaction in both alkaline and acidic media. *Nano Lett.*, 20: 5639–5645.

[11]   Cui, X., L. Gao, S. Lei, S. Liang, J. Zhang, C. D. Sewell, W. Xue, Q. Liu, Z. Lin and Y. Yang. 2021. Simultaneously crafting single-atomic Fe sites and graphitic layer-wrapped Fe_3C nanoparticles encapsulated within mesoporous carbon tubes for oxygen reduction. *Adv. Funct. Mater.*, 31: 2009197.

[12]   Wang, T., Y. He, Y. Liu, F. Guo, X. Li, H. Chen, H. Li and Z. Lin. 2020. A ZIF-triggered rapid polymerization of dopamine renders Co/N-codoped cage-in-cage porous carbon for highly efficient oxygen reduction and evolution. *Nano Energy*, 79: 105487.

[13]   Yang, Y., X. Cui, D. Gao, H. He, Y. Ou, M. Zhou, Q. Lai, X. Wei, P. Xiao and Y. Zhang. 2020. Trimetallic CoFeCr hydroxide electrocatalysts synthesized at a low temperature for accelerating water oxidation via tuning the electronic structure of active sites. *Sustain. Energy Fuels*, 4: 3647–3653.

[14]   Gao, D., J. Guo, X. Cui, L. Yang, Y. Yang, H. He, P. Xiao and Y. Zhang. 2017. Three-dimensional dendritic structures of NiCoMo as efficient electrocatalysts for the hydrogen evolution reaction. *Appl. Mater. Interfaces*, 9: 22420–22431.

[15]   Cui, X., W. Guo, M. Zhou, Y. Yang, Y. Li, P. Xiao, Y. Zhang and X. Zhang. 2015. Promoting effect of Co in Ni_mCo_n (m + n = 4) bimetallic electrocatalysts for methanol oxidation reaction. *ACS Appl. Mater. Interfaces*, 7: 493–503.

[16] Yang, Y., M. Zhou, W. Guo, X. Cui, Y. Li, F. Liu, P. Xiao and Y. Zhang. 2015. NiCoO$_2$ nanowires grown on carbon fiber paper for highly efficient water oxidation. *Electrochim. Acta*, 174: 246–253.

[17] Ma, R., X. Cui, Y. Wang, Z. Xiao, R. Luo, L. Gao, Z. Wei and Y. Yang. 2022. Pyrolysis-free synthesized single-atom cobalt catalysts for efficient oxygen reduction. *J. Mater. Chem. A*, DOI: 10.1039/D1TA08412G.

[18] Cui, X., L. Gao, R. Ma, Z. Wei, C.-H. Lu, Z. Li and Y. Yang. 2021. Pyrolysis-free covalent organic frameworks-based materials for efficient oxygen electrocatalysis. *J. Mater. Chem. A*, 9: 20985–21004.

[19] Liu, F., P. Xiao, W. Q. Tian, M. Zhou, Y. Li, X. Cui, Y. Zhang and X. Zhou. 2015. Hydrogenation of Pt/TiO$_2${101} nanobelts: A driving force for the improvement of methanol catalysis. *Phys. Chem. Chem. Phys.*, 17: 28626–28634.

[20] Zhao, X., Y. Yang, Y. Li, X. Cui, Y. Zhang and P. Xiao. 2016. NiCo-selenide as a novel catalyst for water oxidation. *J. Mater. Sci.*, 51: 3724–3734.

[21] Cui, X., S. Lei, A. C. Wang, L. Gao, Q. Zhang, Yi. Yang and Z. Lin. 2020. Emerging covalent organic frameworks tailored materials for electrocatalysis. *Nano Energy*, 70: 104525.

[22] Zhao, C.-X., B.-Q. Li, J.-N. Liu and Q. Zhang. 2021. Intrinsic electrocatalytic activity regulation of M-N-C single-atom catalysts for the oxygen reduction reaction. *Angew. Chem. Int. Ed.*, 60: 4448–4463.

[23] Liu, L. and A. Corma. 2018. Metal catalysts for heterogeneous catalysis: From single atoms to nanoclusters and nanoparticles. *Chem. Rev.*, 118: 4981–5079.

[24] Qiao, B., A. Wang, X. Yang, L. F. Allard, Z. Jiang, Y. Cui, J. Liu, J. Li and T. Zhang. 2011. Single-atom catalysis of CO oxidation using Pt1/FeOx. *Nat. Chem.*, 3: 634–641.

[25] Yang, X.-F., A. Wang, B. Qiao, J. Li, J. Liu and T. Zhang. 2013. Single-atom catalysts: a new frontier in heterogeneous catalysis. *Acc. Chem. Res.*, 46: 1740–1748.

[26] Wang, Y., H. Su, Y. He, L. Li, S. Zhu, H. Shen, P. Xie, X. Fu, G. Zhou, C. Feng, D. Zhao, F. Xiao, X. Zhu, Y. Zeng, M. Shao, S. Chen, G. Wu, J. Zeng and C. Wang. 2020. Advanced electrocatalysts with single-metal-atom active sites. *Chem. Rev.*, 120: 12217–12314.

[27] Li, X., X. Yang, Y. Huang, T. Zhang and B. Liu. 2019. Supported noble-metal single atoms for heterogeneous catalysis. *Adv. Mater.*, 31: 1902031.

[28] Shi, Y., C. Zhao, H. Wei, Ji. Guo, S. Liang, A. Wang, T. Zhang, J. Liu and T. Ma. 2014. Single-atom catalysis in mesoporous photovoltaics: The principle of utility maximization. *Adv. Mater.*, 26: 8147–8153.

[29] Ding, K., A. Gulec, A. M. Johnson, N. M. Schwitzer, G. D. Stucky, L. D. Marks and P. C. Stair. 2015. Identification of active sites in CO oxidation and water-gas shift over supported Pt catalysts. *Science*, 350: 189–192.

[30] Bayatsarmadi, B., Y. Zheng, A. Vasileff and S.-Z. Qiao. 2017. Recent advances in atomic metal doping of carbon-based nanomaterials for energy conversion. *Small*, 13: 1700191.

[31] Wang, A., J. Li and T. Zhang. 2018. Heterogeneous single-atom catalysis. *Nat. Rev. Chem.*, 2: 65–81.

[32] Gao, G., Y. Jiao, E. R. Waclawik and A. Du. 2016. Single atom (Pd/Pt) supported on graphitic carbon nitride as an efficient photocatalyst for visible-light reduction of carbon dioxide. *J. Am. Chem. Soc.*, 138: 6292–6297.

[33] Liu, J. 2017. Catalysis by supported single metal atoms. *ACS Catal.*, 7: 34–59.

[34] Wang, Y., J. Mao, X. Meng, L. Yu, D. Deng and X. Bao. 2018. Catalysis with two-dimensional materials confining single atoms: concept, design, and applications. *Chem. Rev.*, 119: 1806–1854.

[35] Zhu, C., S. Fu, Q. Shi, D. Du and Y. Lin. 2017. Single-atom electrocatalysts. *Angew. Chem. Int. Ed.*, 56: 13944–13960.

[36] Yang, M., L. F. Allard and M. Flytzani-Stephanopoulos. 2013. Atomically dispersed Au-(OH)$_x$ species bound on titania catalyze the low-temperature water-gas shift reaction. *J. Am. Chem. Soc.*, 135: 3768–3771.

[37] Flytzani-Stephanopoulos, M. 2014. Gold atoms stabilized on various supports catalyze the water-gas shift reaction. *Acc. Chem. Res.*, 47: 783–792.

[38] Zhang, Y., L. Guo, L. Tao, Y. Lu and S. Wang. 2019. Defect-based single-atom electrocatalysts. *Small Methods*, 3: 1800406.

[39] Zhang, B.-W., Y.-X. Wang, S.-L. Chou, H.-K. Liu and S.-X. Dou. 2019. Fabrication of superior single-atom catalysts toward diverse electrochemical reactions. *Small Methods*, 3: 1800497.

[40] Sun, T., S. Mitchell, J. Li, P. Lyu, X. Wu, J. Pérez-Ramírez and J. Lu. 2021. Design of local atomic environments in single-atom electrocatalysts for renewable energy conversions. *Adv. Mater.*, 33: 2003075.

[41] Zhang, L., K. D. Davis and X. Sun. 2019. Pt-based electrocatalysts with high atom utilization efficiency: From nanostructures to single atoms. *Energy Environ. Sci.*, 12: 492–517.

[42] Zhu, C., Q. Shi, S. Feng, D. Du and Y. Lin. 2018. Single-atom catalysts for electrochemical water splitting. *ACS Energy Lett.*, 3: 1713–1721.

[43] Su, X., X.-F. Yang, Y. Huang, B. Liu and T. Zhang. 2019. Single-atom catalysis toward efficient CO_2 conversion to CO and formate products. *Acc. Chem. Res.*, 52: 656–664.

[44] Mitchell, S., E. Vorobyeva and J. Pérez-Ramírez. 2018. The multifaceted reactivity of single-atom heterogeneous catalysts. *Angew. Chem., Int. Ed.*, 57: 15316–15329.

[45] Fei, H., J. Dong, D. Chen, T. Hu, X. Duan, I. Shakir, Y. Huang, X. Duan. 2019. Single atom electrocatalysts supported on graphene or graphene-like carbons. *Chem. Soc. Rev.*, 48: 5207–5241.

[46] Liang, Z., C. Qu, D. Xia, R. Zou and Q. Xu. 2018. Atomically dispersed metal sites in MOF-based materials for electrocatalytic and photocatalytic energy conversion. *Angew. Chem. Int. Ed.*, 57: 9604–9633.

[47] Chen, Y., S. Ji, C. Chen, Q. Peng, D. Wang and Y. Li. 2018. Single-atom catalysts: Synthetic strategies and electrochemical applications. *Joule*, 2: 1242–1264.

[48] Li, X., X. Yang, J. Zhang, Y. Huang and B. Liu. 2019. *In situ*/operando techniques for characterization of single-atom catalysts. *ACS Catal.*, 9: 2521–2531.

[49] Tang, Y., S. Zhao, B. Long, J.-C. Liu and J. Li. 2016. On the nature of support effects of metal dioxides MO_2 (M = Ti, Zr, Hf, Ce, Th) in single-atom gold catalysts: importance of quantum primogenic effect. *J. Phys. Chem. C*, 120: 17514–17526.

[50] Yang, S., Y. J. Tak, J. Kim, A. Soon and H. Lee. 2017. Support effects in single-atom platinum catalysts for electrochemical oxygen reduction. *ACS Catal.*, 7: 1301–1307.

[51] Liu, M., L. Wang, K. Zhao, S. Shi, Q. Shao, L. Zhang, X. Sun, Y. Zhao and J. Zhang. 2019. Atomically dispersed metal catalysts for the oxygen reduction reaction: Synthesis, characterization, reaction mechanisms and electrochemical energy applications. *Energy Environ. Sci.*, 12: 2890–2923.

[52] Li, Z., D. Wang, Y. Wu and Y. Li. 2018. Recent advances in the precise control of isolated single-site catalysts by chemical methods. *Natl. Sci. Rev.*, 5: 673–689.

[53] Zhang, Q. and J. Guan. 2020. Single-atom catalysts for electrocatalytic applications. *Adv. Funct. Mater.*, 30: 2000768.

[54] Ma, L., G. Zhu, D. Wang, H. Chen, Y. Lv, Y. Zhang, X. He and H. Pang. 2020. Emerging metal single atoms in electrocatalysts and batteries. *Adv. Funct. Mater.*, 30: 2003870.

[55] Liu, P., Y. Zhao, R. Qin, S. Mo, G. Chen, L. Gu, D. M. Chevrier, P. Zhang, Q. Guo, D. Zang, B. Wu, G. Fu, N. Zheng. 2016. Photochemical route for synthesizing atomically dispersed palladium catalysts. *Science*, 352: 797–800.

[56] Hu, K., L. Tao, D. Liu, J. Huo and S. Wang. 2016. Sulfur-doped Fe/N/C nanosheets as highly efficient electrocatalysts for oxygen reduction reaction. *ACS Appl. Mater. Interfaces*, 8: 19379–19385.

[57] Wang, X. X., D. A. Cullen, Y.-T. Pan, S. Hwang, M. Wang, Z. Feng, J. Wang, M. H. Engelhard, H. Zhang, Y. He, Y. Shao, D. Su, K. L. More, J. S. Spendelow and G. Wu. 2018. Nitrogen-coordinated single cobalt atom catalysts for oxygen reduction in proton exchange membrane fuel cells. *Adv. Mater.*, 30: 1706758.

[58] Kwak, D.-H., S.-B. Han, Y.-W. Lee, H.-S. Park, I.-A. Choi, K.-B. Ma, M.-C. Kim, S.-J. Kim, D.-H. Kim, J.-I. Sohn and K.-W. Park. 2017. Fe/N/S-doped mesoporous carbon nanostructures as electrocatalysts for oxygen reduction reaction in acid medium. *Appl. Catal. B: Environ.*, 203: 889–898.

[59] Peng, Y., B. Lu and S. Chen. 2018. Carbon-supported single atom catalysts for electrochemical energy conversion and storage. *Adv. Mater.*, 30: 1801995.

[60] Yin, P., T. Yao, Y. Wu, L. Zheng, Y. Lin, W. Liu, H. Ju, J. Zhu, X. Hong, Z. Deng, G. Zhou, S. Wei and Y. Li. 2016. Single cobalt atoms with precise N-coordination as superior oxygen reduction reaction catalysts. *Angew. Chem., Int. Ed.*, 55: 10800–10805.

[61]   Zhang, Z., X. Gao, M. Dou, J. Ji and F. Wang. 2017. Biomass derived N-doped porous carbon supported single Fe atoms as superior electrocatalysts for oxygen reduction. *Small*, 13: 1604290.

[62]   Cui, L., L. Cui, Z. Li, J. Zhang, H. Wang, S. Lu and Y. Xiang. 2019. A copper single-atom catalyst towards efficient and durable oxygen reduction for fuel cells. *J. Mater. Chem. A*, 7: 16690–16695.

[63]   Zhang, C., J. Sha, H. Fei, M. Liu, S. Yazdi, J. Zhang, Q. Zhong, X. Zou, N. Zhao, H. Yu, Z. Jiang, E. Ringe, B. I. Yakobson, J. Dong, D. Chen and J. M. Tour. 2017. Single-atomic ruthenium catalytic sites on nitrogen-doped graphene for oxygen reduction reaction in acidic medium. *ACS Nano*, 11: 6930–6941.

[64]   Chen, Y., S. Ji, Y. Wang, J. Dong, W. Chen, Z. Li, R. Shen, L. Zheng, Z. Zhuang, D. Wang and Y. Li. 2017. Isolated single iron atoms anchored on N-doped porous carbon as an efficient electrocatalyst for the oxygen reduction reaction. *Angew. Chem. Int. Ed.*, 56: 6937–6941.

[65]   Zheng, T., K. Jiang, N. Ta, Y. Hu, J. Zeng, J. Liu and H. Wang. 2019. Large-scale and highly selective CO_2 electrocatalytic reduction on nickel single-atom catalyst. *Joule*, 3: 265–278.

[66]   Flytzani-Stephanopoulos1, M. and B. C. Gates. 2012. Atomically dispersed supported metal catalysts. *Annu. Rev. Chem. Biomol. Eng.*, 3: 545–574.

[67]   Beniya, A. and S. Higashi. 2019. Towards dense single-atom catalysts for future automotive applications. *Nat. Catal.*, 2: 590–602.

[68]   Cui, X., W. Li, P. Ryabchuk, K. Junge and M. Beller. 2018. Bridging homogeneous and heterogeneous catalysis by heterogeneous single-metal-site catalysts. *Nat. Catal.*, 1: 385–397.

[69]   Huang, F., Y. Deng, Y. Chen, X. Cai, M. Peng, Z. Jia, J. Xie, D. Xiao, X. Wen, N. Wang, Z. Jiang, H. Liu and D. Ma. 2019. Anchoring Cu_1 species over nanodiamond-graphene for semi-hydrogenation of acetylene. *Nat. Commun.*, 10: 4431.

[70]   Zhang, L., Y. Jia, G. Gao, X. Yan, N. Chen, J. Chen, M. T. Soo, B. Wood, D. Yang, A. Du and X. Yao. 2018. Graphene defects trap atomic Ni species for hydrogen and oxygen evolution reactions. *Chem.*, 4: 285–297.

[71]   Zhang, X., J. Guo, P. Guan, C. Liu, H. Huang, F. Xue, X. Dong, S. J. Pennycook and M. F. Chisholm. 2013. Catalytically active single-atom niobium in graphitic layers. *Nat. Commun.*, 4: 1924.

[72]   Huang, F., Y. Deng, Y. Chen, X. Cai, M. Peng, Z. Jia, P. Ren, D. Xiao, X. Wen, N. Wang, H. Liu and D. Ma. 2018. Atomically dispersed Pd on nanodiamond/graphene hybrid for selective hydrogenation of acetylene. *J. Am. Chem. Soc.*, 140: 13142–13146.

[73]   Cheng, N., S. Stambula, D. Wang, M. N. Banis, J. Liu, A. Riese, B. Xiao, R. Li, T.-K. Sham, L.-M. Liu, G. A. Botton and X. Sun. 2016. Platinum single-atom and cluster catalysis of the hydrogen evolution reaction. *Nat. Commun.*, 7: 13638.

[74]   Huang, X., Y. Xia, Y. Cao, X. Zheng, H. Pan, J. Zhu, C. Ma, H. Wang, J. Li, R. You, S. Wei, W. Huang and J. Lu. 2017. Enhancing both selectivity and coking-resistance of a single-atom Pd_1/C_3N_4 catalyst for acetylene hydrogenation. *Nano Res.*, 10: 1302–1312.

[75]   Yan, H., H. Cheng, H. Yi, Y. Lin, T. Yao, C. Wang, J. Li, S. Wei and J. Lu. 2015. Single-atom Pd_1/graphene catalyst achieved by atomic layer deposition: remarkable performance in selective hydrogenation of 1,3-butadiene. *J. Am. Chem. Soc.*, 137: 10484–10487.

[76]   Stambula, S., N. Gauquelin, M. Bugnet, S. Gorantla, S. Turner, S. Sun, J. Liu, G. Zhang, X. Sun and G. A. Botton. 2014. Chemical structure of nitrogen-doped graphene with single platinum atoms and atomic clusters as a platform for the PEMFC electrode. *J. Phys. Chem. C*, 118: 3890–3900.

[77]   Sun, S., G. Zhang, N. Gauquelin, N. Chen, J. Zhou, S. Yang, W. Chen, X. Meng, D. Geng, M. N. Banis, R. Li, S. Ye, S. Knights, G. A. Botton, T.-K. Sham and X. Sun. 2013. Single-atom catalysis using Pt/graphene achieved through atomic layer deposition. *Sci. Rep.*, 3: 1775.

[78]   O'Neill, B. J., D. H. K. Jackson, J. Lee, C. Canlas, P. C. Stair, C. L. Marshall, J. W. Elam, T. F. Kuech, J. A. Dumesic and G. W. Huber. 2015. Catalyst design with atomic layer deposition. *ACS Catal.*, 5: 1804–1825.

[79]   Piernavieja-Hermida, M., Z. Lu, A. White, K.-B. Low, T. Wu, J. W. Elam, Z. Wu, and Y. Lei. 2016. Towards ALD thin film stabilized single-atom Pd_1 catalysts. *Nanoscale*, 8: 15348–15356.

[80]   Grillo, F., H. V. Bui, D. L. Zara, A. A. I. Aarnink, A. Y. Kovalgin, P. Kooyman, M. T. Kreutzer, and J. R. van Ommen. 2018. From single atoms to nanoparticles: Autocatalysis and metal aggregation in atomic layer deposition of Pt on TiO_2 nanopowder. *Small*, 14: 1800765.

[81] Zhao, J., Q. Deng, A. Bachmatiuk, G. Sandeep, A. Popov, J. Eckert and M. H. Rummeli. 2014. Free-standing single-atom-thick iron membranes suspended in graphene pores. *Science*, 343: 1228–1232.

[82] Ta, H. Q., L. Zhao, W. Yin, D. Pohl, B. Rellinghaus, T. Gemming, B. Trzebicka, J. Palisaitis, G. Jing, P. O. A. Persson, Z. Liu, A. Bachmatiuk and M. H. Rummeli. 2018. Single Cr atom catalytic growth of graphene. *Nano Res.*, 11: 2405–2411.

[83] Tavakkoli, M., N. Holmberg, R. Kronberg, H. Jiang, J. Sainio, E. I. Kauppinen, T. Kallio and K. Laasonen. 2017. Electrochemical activation of single-walled carbon nanotubes with pseudo-atomic-scale platinum for the hydrogen evolution reaction. *ACS Catal.*, 7: 3121–3130.

[84] Zhang, L., L. Han, H. Liu, X. Liu and J. Luo. 2017. Potential-cycling synthesis of single platinum atoms for efficient hydrogen evolution in neutral media. *Angew. Chem., Int. Ed.*, 56: 13694–13698.

[85] Zhang, H., K. Kawashima, M. Okumura and N. Toshima. 2014. Colloidal Au single-atom catalysts embedded on Pd nanoclusters. *J. Mater. Chem. A*, 2: 13498–13508.

[86] Shao, X., X. Yang, J. Xu, S. Liu, S. Miao, X. Liu, X. Su, H. Duan, Y. Huang and T. Zhang. 2019. Iridium single-atom catalyst performing a quasi-homogeneous hydrogenation transformation of CO_2 to formate. *Chem*, 5: 693–705.

[87] Wei, H., H. Wu, K. Huang, B. Ge, J. Ma, J. Lang, D. Zu, M. Lei, Y. Yao, W. Guo and H. Wu. 2019. Ultralow-temperature photochemical synthesis of atomically dispersed Pt catalysts for the hydrogen evolution reaction. *Chem. Sci.*, 10: 2830–2836.

[88] Tavakkoli, M., E. Flahaut, P. Peljo, J. Sainio, F. Davodi, E. V. Lobiak, K. Mustonen and E. I. Kauppinen. 2020. Mesoporous single-atom-doped graphene-carbon nanotube hybrid: Synthesis and tunable electrocatalytic activity for oxygen evolution and reduction reactions. *ACS Catal.*, 10(8): 4647–4658.

[89] Zhang, Z., C. Feng, C. Liu, M. Zuo, L. Qin, X. Yan, Y. Xing, H. Li, R. Si, S. Zhou and J. Zeng. 2020. Electrochemical deposition as a universal route for fabricating single-atom catalysts. *Nat. Commun.*, 11: 1215.

[90] Li, H., H. Zhu, Z. Zhuang, S. Lu, F. Duan and M. Du. 2020. Single-atom catalysts for electrochemical clean energy conversion: recent progress and perspectives. *Sustain. Energy Fuels*, 4: 996–1011.

[91] Kou, Z., W. Zang, P. Wang, X. Li and J. Wang. 2020. Single atom catalysts: A surface heterocompound perspective. *Nanoscale Horiz.*, 5: 757–764.

[92] Chang, T.-Y., Y. Tanaka, R. Ishikawa, K. Toyoura, K. Matsunaga, Y. Ikuhara and N. Shibata. 2014. Direct imaging of Pt single atoms adsorbed on TiO_2 (110) surfaces. *Nano Lett.*, 14: 134–138.

[93] Deng, D., X. Chen, L. Yu, X. Wu, Q. Liu, Y. Liu, H. Yang, H. Tian, Y. Hu, P. Du, R. Si, J. Wang, X. Cui, H. Li, J. Xiao, T. Xu, J. Deng, F. Yang, P. N. Duchesne, P. Zhang, J. Zhou, L. Sun, J. Li, X. Pan and X. Bao. 2015. A single iron site confined in a graphene matrix for the catalytic oxidation of benzene at room temperature. *Sci. Adv.*, 1: e1500462.

[94] Ogino, I. 2017. X-ray absorption spectroscopy for single-atom catalysts: critical importance and persistent challenges. *Chin. J. Catal.*, 38: 1481–1488.

[95] Koningsberger, D. C. and B. C. Gates. 1992. Nature of the metal-support interface in supported metal catalysts: Results from X-ray absorption spectroscopy. *Catal. Lett.*, 14: 271–277.

[96] Zang, W., T. Yang, H. Zou, S. Xi, H. Zhang, X. Liu, Z. Kou, Y. Du, Y. P. Feng, L. Shen, L. Duan, J. Wang and S. J. Pennycook. 2019. Copper single atoms anchored in porous nitrogen-doped carbon as efficient pH-universal catalysts for the nitrogen reduction reaction. *ACS Catal.*, 9: 10166–10173.

[97] Ma, Y., T. Yang, H. Zou, W. Zang, Z. Kou, L. Mao, Y. Feng, L. Shen, S. J. Pennycook, L. Duan, X. Li and J. Wang. 2020. Synergizing Mo single atoms and Mo_2C nanoparticles on CNTs synchronizes selectivity and activity of electrocatalytic N_2 reduction to ammonia. *Adv. Mater.*, 32: 2002177.

[98] Kou, Z., W. Zang, W. Pei, L. Zheng, S. Zhou, S. Zhang, L. Zhang and J. Wang. 2020. A sacrificial Zn strategy enables anchoring of metal single atoms on the exposed surface of holey 2D molybdenum carbide nanosheets for efficient electrocatalysis. *J. Mater. Chem. A*, 8: 3071–3082.

[99] Iwase, K., S. Nakanishi, M. Miyayama and K. Kamiya. 2020. Rational molecular design of electrocatalysts based on single-atom modified covalent organic frameworks for efficient oxygen reduction reaction. *ACS Appl. Energy Mater.*, 3: 1644–1652.

[100] Kattel, S., P. Atanassov and B. Kiefer. 2012. Density functional theory study of Ni-N$_x$/C electrocatalyst for oxygen reduction in alkaline and acidic media. *J. Phys. Chem. C*, 116: 17378–17383.

[101] Iwase, K., K. Kamiya, M. Miyayama, K. Hashimoto and S. Nakanishi. 2018. Sulfur-linked covalent triazine frameworks doped with coordinatively unsaturated Cu(I) as electrocatalysts for oxygen reduction. *Chem. Electro. Chem.*, 5: 805–810.

[102] Liu, M., Z. Zhao, X. Duan and Y. Huang. 2019. Nanoscale structure design for high-performance Pt-based ORR catalysts. *Adv. Mater.*, 31: 1802234.

[103] Pegis, M. L., C. F. Wise, D. J. Martin and J. M. Mayer. 2018. Oxygen reduction by homogeneous molecular catalysts and electrocatalysts. *Chem. Rev.*, 118: 2340–2391.

[104] Zhang, H. and P. K. Shen. 2012. Recent development of polymer electrolyte membranes for fuel cells. *Chem. Rev.*, 112: 2780–2832.

[105] Choi, C. H., M. Kim, H. C. Kwon, S. J. Cho, S. Yun, H.-T. Kim, K. J. J. Mayrhofer, H. Kim and M. Choi. 2016. Tuning selectivity of electrochemical reactions by atomically dispersed platinum catalyst. *Nat. Commun.*, 7: 10922.

[106] Liu, J., M. Jiao, B. Mei, Y. Tong, Y. Li, M. Ruan, P. Song, G. Sun, L. Jiang, Y. Wang, Z. Jiang, L. Gu, Z. Zhou and W. Xu. 2019. Carbon-supported divacancy-anchored platinum single-atom electrocatalysts with superhigh Pt utilization for the oxygen reduction reaction. *Angew. Chem., Int. Ed.*, 58: 1163–1167.

[107] Liu, J., M. Jiao, L. Lu, H. M. Barkholtz, Y. Li, Y. Wang, L. Jiang, Z. Wu, D.-J. Liu, L. Zhuang, C. Ma, J. Zeng, B. Zhang, D. Su, P. Song, W. Xing, W. Xu, Y. Wang, Z. Jiang and G. Sun. 2017. High performance platinum single atom electrocatalyst for oxygen reduction reaction. *Nat. Commun.*, 8: 15938.

[108] Li, T., J. Liu, Y. Song and F. Wang. 2018. Photochemical solid-phase synthesis of platinum single atoms on nitrogen-doped carbon with high loading as bifunctional catalysts for hydrogen evolution and oxygen reduction reactions. *ACS Catal.*, 8: 8450–8458.

[109] Miao, Z., X. Wang, M. -C. Tsai, Q. Jin, J. Liang, F. Ma, T. Wang, S. Zheng, B.-J. Hwang, Y. Huang, S. Guo and Q. Li. 2018. Atomically dispersed Fe-N$_x$/C electrocatalyst boosts oxygen catalysis via a new metal-organic polymer supramolecule strategy. *Adv. Energy Mater.*, 8: 1801226.

[110] Liu, S., Z. Wang, S. Zhou, F. Yu, M. Yu, C.-Y. Chiang, W. Zhou, J. Zhao and J. Qiu. 2017. Metal-organic-framework-derived hybrid carbon nanocages as a bifunctional electrocatalyst for oxygen reduction and evolution. *Adv. Mater.*, 29: 1700874.

[111] Yang, Z., Y. Wang, M. Zhu, Z. Li, W. Chen, W. Wei, T. Yuan, Y. Qu, Q. Xu, C. Zhao, X. Wang, P. Li, Y. Li, Y. Wu and Y. Li. 2019. Boosting oxygen reduction catalysis with Fe-N$_4$ sites decorated porous carbons toward fuel cells. *ACS Catal.*, 9: 2158–2163.

[112] Zhang, H., H. T. Chung, D. A. Cullen, S. Wagner, U. I. Kramm, K. L. More, P. Zelenay and G. Wu. 2019. High-performance fuel cell cathodes exclusively containing atomically dispersed iron active sites. *Energy Environ. Sci.*, 12: 2548–2558.

[113] Pan, Y., S. Liu, K. Sun, X. Chen, B. Wang, K. Wu, X. Cao, W.-C. Cheong, R. Shen, A. Han, Z. Chen, L. Zheng, J. Luo, Y. Lin, Y. Liu, D. Wang, Q. Peng, Q. Zhang, C. Chen and Y. Li. 2018. A bimetallic Zn/Fe polyphthalocyanine-derived single-atom Fe-N$_4$ catalytic site: A superior trifunctional catalyst for overall water splitting and Zn-air batteries. *Angew. Chem., Int. Ed.*, 57: 8614–8618.

[114] Ma, L., S. Chen, Z. Pei, Y. Huang, G. Liang, F. Mo, Q. Yang, J. Su, Y. Gao, J. A. Zapien and C. Zhi. 2018. Single-site active iron-based bifunctional oxygen catalyst for a compressible and rechargeable zinc-air battery. *ACS Nano*, 12: 1949–1958.

[115] He, Y., S. Hwang, D. A. Cullen, M. A. Uddin, L. Langhorst, B. Li, S. Karakalos, A. J. Kropf, E. C. Wegener, J. Sokolowski, M. Chen, D. Myers, D. Su, K. L. More, G. Wang, S. Litster and G. Wu. 2019. Highly active atomically dispersed CoN$_4$ fuel cell cathode catalysts derived from surfactant-assisted MOFs: Carbon-shell confinement strategy. *Energy Environ. Sci.*, 12: 250–260.

[116] Bai, L., Z. Duan, X. Wen, R. Si and J. Guan. 2019. Atomically dispersed manganese-based catalysts for efficient catalysis of oxygen reduction reaction. *Appl. Catal. B: Environ.*, 257: 117930.

[117] Yang, Y., K. Mao, S. Gao, H. Huang, G. Xia, Z. Lin, P. Jiang, C. Wang, H. Wang and Q. Chen. 2018. O-, N-atoms-coordinated Mn cofactors within a graphene framework as bioinspired oxygen reduction reaction electrocatalysts. *Adv. Mater.*, 30: 1801732.

[118] Li, F., G.-F. Han, H.-J. Noh, S.-J. Kim, Y. Lu, H. Y. Jeong, Z. Fu and J.-B. Baek. 2018. Boosting oxygen reduction catalysis with abundant copper single atom active sites. *Energy Environ. Sci.*, 11: 2263–2269.

[119] Li, J., S. Chen, N. Yang, M. Deng, S. Ibraheem, J. Deng, J. Li, L. Li and Z. Wei. 2019. Ultrahigh-loading zinc single-atom catalyst for highly efficient oxygen reduction in both acidic and alkaline media, *Angew. Chem. Int. Ed.*, 58: 7035–7039.

[120] Song, F., L. Bai, A. Moysiadou, S. Lee, C. Hu, L. Liardet and X. Hu. 2018. Transition metal oxides as electrocatalysts for the oxygen evolution reaction in alkaline solutions: An application-inspired renaissance. *J. Am. Chem. Soc.*, 140: 7748–7759.

[121] Vij, V., S. Sultan, A. M. Harzandi, A. Meena, J. N. Tiwari, W.-G. Lee, T. Yoon and K. S. Kim. 2017. Nickel-based electrocatalysts for energy-related applications: oxygen reduction, oxygen evolution, and hydrogen evolution reactions. *ACS Catal.*, 7: 7196–7225.

[122] Tan, P., B. Chen, H. Xu, H. Zhang, W. Cai, M. Ni, M. Liu and Z. Shao. 2017. Flexible Zn- and Li-air batteries: recent advances, challenges, and future perspectives. *Energy Environ. Sci.*, 10: 2056–2080.

[123] Wang, D., Q. Li, C. Han, Q. Lu, Z. Xing and X. Yang. 2019. Atomic and electronic modulation of self-supported nickel-vanadium layered double hydroxide to accelerate water splitting kinetics. *Nat. Commun.*, 10: 3899.

[124] Zhuang, L., Y. Jia, H. Liu, X. Wang, R. K. Hocking, H. Liu, J. Chen, L. Ge, L. Zhang, M. Li, C.-L. Dong, Y.-C. Huang, S. Shen, D. Yang, Z. Zhu and X. Yao. 2019. Defect-induced Pt-Co-Se coordinated sites with highly asymmetrical electronic distribution for boosting oxygen-involving electrocatalysis. *Adv. Mater.*, 31: 1805581.

[125] Zhang, J., J. Liu, L. Xi, Y. Yu, N. Chen, S. Sun, W. Wang, K. M. Lange and B. Zhang. 2018. Single-atom Au/NiFe layered double hydroxide electrocatalyst: Probing the origin of activity for oxygen evolution reaction. *J. Am. Chem. Soc.*, 140: 3876–3879.

[126] Hou, Y., M. Qiu, M. G. Kim, P. Liu, G. Nam, T. Zhang, X. Zhuang, B. Yang, J. Cho, M. Chen, C. Yuan, L. Lei and X. Feng. 2019. Atomically dispersed nickel-nitrogen-sulfur species anchored on porous carbon nanosheets for efficient water oxidation. *Nat. Commun.*, 10: 1392.

[127] Fei, H., J. Dong, Y. Feng, C. S. Allen, C. Wan, B. Volosskiy, M. Li, Z. Zhao, Y. Wang, H. Sun, P. An, W. Chen, Z. Guo, C. Lee, D. Chen, I. Shakir, M. Liu, T. Hu, Y. Li, A. I. Kirkland, X. Duan, Y. Huang. 2018. General synthesis and definitive structural identification of MN_4C_4 single-atom catalysts with tunable electrocatalytic activities. *Nat. Catal.*, 1: 63–72.

[128] Han, X., X. Ling, D. Yu, D. Xie, L. Li, S. Peng, C. Zhong, N. Zhao, Y. Deng and W. Hu. 2019. Atomically dispersed binary Co-Ni sites in nitrogen-doped hollow carbon nanocubes for reversible oxygen reduction and evolution. *Adv. Mater.*, 31: 1905622.

[129] Zhang, Z., C. Liu, C. Feng, P. Gao, Y. Liu, F. Ren, Y. Zhu, C. Cao, W. Yan, R. Si, S. Zhou and J. Zeng. 2019. Breaking the local symmetry of $LiCoO_2$ via atomic doping for efficient oxygen evolution. *Nano Lett.*, 19: 8774–8779.

[130] Yan, J., L. Kong, Y. Ji, J. White, Y. Li, J. Zhang, P. An, S. Liu, S.-T. Lee and T. Ma. 2019. Single atom tungsten doped ultrathin α-$Ni(OH)_2$ for enhanced electrocatalytic water oxidation. *Nat. Commun.*, 10: 2149.

[131] Guan, J., Z. Duan, F. Zhang, S. D. Kelly, R. Si, M. Dupuis, Q. Huang, J. Q. Chen, C. Tang and C. Li. 2018. Water oxidation on a mononuclear manganese heterogeneous catalyst. *Nat. Catal.*, 1: 870–877.

[132] Zheng, Y., Y. Jiao, M. Jaroniec and S. Z. Qiao. 2015. Advancing the electrochemistry of the hydrogen-evolution reaction through combining experiment and theory. *Angew. Chem. Int. Ed.*, 54: 52–65.

[133] Jiang, W., X. Zou, H. Du, L. Gan, C. Xu, F. Kang, W. Duan and J. Li. 2018. Universal descriptor for large-scale screening of high-performance MXene-based materials for energy storage and conversion. *Chem. Mater.*, 30: 2687–2693.

[134] Zou, X. and Y. Zhang. 2015. Noble metal-free hydrogen evolution catalysts for water splitting. *Chem. Soc. Rev.*, 44: 5148–5180.

[135] Seh, Z. W., J. Kibsgaard, C. F. Dickens, I. Chorkendorff, J. K. Nørskov and T. F. Jaramillo. 2017. Combining theory and experiment in electrocatalysis: Insights into materials design, *Science*, 355: eaad4998.

[136] Subbaraman, R., D. Tripkovic, D. Strmcnik, K.-C. Chang, M. Uchimura, A. P. Paulikas, V. Stamenkovic and N. M. Markovic. 2011. Enhancing hydrogen evolution activity in water splitting by tailoring Li$^+$-Ni(OH)$_2$-Pt interfaces. *Science*, 334: 1256–1260.

[137] Nørskov, J. K., T. Bligaard, A. Logadottir, J. R. Kitchin, J. G. Chen, S. Pandelov and U. Stimming. 2005. Trends in the exchange current for hydrogen evolution. *J. Electrochem. Soc.*, 152: J23–J26.

[138] Huang, X., Z. Zeng, S. Bao, M. Wang, X. Qi, Z. Fan and H. Zhang. 2013. Solution-phase epitaxial growth of noble metal nanostructures on dispersible single-layer molybdenum disulfide nanosheets. *Nat. Commun.*, 4: 1444.

[139] Wang, P., X. Zhang, J. Zhang, S. Wan, S. Guo, G. Lu, J. Yao and X. Huang. 2017. Precise tuning in platinum-nickel/nickel sulfide interface nanowires for synergistic hydrogen evolution catalysis. *Nat. Commun.*, 8: 14580.

[140] Liu, D., X. Li, S. Chen, H. Yan, C. Wang, C. Wu, Y. A. Haleem, S. Duan, J. Lu, B. Ge, P. M. Ajayan, Y. Luo, J. Jiang and L. Song. 2019. Atomically dispersed platinum supported on curved carbon supports for efficient electrocatalytic hydrogen evolution. *Nat. Energy*, 4: 512–518.

[141] Lu, B., L. Guo, F. Wu, Y. Peng, J. E. Lu, T. J. Smart, N. Wang, Y. Z. Finfrock, D. Morris, P. Zhang, N. Li, P. Gao, Y. Ping and S. Chen. 2019. Ruthenium atomically dispersed in carbon outperforms platinum toward hydrogen evolution in alkaline media. *Nat. Commun.*, 10: 631.

[142] Pu, Z., J. Zhao, I. S. Amiinu, W. Li, M. Wang, D. He and S. Mu. 2019. A universal synthesis strategy for P-rich noble metal diphosphide-based electrocatalysts for the hydrogen evolution reaction. *Energy Environ. Sci.*, 12: 952–957.

[143] Gao, X., Y. Zhou, Y. Tan, S. Liu, Z. Cheng and Z. Shen. 2019. Graphyne doped with transition-metal single atoms as effective bifunctional electrocatalysts for water splitting. *Appl. Surf. Sci.*, 492: 8–15.

[144] Gao, X., Y. Zhou, S. Liu, Z. Cheng, Y. Tan and Z. Shen. 2020. Single cobalt atom anchored on N-doped graphyne for boosting the overall water splitting. *Appl. Surf. Sci.*, 502: 144155.

[145] Hossain, M. D., Z. Liu, M. Zhuang, X. Yan, G.-L. Xu, C. A. Gadre, A. Tyagi, I. H. Abidi, C.-J. Sun, H. Wong, A. Guda, Y. Hao, X. Pan, K. Amine and Z. Luo. 2019. Rational design of graphene-supported single atom catalysts for hydrogen evolution reaction. *Adv. Energy Mater.*, 9: 1803689.

[146] Yi, J.-D., R. Xu, G.-L. Chai, T. Zhang, K. Zang, B. Nan, H. Lin, Y.-L. Liang, J. Lv, J. Luo, R. Si, Y.-B. Huang and R. Cao. 2019. Cobalt single-atoms anchored on porphyrinic triazine-based frameworks as bifunctional electrocatalysts for oxygen reduction and hydrogen evolution reactions. *J. Mater. Chem. A*, 7: 1252–1259.

[147] Chen, W., J. Pei, C.-T. He, J. Wan, H. Ren, Y. Zhu, Y. Wang, J. Dong, S. Tian, W.-C. Cheong, S. Lu, L. Zheng, X. Zheng, W. Yan, Z. Zhuang, C. Chen, Q. Peng, D. Wang and Y. Li. 2017. Rational design of single molybdenum atoms anchored on N-doped carbon for effective hydrogen evolution reaction. *Angew. Chem. Int. Ed.*, 56: 16086–16090.

[148] Chen, W., J. Pei, C.-T. He, J. Wan, H. Ren, Y. Wang, J. Dong, K. Wu, W.-C. Cheong, J. Mao, X. Zheng, W. Yan, Z. Zhuang, C. Chen, Q. Peng, D. Wang and Y. Li. 2018. Single tungsten atoms supported on MOF-derived N-doped carbon for robust electrochemical hydrogen evolution. *Adv. Mater.*, 30: 1800396.

[149] Ji, L., P. Yan, C. Zhu, C. Ma, W. Wu, C. Wei, Y. Shen, S. Chu, J. Wang, Y. Du, J. Chen, X. Yang and Q. Xu. 2019. One-pot synthesis of porous 1T-phase MoS$_2$ integrated with single-atom Cu doping for enhancing electrocatalytic hydrogen evolution reaction. *Appl. Catal. B: Environ.*, 251: 87–93.

[150] Xue, Y., B. Huang, Y. Yi, Y. Guo, Z. Zuo, Y. Li, Z. Jia, H. Liu and Y. Li. 2018. Anchoring zero valence single atoms of nickel and iron on graphdiyne for hydrogen evolution. *Nat. Commun.*, 9: 1460.

[151] Gusmao, R., M. Vesely and Z. Sofer. 2020. Recent developments on the single atom supported at 2D materials beyond graphene as catalysts. *ACS Catal.*, 10: 9634–9648.

[152] Olah, G. A., G. K. S. Prakash and A. Goeppert. 2011. Anthropogenic chemical carbon cycle for a sustainable future. *J. Am. Chem. Soc.*, 133: 12881–12898.

[153] Den Elzen, M. G. J. and D. P. Van Vuuren. 2007. Peaking profiles for achieving long-term temperature targets with more likelihood at lower costs. *Proc. Natl. Acad. Sci. U. S. A.*, 104: 17931–17936.

[154] Qiao, J., Y. Liu, F. Hong and J. Zhang. 2014. A review of catalysts for the electroreduction of carbon dioxide to produce low-carbon fuels. *Chem. Soc. Rev.*, 43: 631–675.

[155] Kumar, B., J. P. Brian, V. Atla, S. Kumari, K. A. Bertram, R. T. White and J. M. Spurgeon. 2016. New trends in the development of heterogeneous catalysts for electrochemical CO_2 reduction. *Catal. Today*, 270: 19–30.

[156] Zhang, L., Z.-J. Zhao and J. Gong. 2017. Nanostructured materials for heterogeneous electrocatalytic CO_2 reduction and their related reaction mechanisms. *Angew. Chem. Int. Ed.*, 56: 11326–11353.

[157] Varela, A. S., W. Ju, A. Bagger, P. Franco, J. Rossmeisl and P. Strasser. 2019. Electrochemical reduction of CO_2 on metal-nitrogen-doped carbon catalysts. *ACS Catal.*, 9: 7270–7284.

[158] Jiang, K., S. Siahrostami, T. Zheng, Y. Hu, S. Hwang, E. Stavitski, Y. Peng, J. Dynes, M. Gangisetty, D. Su, K. Attenkoferf and H. Wang. 2018. Isolated Ni single atoms in graphene nanosheets for high-performance CO_2 reduction. *Energy Environ. Sci.*, 11: 893–903.

[159] Fan, Q., P. Hou, C. Choi, T.-S. Wu, S. Hong, F. Li, Y.-L. Soo, P. Kang, Y. Jung and Z. Sun. 2020. Activation of Ni particles into single Ni-N atoms for efficient electrochemical reduction of CO_2. *Adv. Energy Mater.*, 10: 1903068.

[160] Yang, J., Z. Qiu, C. Zhao, W. Wei, Z. Li, Y. Qu, J. Dong, J. Luo, Z. Li and Y. Wu. 2018. *In situ* thermal atomization to convert supported nickel nanoparticles into surface-bound nickel single-atom catalysts. *Angew. Chem. Int. Ed.*, 57: 14095–14100.

[161] Yang, H., L. Shang, Q. Zhang, R. Shi, G. I. N. Waterhouse, L. Gu and T. Zhang. 2019. A universal ligand mediated method for large scale synthesis of transition metal single atom catalysts. *Nat. Commun.*, 10: 4585.

[162] Pan, Y., R. Lin, Y. Chen, S. Liu, W. Zhu, X. Cao, W. Chen, K. Wu, W.-C. Cheong, Y. Wang, L. Zheng, J. Luo, Y. Lin, Y. Liu, C. Liu, J. Li, Q. Lu, X. Chen, D. Wang, Q. Peng, C. Chen and Y. Li. 2018. Design of single-atom Co-N_5 catalytic site: A robust electrocatalyst for CO_2 reduction with nearly 100% CO selectivity and remarkable stability. *J. Am. Chem. Soc.*, 140: 4218–4221.

[163] Zhang, C., S. Yang, J. Wu, M. Liu, S. Yazdi, M. Ren, J. Sha, J. Zhong, K. Nie, A. S. Jalilov, Z. Li, H. Li, B. I. Yakobson, Q. Wu, E. Ringe, H. Xu, P. M. Ajayan and J. M. Tour. 2018. Electrochemical CO_2 reduction with atomic iron-dispersed on nitrogen-doped graphene. *Adv. Energy Mater.*, 8: 1703487.

[164] Zhao, Z. and G. Lu. 2019. Cu-based single-atom catalysts boost electroreduction of CO_2 to CH_3OH: First-principles predictions. *J. Phys. Chem. C*, 123: 4380–4387.

[165] Zheng, W., J. Yang, H. Chen, Y. Hou, Q. Wang, M. Gu, F. He, Y. Xia, Z. Xia, Z. Li, B. Yang, L. Lei, C. Yuan, Q. He, M. Qiu and X. Feng. 2020. Atomically defined undercoordinated active sites for highly efficient CO_2 electroreduction. *Adv. Funct. Mater.*, 30: 1907658.

[166] Huan, T. N., N. Ranjbar, G. Rousse, M. Sougrati, A. Zitolo, V. Mougel, F. Jaouen and M. Fontecave. 2017. Electrochemical reduction of CO_2 catalyzed by Fe-N-C materials: A structure-selectivity study. *ACS Catal.*, 7: 1520–1525.

Index

About the Editors

Zhiqun Lin is a Professor in the School of Materials Science and Engineering at the Georgia Institute of Technology. He received his Ph.D. in Polymer Science and Engineering from the University of Massachusetts, Amherst in 2002. His research interests include perovskite solar cells, polymer solar cells, dye-sensitized solar cells, photocatalysis, hydrogen generation, lithium ion batteries, quantum dots (rods), conjugated polymers, block copolymers, polymer blends, hierarchical structure formation and assembly, surface and interfacial properties, multifunctional nanocrystals, and Janus nanostructures.

Xueqin Liu is an Associate Professor in the Faculty of Materials Science and Chemistry, China University of Geosciences (Wuhan). He received his Ph.D. in Materials Science and Engineering from China University of Geosciences (Wuhan) in 2016. He was selected to be a community board member for Nanoscale Horizons. His research interests include the structure modification of catalytic materials and their application in solar energy conversion and environment remediation, such as hydrogen generation, CO_2 reduction and N_2 reduction.

Zhen Li is a Professor in the Faculty of Materials Science and Chemistry, China University of Geosciences (Wuhan). She received her Ph.D. in mineral-petrological materials science from China University of Geosciences (Wuhan) in 2004. Her research is devoted to the development and application of graphite materials, the growth and application of multifunctional nanocrystals, the comprehensive utilization of non-metallic minerals and the preparation of functional nanocomposites. She has published more than 100 peer reviewed journal articles.

Yanqiu Huang is a Professor in the Faculty of Materials Science and Chemistry, China University of Geosciences (Wuhan). He received her PhD degree in Microelectronics and Solid State Electronics from Huazhong University of Science and Technology in 2002. His research interests include piezoelectric and ferroelectric materials as well as graphite and graphite based composite electrode materials for the lithium-ion battery. He has published more than 60 peer reviewed journal articles.

For Product Safety Concerns and Information please contact our EU
representative GPSR@taylorandfrancis.com
Taylor & Francis Verlag GmbH, Kaufingerstraße 24, 80331 München, Germany

www.ingramcontent.com/pod-product-compliance
Lightning Source LLC
Chambersburg PA
CBHW060348220326
41598CB00023B/2840

9 781032 382593